11/25/96

ELECTROMECHANICAL MOTION DEVICES

McGraw-Hill Series in Electrical Engineering

Consulting Editor

Stephen W. Director, *Carnegie-Mellon University*

Circuits and Systems
Communications and Signal Processing
Control Theory
Electronics and Electronic Circuits
Power and Energy
Electromagnetics
Computer Engineering
Introductory
Radar and Antennas
VLSI

Previous Consulting Editors

Ronald N. Bracewell, Colin Cherry, James F. Gibbons, Willis W. Harman, Hubert Heffner, Edward W. Herold, John G. Linvill, Simon Ramo, Ronald A. Rohrer, Anthony E. Siegman, Charles Susskind, Frederick E. Terman, John G. Truxal, Ernst Weber, and John R. Whinnery

Power and Energy

Consulting Editor

Stephen W. Director, *Carnegie-Mellon University*

Chapman: *Electric Machinery Fundamentals*
Elgerd: *Electric Energy Systems Theory: An Introduction*
Fitzgerald, Kingsley, and Umans: *Electric Machinery*
Gonen: *Electric Power Distribution System Engineering*
Hu and White: *Solar Cells: From Basic to Advanced Systems*
Krause: *Analysis of Electric Machinery*
Krause and Wasynczuk: *Electromechanical Motion Devices*
Stevenson: *Elements of Power System Analysis*

Also Available from McGraw-Hill

Schaum's Outline Series in Electronic & Electrical Engineering

Each outline includes basic theory, definitions, and hundreds of solved problems and supplementary problems with answers.

Current List Includes:

Basic Circuit Analysis
Basic Electrical Engineering
Basic Electricity
Basic Mathematics for Electricity and Electronics
Electric Circuits, 2d edition
Electric Machines and Electromechanics
Electromagnetics
Electronic Circuits
Electronic Communication
Electronics Technology
Feedback and Control Systems
Transmission Lines
State Space & Linear Systems

Schaum's Solved Problems Series

Each title in this series is a complete and expert source of solved problems containing thousands of problems with worked out solutions.

Current List Includes:

3000 Solved Problems in Electric Circuits

Available at Your College Bookstore

ELECTROMECHANICAL MOTION DEVICES

Paul C. Krause

Professor of Electrical Engineering
School of Electrical Engineering
Purdue University

Oleg Wasynczuk

Professor of Electrical Engineering
School of Electrical Engineering
Purdue University

McGraw-Hill Book Company

New York St. Louis San Francisco Auckland Bogotá Caracas
Colorado Springs Hamburg Lisbon London Madrid Mexico Milan
Montreal New Delhi Oklahoma City Panama Paris San Juan
São Paulo Singapore Sydney Tokyo Toronto

This book was set in Times Roman.
The editors were Alar E. Elken and John M. Morriss;
the production supervisor was Denise L. Puryear.
The cover was designed by Amy E. Becker.
Project supervision was done by The Total Book.
R. R. Donnelley & Sons Company was printer and binder.

ELECTROMECHANICAL MOTION DEVICES

234567890 DOC DOC 89432109

ISBN 0-07-035494-4

Library of Congress Cataloging-in-Publication Data

Krause, Paul C.
Electromechanical motion devices / Paul C. Krause and
Oleg Wasynczuk.
p.   cm.
Includes bibliographies and index.
ISBN 0-07-035494-4.   ISBN 0-07-035524-X (solutions manual)
1. Electromechanical devices.   2. Magnetic circuits.
I. Wasynczuk, Oleg.   II. Title.
TK153.K73 1989
621.8—dc19   88-13319

ABOUT THE AUTHORS

Paul C. Krause is a Professor of electrical engineering at Purdue University, where he has taught since 1970. He has also held teaching positions at the University of Kansas and the University of Wisconsin. Dr. Krause is a member of Sigma Tau, Eta Kappa Nu, Pi Tau Sigma, Phi Theta Kappa, and a Fellow of IEEE. He has authored or co-authored over 100 technical papers and a textbook entitled *Analysis of Electric Machinery* published by McGraw-Hill in 1986. He received B.S. degrees in electrical and mechanical engineering and an M.S. degree in electrical engineering from the University of Nebraska and a Ph.D. degree in electrical engineering from the University of Kansas.

Oleg Wasynczuk is a Professor of electrical engineering at Purdue University, where he has taught since 1979. Dr. Wasynczuk is a member of Eta Kappa Nu, Tau Beta Pi, Phi Kappa Phi, and a Senior Member of IEEE. He has authored or co-authored over 30 technical papers. He received a B.S. degree in electrical engineering from Bradley University and M.S. and Ph.D. degrees in electrical engineering from Purdue University.

TO OUR FAMILIES

CONTENTS

PREFACE

In the past, the course in electric machines taught at most engineering schools has been oriented toward the electric power area. The advent of the computer and the ever increasing interest in low-power automated systems are forces causing many of the university professors responsible for the undergraduate electromechanical energy conversion course to search for a text which covers a broader spectrum of material than just electric machines in power systems. This textbook is an attempt to provide for this need. It is written for the junior engineering student in electrical, mechanical, or industrial engineering who has completed the basic courses in physics and an introductory course in electric circuits.

The purpose is to provide a basic knowledge of electromechanical motion devices for students interested in either robotics, controls, or computers as well as for students interested in the energy or power systems area. To achieve this goal, considerable attention is given to electromechanical rotational devices commonly used in low-power automated control systems. The permanent magnet dc machine, the 2-phase induction, the brushless dc, and the stepper motor are examples. The material is arranged so that the instructor can select topics which best fit the particular course requirements without being forced to cover material of secondary or passing interest. In this regard, Chapters 1 and 2 contain basic, introductory material and should be covered, for the most part, by all students. This material provides sufficient background to begin study of either Chapter 3 on dc machines or Chapter 4. Chapter 3 is not a prerequisite for any other chapter. In Chapter 4, features which are common to rotating magnetic field electromechanical devices (ac machines) are addressed. Once Chapter 4 has been completed, the student has the background to proceed to any one of the next four chapters. It is not necessary to study one of these chapters before the other. None is prerequisite for the other. In other words, the student may study the induction, synchronous, brushless dc, or stepper motor in any order whatsoever. The final chapter on single-phase and 2-phase servomotors is an extension of Chapter 5 and has Chapter 5 as a prerequisite.

The material in Chapters 3 and 5 through 7 is arranged so that the further one proceeds in these chapters, the more involved the material becomes; however, this advanced material is not prerequisite for the following chapters. For example, attention is focused on 2-phase devices; however, in the latter sections of Chapters 5 through 7, the 3-phase device is considered. Although the core material requires only physics and an introductory course in electric circuits, some material is presented for the student who is taking more advanced courses in the control and/or the electric circuits areas. For example, in the latter sections in the chapters on dc and brushless dc motors, the formulation of time-domain transfer functions and state equations is presented. Moreover, in the chapter on brushless dc motors, inverter drive systems are covered briefly. Students who are taking a course in switching circuits will be able to appreciate this material more than those who are not. Regardless of one's background, the computer traces illustrating the performance of the inverter are instructive. The Laplace transformation method is not a prerequisite for any of the material in this text; however, some attention is given to this method of solving linear differential equations in the final section of Chapter 3.

Perhaps two scenarios bracket the possible classroom uses of this text. To emphasize the electric power area, Chapters 1 through 6 would be covered nearly completely with some coverage of single-phase machines from Chapter 9. To emphasize the control and automation area, Chapters 1 through 4 would be covered but with only a very brief discussion of transformer connections in Chapter 1. This scenario would be completed with coverage of only brushless dc and stepper motors, Chapters 7 and 8, respectively. Induction and synchronous machines would be omitted, Chapters 5, 6, and 9. Actually, the organization of the material by most instructors will probably fall somewhere in between these two scenarios.

Throughout the text, the more important equations are "boxed in" and many numerical examples are set out. Computer traces are given for the purpose of illustrating the dynamic performance of the electromechanical devices analyzed. However, a background is not required, whatsoever, in either simulation or computer programming in order to understand or to make use of these computer traces. Also, at the end of each section, short problems (SPs) are given with answers. Some of these problems can be answered without pencil and paper; others may take 5 to 10 minutes to solve. Also, the more involved or lengthy problems at the end of each chapter are denoted by an asterisk to aid the instructor in assigning problems. A *Solutions Manual* is provided wherein the solutions to all of the SPs and all of the problems at the end of each chapter are given in detail.

Linda Stovall typed the original manuscript of the *Solutions Manual*. We are convinced that she is the best there is and we are glad that we have had the opportunity to work with her. We are also grateful to the following reviewers for their many helpful comments and suggestions: A. W. Dipert, University of Illinois at Urbana, Champaign; George Gela, Ohio State University; Alvin Laday, Iowa State University; George Vachtsevanos, Georgia Institute of Technology; and Alan K. Wallace, Oregon State University.

Paul C. Krause
Oleg Wasynczuk

ELECTROMECHANICAL MOTION DEVICES

CHAPTER
1

MAGNETIC AND MAGNETICALLY COUPLED CIRCUITS

1.1 INTRODUCTION

Before diving into the analysis of electromechanical motion devices, it is helpful to review briefly some of our previous work in physics and in basic electric circuit analysis. In particular, the analysis of magnetic circuits, the basic properties of magnetic materials, and the derivation of equivalent circuits of stationary, magnetically coupled devices are topics presented in this chapter. This material will be a review for most, since it is covered either in a sophomore physics course for engineers or in introductory electrical engineering courses in circuit theory. Nevertheless, reviewing this material and establishing concepts and terms for later use sets the appropriate stage for our study of electromechanical motion devices.

Perhaps the most important new concept presented in this chapter is the fact that in all electromechanical devices, mechanical motion must occur, either translational or rotational, and this motion is reflected into the electric system either as a change of flux linkages in the case of an electromagnetic system or as a change of charge in the case of an electrostatic system. We will deal

1

primarily with electromagnetic systems. If the magnetic system is linear, then the change in flux linkages results owing to a change in the inductance. In other words, we will find that the inductances of the electric circuits associated with electromechanical motion devices are functions of the mechanical motion. In this chapter, we shall learn to express the self- and mutual inductances for simple translational and rotational electromechanical devices, and to handle these changing inductances in the voltage equations describing the electric circuits associated with the electromechanical system.

Throughout this text, we shall "box in" some of the more important equations and give short problems (SP's) with answers, following most sections. If we have done our job, each short problem should take less than ten minutes to solve. Also, it may be appropriate to skip or deemphasize some material in this chapter. For example, those interested primarily in electromechanical motion devices will find it sufficient to read only the first part of the section on three-phase systems and transformer connections. Others, who are interested in the power system aspects of electromechanical devices, will find it helpful to study this section in detail and work the associated problems at the end of the chapter. At the close of each chapter, we shall take a moment to look back over some of the important aspects of the material which we have just covered and mention what is coming next and how we plan to fit things together as we go along.

1.2   MAGNETIC CIRCUITS

An elementary magnetic circuit is shown in Fig. 1.2-1. This system consists of an electric conductor wound N times about the magnetic member which is generally some type of ferromagnetic material. In this example system, the magnetic member contains an air gap of uniform length between points a and b. We will assume that the magnetic system (circuit) consists only of the magnetic member and the air gap. Recall that Ampere's law states that the line integral of the field intensity **H** about a closed path is equal to the net current

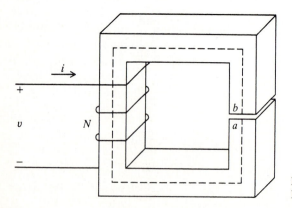

FIGURE 1.2-1
Elementary magnetic system.

enclosed within this closed path of integration. That is,

$$\oint \mathbf{H} \cdot d\mathbf{L} = i_n \qquad (1.2\text{-}1)$$

where i_n is the net current enclosed. Let us apply Ampere's law to the closed path depicted as a dashed line in Fig. 1.2-1. In particular,

$$\int_a^b H_i \, dL + \int_b^a H_g \, dL = Ni \qquad (1.2\text{-}2)$$

where the path of integration is assumed to be in the clockwise direction. This equation requires some explanation. First, we are assuming that the field intensity exists only in the direction of the given path, hence we have dropped the vector notation. The subscript i denotes the field intensity (H_i) in the ferromagnetic material (iron or steel) and g denotes the field intensity (H_g) in the air gap. The path of integration is taken as the mean length about the magnetic member for purposes we shall explain later. The right-hand side of (1.2-2) represents the net current enclosed. In particular, we have enclosed the current i, N times. This has the units of amperes but is commonly referred to as ampere-turns (At) or magnetomotive force (mmf). We will find that the mmf in magnetic circuits is analogous to the electromotive force (emf) in electric circuits. Note that the current enclosed is positive in (1.2-2) if the current i is positive. The sign of the right-hand side of (1.2-2) may be determined by the so-called "cork-screw" rule. That is, the current enclosed is positive if its assumed positive direction is in the same direction as the advance of a right-hand screw if it were turned in the direction of the path of integration, which in Fig. 1.2-1 is clockwise. Before continuing, it should be mentioned that we refer to $\mathbf{H}$ as the field intensity; however, some authors prefer to call $\mathbf{H}$ the field strength.

If we carry out the line integration, (1.2-2) can be written

$$H_i l_i + H_g l_g = Ni \qquad (1.2\text{-}3)$$

where l_i is the mean length of the magnetic material and l_g is the length across the air gap. Now, we have some explaining to do. We have assumed that the magnetic circuit consists only of the ferromagnetic material and the air gap, and that the magnetic field intensity is always in the direction of the path of integration or, in other words, perpendicular to a cross section of the magnetic material taken in the same sense as the air gap is cut through the material. The assumed direction of the magnetic field intensity is valid except in the vicinity of the corners. The direction of the field intensity changes gradually rather than abruptly at the corners. Nevertheless, the "mean length approximation" is widely used as an adequate means of analyzing this type of magnetic circuit.

Let us now take a cross section of the magnetic material as shown in Fig. 1.2-2. From our study of physics, we know that for linear, isotropic magnetic

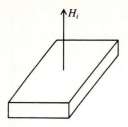

FIGURE 1.2-2
Cross section of magnetic material.

materials the flux density **B** is related to the field intensity as

$$\boxed{\mathbf{B} = \mu\mathbf{H}}$$ (1.2-4)

where μ is the permeability of the medium. Hence, we can write (1.2-3) in terms of flux density as

$$\frac{B_i}{\mu_i}l_i + \frac{B_g}{\mu_g}l_g = Ni$$ (1.2-5)

The surface integral of the flux density is equal to the flux Φ, thus

$$\boxed{\Phi = \int_A \mathbf{B}\cdot d\mathbf{S}}$$ (1.2-6)

If we assume that the flux density is uniformly distributed over the cross-sectional area, then

$$\Phi_i = B_i A_i$$ (1.2-7)

where Φ_i is the total flux in the magnetic material and A_i is the associated cross-sectional area. In the air gap

$$\Phi_g = B_g A_g$$ (1.2-8)

where A_g is the cross-sectional area of the gap. From physics, it is known that the streamlines of flux density **B** are closed, hence the flux in the air gap is equal to the flux in the core. That is, $\Phi_i = \Phi_g$ and, if the air gap is small, $A_i \cong A_g$. However, the effective area of the air gap is larger than that of the magnetic material, since the flux will tend to "balloon or spread out" (fringing effect), covering a maximum area midway across the air gap. Generally, this is taken into account by assuming that $A_g = kA_i$, where $k > 1$ is determined primarily by the length of the air gap. Although we shall keep this in mind, it is sufficient for our purposes to assume $A_g = A_i$. If we let $\Phi_i = \Phi_g = \Phi$ and substitute (1.2-7) and (1.2-8) into (1.2-5), we obtain

$$\frac{l_i}{\mu_i A_i}\Phi + \frac{l_g}{\mu_g A_g}\Phi = Ni$$ (1.2-9)

The analogy to Ohm's law is at hand. Ni (mmf) is analogous to the voltage (emf), and the flux Φ is analogous to the current. We can complete this analogy if we recall that the resistance of a conductor is proportional to its length and inversely proportional to its conductivity and cross-sectional area. Similarly, $l_i/\mu_i A_i$ and $l_g/\mu_g A_g$ are the reluctances of the magnetic material and air gap, respectively. Generally the permeability is expressed in terms of relative permeability as

$$\mu_i = \mu_{ri}\mu_0 \tag{1.2-10}$$

$$\mu_g = \mu_{rg}\mu_0 \tag{1.2-11}$$

where μ_0 is the permeability of free space ($4\pi \times 10^{-7}$ Wb/A·m or $4\pi \times 10^{-7}$ H/m, since Wb/A is a henry) and μ_{ri} and μ_{rg} are the relative permeability of the magnetic material and the air gap, respectively. For all practical purposes, $\mu_{rg} = 1$; however, μ_{ri} may be as large as 500 to 4000 depending upon the type of ferromagnetic material. We will use $\mathscr{R}$ to denote reluctance so as to distinguish reluctance from resistance, which will be denoted by r or R. We can now write (1.2-9) as

$$(\mathscr{R}_i + \mathscr{R}_g)\Phi = Ni \tag{1.2-12}$$

where $\mathscr{R}_i$ and $\mathscr{R}_g$ are the reluctance of the iron and air gap, respectively.

Example 1A. A magnetic system is shown in Fig. 1A-1. The total number of turns is 100, the relative permeability of the iron is 1000, and the current is 10 A. Calculate the total flux in the center leg.

Let us draw the electric circuit analog of this magnetic system for which we will need to calculate the reluctance of the various paths.

$$\mathscr{R}_{ab} = \frac{l_{ab}}{\mu_{ri}\mu_0 A_i}$$

$$= \frac{0.22}{(1000)(4\pi \times 10^{-7})(0.04)^2} = 109{,}419 \text{ H}^{-1} \tag{1A-1}$$

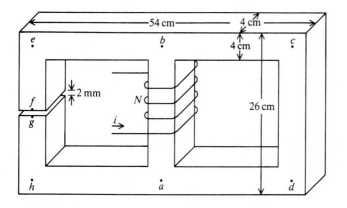

FIGURE 1A-1
Single-winding magnetic system.

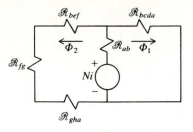

FIGURE 1A-2
Electric circuit analog of Fig. 1A-1.

Similarly,

$$\mathcal{R}_{bcda} = \frac{0.25 + 0.22 + 0.25}{(1000)(4\pi \times 10^{-7})(0.04)^2} = 358{,}099 \text{ H}^{-1} \qquad (1A\text{-}2)$$

Neglecting the air gap length,

$$\mathcal{R}_{bef} = \mathcal{R}_{gha} = \tfrac{1}{2}\mathcal{R}_{bcda} = 179{,}049 \text{ H}^{-1} \qquad (1A\text{-}3)$$

The reluctance of the air gap is

$$\mathcal{R}_{fg} = \frac{0.002}{(4\pi \times 10^{-7})(0.04)^2} = 994{,}718 \text{ H}^{-1} \qquad (1A\text{-}4)$$

The electric circuit analog is given in Fig. 1A-2. The polarity of the mmf is determined by the right-hand rule. That is, if we grasp one of the turns of the winding with our right hand with the thumb pointed in the direction of positive current, then our fingers will point in the direction of positive flux which flows in the direction of an mmf rise. Or if we grasp the winding (center leg) with the fingers of our right hand in the direction of positive current, then our thumb will be in the direction of positive flux and in the direction of a rise in mmf.

We can now apply dc circuit theory to solve for the total flux, $\Phi_1 + \Phi_2$, flowing in the center leg. For example, we can use loop equations or, as we will do here, reduce the series-parallel circuit to an equivalent reluctance. The equivalent reluctance of the parallel combination is

$$\mathcal{R}_{eq} = \frac{(\mathcal{R}_{bcda})(\mathcal{R}_{bef} + \mathcal{R}_{fg} + \mathcal{R}_{gha})}{\mathcal{R}_{bcda} + \mathcal{R}_{bef} + \mathcal{R}_{fg} + \mathcal{R}_{gha}}$$

$$= \frac{(358{,}099)(179{,}049 + 994{,}718 + 179{,}049)}{358{,}099 + 179{,}049 + 994{,}718 + 179{,}049}$$

$$= \frac{(358{,}099)(1{,}352{,}816)}{1{,}710{,}915} = 283{,}148 \text{ H}^{-1} \qquad (1A\text{-}5)$$

$$\Phi_1 + \Phi_2 = \frac{Ni}{\mathcal{R}_{ab} + \mathcal{R}_{eq}}$$

$$= \frac{(100)(10)}{109{,}419 + 283{,}148} = 2.547 \times 10^{-3} \text{ Wb} \qquad (1A\text{-}6)$$

Example 1B. Consider the magnetic system shown in Fig. 1B-1 [1]. The windings are supplied from ac sources and, in the steady state $I_1 = \sqrt{2}\cos \omega_e t$ and $I_2 = \sqrt{2}\,0.3\cos(\omega_e t + 45°)$, where capital letters are used to denote steady-state conditions. (Although we may violate this rule once in a while, capital letters will be used to denote steady-state voltages, currents, and flux linkages.) $N_1 = 150$ turns, $N_2 = 90$ turns, and $\mu_r = 3000$. Calculate the flux in the center leg.

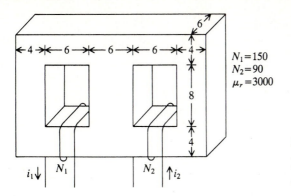

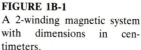

FIGURE 1B-1
A 2-winding magnetic system with dimensions in centimeters.

The electric circuit analog is given in Fig. 1B-2. The reluctance $\mathcal{R}_x$ is the reluctance of the center leg and $\mathcal{R}_y$ is the reluctance of one of the two parallel paths from the top of the center leg through an outside leg to the bottom of the center leg. In particular,

$$\mathcal{R}_y = \frac{2(0.03 + 0.06 + 0.02) + 0.12}{(3000)(4\pi \times 10^{-7})(0.06)(0.04)} = 37{,}578 \text{ H}^{-1} \tag{1B-1}$$

$$\mathcal{R}_x = \frac{0.12}{(3000)(4\pi \times 10^{-7})(0.06)^2} = 8842 \text{ H}^{-1} \tag{1B-2}$$

Since the currents are sinusoidal, the mmf's will be sinusoidal. Thus, it is convenient to use phasors to solve for Φ_1 and Φ_2. Phasors, which will be denoted by a raised tilde, are reviewed in Appendix B. The loop equations are

$$\text{m}\tilde{\text{m}}\text{f}_1 = \mathcal{R}_y\tilde{\Phi}_1 + \mathcal{R}_x(\tilde{\Phi}_1 - \tilde{\Phi}_2) \tag{1B-3}$$

$$\text{m}\tilde{\text{m}}\text{f}_2 = \mathcal{R}_x(\tilde{\Phi}_2 - \tilde{\Phi}_1) + \mathcal{R}_y\tilde{\Phi}_2 \tag{1B-4}$$

which may be written in matrix form as

$$\begin{bmatrix} \text{m}\tilde{\text{m}}\text{f}_1 \\ \text{m}\tilde{\text{m}}\text{f}_2 \end{bmatrix} = \begin{bmatrix} \mathcal{R}_x + \mathcal{R}_y & -\mathcal{R}_x \\ -\mathcal{R}_x & \mathcal{R}_x + \mathcal{R}_y \end{bmatrix} \begin{bmatrix} \tilde{\Phi}_1 \\ \tilde{\Phi}_2 \end{bmatrix} \tag{1B-5}$$

A review of matrix algebra is given in Appendix D. Now,

$$\text{m}\tilde{\text{m}}\text{f}_1 = N_1\tilde{I}_1 = (150)(1\underline{/0^\circ}) = 150\underline{/0^\circ} \text{ At} \tag{1B-6}$$

$$\text{m}\tilde{\text{m}}\text{f}_2 = N_2\tilde{I}_2 = (90)(0.3\underline{/45^\circ}) = 27\underline{/45^\circ} \text{ At} \tag{1B-7}$$

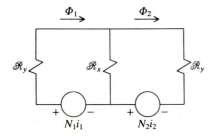

FIGURE 1B-2
Electric circuit analog of Fig. 1B-1.

Solving (1B-5) yields

$$\tilde{\Phi}_1 = (3.434 + j0.081) \times 10^{-3} \text{ Wb} \qquad (1B-8)$$

$$\tilde{\Phi}_2 = (1.065 + j0.427) \times 10^{-3} \text{ Wb} \qquad (1B-9)$$

The flux flowing down through the center leg is

$$\tilde{\Phi}_1 - \tilde{\Phi}_2 = (2.369 - j0.346) \times 10^{-3}$$

$$= 2.39 \times 10^{-3} \underline{/-8.3°} \text{ Wb} \qquad (1B-10)$$

SP1.2-1. Calculate Φ_1 in Example 1A. [$\Phi_1 = 2.014 \times 10^{-3}$ Wb]

SP1.2-2. Calculate $\tilde{\Phi}_1 + \tilde{\Phi}_2$ in Example 1A when $I = \sqrt{2}\, 10 \cos(\omega_e t - 30°)$. [$\tilde{\Phi}_1 + \tilde{\Phi}_2 = 2.547 \times 10^{-3}$ $\underline{/-30°}$ Wb, rms]

SP1.2-3. Remove the center leg of the magnetic system shown in Fig. 1B-1. Calculate the total flux when $I_1 = 9$ A and $I_2 = -15$ A. [Zero]

1.3 PROPERTIES OF MAGNETIC MATERIALS

We may be aware from our study of physics that, when ferromagnetic materials such as iron, nickel, cobalt or alloys of these elements, such as various types of steels, are placed in a magnetic field, the flux produced is markedly larger (500 to 4000 times, for example) than that which would be produced when a nonmagnetic material is subjected to the same magnetic field. We must take some time to review briefly the basic properties of ferromagnetic materials and to establish terminology for later use.

Let us begin by considering the relationship between B and H shown in Fig. 1.3-1 which is typical of silicon steel used in transformers. We will assume

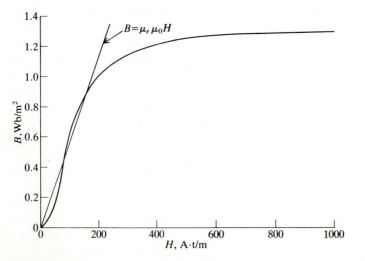

FIGURE 1.3-1
B-H curve for typical silicon steel used in transformers.

that the ferromagnetic core is initially completely demagnetized (both B and H are zero). As we apply an external H field by increasing the current in a winding wound around the core, the flux density B also increases but non-linearly, as shown in Fig. 1.3-1. After H reaches a value of approximately 150 A/m, the flux density rises more slowly and the material begins to saturate when H is several hundred amperes per meter.

In ferromagnetic materials, the combination of the magnetic moments produced by the electrons orbiting the nucleus of an atom and the electron itself spinning on its axis produce a net magnetic moment of the atom which is not canceled by an opposing magnetic moment of a neighboring atom. Microscopically, ferromagnetic materials have been found to be divided into magnetic domains wherein all magnetic moments (dipoles) are aligned. Although the magnetic moments are all aligned within a magnetic domain, the direction of this alignment will differ from one domain to another.

When a ferromagnetic material is subjected to an external magnetic field, those domains, which originally tend to be aligned with the applied magnetic field, grow at the expense of those domains with magnetic moments which are less aligned. Thereby, the flux is increased from that which would occur with a nonmagnetic material. This is known as *domain wall motion* [2]. As the strength of the magnetic field increases, the aligned domains continue to grow in nearly a linear fashion. Thus a nearly linear B-H curve results ($B \cong \mu_r \mu_0 H$) until the ability of the aligned domains to take from the unaligned domains starts to slow. This gives rise to the knee of the B-H curve and saturation is beginning. At this point, the displacements of the domain walls are complete. That is, there are no longer unaligned domains from which to take. However, the remaining domains may still not be in perfect alignment with the external H field. A further increase in H will cause a rotation of the atomic dipole moments within the remaining domains toward a more perfect alignment. However, the marginal increase in B due to rotation is less than the original increase in B due to domain wall motion, resulting in a decrease in slope of the B-H curve. The magnetic material is said to be completely saturated when the remaining domains are perfectly aligned. In this case, the slope of the B-H curve becomes μ_0 [2]. If it is assumed that the magnetic flux is uniform through most of the magnetic material, then B is proportional to Φ and H is proportional to mmf. Hence, a plot of flux versus current is of the same shape as the B-H curve.

A transformer is generally designed so that some saturation occurs during normal operation. Electric machines are also designed similarly in that a machine generally operates slightly in the saturated region during normal, rated operating conditions. Since saturation causes the coefficients of the differential equations describing the behavior of an electromagnetic device to be functions of the winding currents, a transient analysis is difficult without the aid of a computer. However, it is not our purpose to set forth methods of analyzing nonlinear magnetic systems.

In the previous discussion, we have assumed that the ferromagnetic

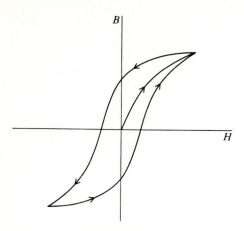

FIGURE 1.3-2
Hysteresis loop.

material is initially demagnetized and that the applied field intensity is gradual-ly increased from zero. However, if a ferromagnetic material is subjected to an alternating field intensity, the resulting *B-H* curve exhibits hysteresis. For example, let us assume that a ferromagnetic material is subjected to an alternating field intensity (alternating current flowing in the winding) and initially the flux density and field intensity are both zero. As *H* increases from zero, *B* increases along the initial *B-H* curve, as shown in Fig. 1.3-2. However the field intensity varies sinusoidally and, when *H* decreases from a maximum, *B* does not follow back down the original *B-H* curve. After several cycles, the magnetic system will reach a steady-state condition and the plot of *B* versus *H* will form a hysteresis loop or a double-valued function, as shown in Fig. 1.3-2. What is happening is very complex. In simple terms, the growth of aligned domains for an incremental change in *H* in one direction is not equal to the growth of oppositely aligned domains if this change in *H* were suddenly reversed. We could become quite involved by discussing minor hysteresis loops which would occur if, during the sinusoidal variation of *H*, it were suddenly stopped at some nonzero value then reversed, stopped, and reversed again [2]. We shall only mention this phenomenon in passing.

A family of hysteresis loops is shown in Fig. 1.3-3. In each case, the applied *H* is sinusoidal; however, the amplitude of the *H* field is varied to give the family of loops shown in Fig. 1.3-3. A magnetization or *B-H* curve for a given material is obtained by connecting the tips of the hysteresis loops, as shown by the dashed line in Fig. 1.3-3. The locus of the tips of the hysteresis curves is about the same as the original *B-H* curve in Fig. 1.3-1, which corresponds to a gradual increase of *H* in an initially demagnetized material. If *H* were suddenly stopped at zero, the flux density remaining in the ferromag-netic material is called the *residual flux density* (B_r). The negative field intensity necessary to bring this residual flux density to zero is called the *coercive force* (H_c). These two quantities are indicated in Fig. 1.3-3 for the largest hysteresis loop shown.

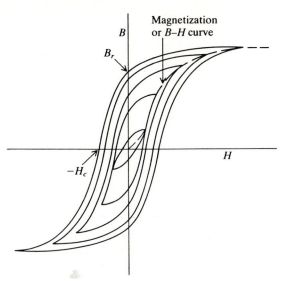

FIGURE 1.3-3
Family of steady-state hysteresis loops.

Energy is required to increase the size of the magnetic domains of the ferromagnetic material. It can be shown that the energy necessary to align alternately the magnetic domains is equal to the area enclosed by the hysteresis loop. This energy causes a rise in the temperature of the magnetic material, and the power associated with this energy loss is called the *hysteresis loss*.

When a solid block of magnetic material such as that shown in Fig. 1.2-1 is subjected to an alternating field intensity, the resulting alternating flux induces current in the solid magnetic material which will circulate in a loop perpendicular to the flux density (**B**) inducing it. These so-called eddy currents have two undesirable effects. First, the mmf established by these circulating currents opposes the mmf produced by the winding, and this opposition is greatest at the center of the material because that tends to be also the center of the current loops. Thus, the flux would tend not to flow through the center of the solid magnetic member, thereby not utilizing the full benefits of the ferromagnetic material. Second, there is an i^2r loss associated with these eddy currents, called *eddy current loss*, which is dissipated as heat. These two adverse effects can be minimized in several ways, but the most common is to build the ferromagnetic core of laminations (thin strips) insulated from each other and oriented in the direction of the magnetic field (**B** or **H**). These thin strips offer a much smaller area in which the eddy currents can flow, hence smaller currents and smaller losses result.

The core losses associated with ferromagnetic materials are the combination of the hysteresis and eddy current losses. Electromagnetic devices are designed to minimize these losses; however, they are always present and are often taken into account in a linear system analysis by assuming that their effects on the electric system can be represented by a resistance.

SP1.3-1. The magnetic circuit of Fig. 1.2-1 is constructed by using silicon sheet steel. Its magnetization curve is given by Fig. 1.3-1. The gap length l_g is 1 mm, the mean core length l_i is 100 cm, $N = 500$, and $A_i = A_g = 25$ cm^2. Determine the current needed to produce a flux Φ of 2.5×10^{-3} Wb. [*Hint:* First establish H_i, H_g, and use (1.2-3).] [$I = 1.99$ A]

1.4 STATIONARY MAGNETICALLY COUPLED CIRCUITS

Magnetically coupled electric circuits are central to the operation of transformers and electromechanical motion devices. In the case of transformers, stationary circuits are magnetically coupled for the purpose of changing the voltage and current levels. In the case of electromechanical devices, circuits in relative motion are magnetically coupled for the purpose of transferring energy between the mechanical and electric systems. Since magnetically coupled circuits play such an important role in energy conversion, it is important to establish the equations which describe their behavior and to express these equations in a form convenient for analysis. Many of these goals may be achieved by considering two stationary electric circuits which are magnetically coupled, as shown in Fig. 1.4-1. The two windings consist of turns N_1 and N_2, and they are wound on a common core which is a ferromagnetic material with a permeability large relative to that of air. The magnetic core is not illustrated in three dimensions.

Before proceeding, a comment or two is in order. Generally, the concept of an ideal transformer is introduced in a basic circuits course. In the ideal case, v_2 in Fig. 1.4-1, is $(N_2/N_1)v_1$ and i_2 is $-(N_1/N_2)i_1$. Only the turns-ratio of the transformer is considered. However, this treatment is often not sufficient

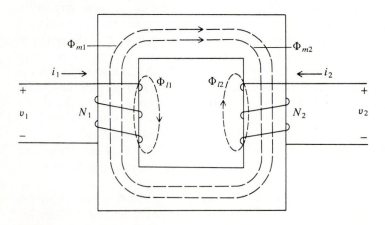

FIGURE 1.4-1
Magnetically coupled circuits.

for a detailed analysis of transformers, and it is seldom appropriate in the analysis of electromechanical motion devices, since an air gap is necessary for motion to occur; hence, the windings are not as tightly coupled as in the case of transformers and the leakage flux must be taken into account.

In general, the flux produced by each winding can be separated into two components—a leakage component denoted with the subscript l and a magnetizing component denoted by the subscript m. Each of these components is depicted by a single streamline with the positive direction determined by applying the right-hand rule to the directions of current flow in the winding. (The right-hand rule was reviewed in Example 1A.) The leakage flux associated with a given winding links only that winding, whereas the magnetizing flux, whether it is due to current in winding 1 or winding 2, links both windings. In some cases, i_2 is selected positive out of the top of winding 2 and a dot is placed at that terminal. Although the "dot notation" is convenient for transformers, it is not used in the case of electromechanical devices.

The flux linking each winding may be expressed as

$$\Phi_1 = \Phi_{l1} + \Phi_{m1} + \Phi_{m2} \tag{1.4-1}$$

$$\Phi_2 = \Phi_{l2} + \Phi_{m2} + \Phi_{m1} \tag{1.4-2}$$

The leakage flux Φ_{l1} is produced by current flowing in winding 1 and it links only the turns of winding 1. Likewise, the leakage flux Φ_{l2} is produced by current flowing in winding 2 and it links only the turns of winding 2. The flux Φ_{m1} is produced by current flowing in winding 1 and it links all turns of windings 1 and 2. Similarly, the magnetizing flux Φ_{m2} is produced by current flowing in winding 2 and it also links all turns of windings 1 and 2. Both Φ_{m1} and Φ_{m2} are called *magnetizing fluxes*. With the selected positive directions of current flow and the manner in which the windings are wound, magnetizing flux produced by positive current flowing in one winding adds to the magnetizing flux produced by positive current flowing in the other winding. For this case, we will find that the mutual inductance is positive.

It is appropriate to point out that this is an idealization of the actual magnetic system. It seems logical that all of the leakage flux will not link all the turns of the winding producing it; hence, Φ_{l1} and Φ_{l2} are "equivalent" leakage fluxes. Similarly, all of the magnetizing flux of one winding may not link all of the turns of the other winding. To acknowledge this practical aspect of the magnetic system, N_1 and N_2 are often considered to be the equivalent number of turns rather than the actual number.

The voltage equations may be expressed as

$$v_1 = r_1 i_1 + \frac{d\lambda_1}{dt} \tag{1.4-3}$$

$$v_2 = r_2 i_2 + \frac{d\lambda_2}{dt} \tag{1.4-4}$$

In matrix form,

$$\begin{bmatrix} v_1 \\ v_2 \end{bmatrix} = \begin{bmatrix} r_1 & 0 \\ 0 & r_2 \end{bmatrix} \begin{bmatrix} i_1 \\ i_2 \end{bmatrix} + \frac{d}{dt} \begin{bmatrix} \lambda_1 \\ \lambda_2 \end{bmatrix} \tag{1.4-5}$$

A review of matrix algebra is given in Appendix D. The resistances r_1 and r_2 and the flux linkages λ_1 and λ_2 are related to windings 1 and 2, respectively. Since it is assumed that Φ_1 links the equivalent turns of winding 1 (N_1) and Φ_2 links the equivalent turns of winding 2 (N_2), the flux linkages may be written as

$$\lambda_1 = N_1 \Phi_1 \tag{1.4-6}$$

$$\lambda_2 = N_2 \Phi_2 \tag{1.4-7}$$

where Φ_1 and Φ_2 are given by (1.4-1) and (1.4-2), respectively.

If we assume that the magnetic system is linear, we may apply Ohm's law for magnetic circuits to express the fluxes. Thus, the fluxes may be written as

$$\Phi_{l1} = \frac{N_1 i_1}{\mathcal{R}_{l1}} \tag{1.4-8}$$

$$\Phi_{m1} = \frac{N_1 i_1}{\mathcal{R}_m} \tag{1.4-9}$$

$$\Phi_{l2} = \frac{N_2 i_2}{\mathcal{R}_{l2}} \tag{1.4-10}$$

$$\Phi_{m2} = \frac{N_2 i_2}{\mathcal{R}_m} \tag{1.4-11}$$

where $\mathcal{R}_{l1}$ and $\mathcal{R}_{l2}$ are the reluctances of the leakage paths and $\mathcal{R}_m$ is the reluctance of the path of magnetizing fluxes. Typically, the reluctances associated with leakage paths are much larger than the reluctance of the magnetizing path. The reluctance associated with an individual leakage path is difficult to determine exactly, and it is usually approximated from test data. On the other hand, the reluctance of the magnetizing path of the core shown in Fig. 1.4-1 may be computed with sufficient accuracy as in Example 1A.

Substituting (1.4-8) through (1.4-11) into (1.4-1) and (1.4-2) yields

$$\Phi_1 = \frac{N_1 i_1}{\mathcal{R}_{l1}} + \frac{N_1 i_1}{\mathcal{R}_m} + \frac{N_2 i_2}{\mathcal{R}_m} \tag{1.4-12}$$

$$\Phi_2 = \frac{N_2 i_2}{\mathcal{R}_{l2}} + \frac{N_2 i_2}{\mathcal{R}_m} + \frac{N_1 i_1}{\mathcal{R}_m} \tag{1.4-13}$$

Substituting (1.4-12) and (1.4-13) into (1.4-6) and (1.4-7) yields

$$\lambda_1 = \frac{N_1^2}{\mathcal{R}_{l1}} i_1 + \frac{N_1^2}{\mathcal{R}_m} i_1 + \frac{N_1 N_2}{\mathcal{R}_m} i_2 \tag{1.4-14}$$

$$\lambda_2 = \frac{N_2^2}{\mathcal{R}_{l2}} i_2 + \frac{N_2^2}{\mathcal{R}_m} i_2 + \frac{N_2 N_1}{\mathcal{R}_m} i_1 \qquad (1.4\text{-}15)$$

When the magnetic system is linear, the flux linkages are generally expressed in terms of inductances and the currents. We see that the coefficients of the first two terms on the right-hand side of (1.4-14) depend upon N_1 and the reluctance of the magnetic system, independent of the existence of winding 2. An analogous statement may be made regarding (1.4-15) with the roles of winding 1 and winding 2 reversed. Hence, the self-inductances are defined as

$$L_{11} = \frac{N_1^2}{\mathcal{R}_{l1}} + \frac{N_1^2}{\mathcal{R}_m} = L_{l1} + L_{m1} \qquad (1.4\text{-}16)$$

$$L_{22} = \frac{N_2^2}{\mathcal{R}_{l2}} + \frac{N_2^2}{\mathcal{R}_m} = L_{l2} + L_{m2} \qquad (1.4\text{-}17)$$

where L_{l1} and L_{l2} are the leakage inductances and L_{m1} and L_{m2} are the magnetizing inductances of windings 1 and 2, respectively. From (1.4-16) and (1.4-17) it follows that the magnetizing inductances may be related as

$$\frac{L_{m2}}{N_2^2} = \frac{L_{m1}}{N_1^2} \qquad (1.4\text{-}18)$$

The mutual inductances are defined as the coefficient of the third term on the right-hand side of (1.4-14) and (1.4-15). In particular,

$$L_{12} = \frac{N_1 N_2}{\mathcal{R}_m} \qquad (1.4\text{-}19)$$

$$L_{21} = \frac{N_2 N_1}{\mathcal{R}_m} \qquad (1.4\text{-}20)$$

We see that $L_{12} = L_{21}$ and, with the assumed positive direction of current flow and the manner in which the windings are wound, the mutual inductances are positive. If, however, the assumed positive directions of current were such that Φ_{m1} opposed Φ_{m2}, then the mutual inductances would be negative.

The mutual inductances may be related to the magnetizing inductances. Comparing (1.4-16) and (1.4-17) with (1.4-19) and (1.4-20), we see that

$$L_{12} = \frac{N_2}{N_1} L_{m1} = \frac{N_1}{N_2} L_{m2} \qquad (1.4\text{-}21)$$

The flux linkages may now be written as

$$\lambda_1 = L_{11} i_1 + L_{12} i_2 \qquad (1.4\text{-}22)$$
$$\lambda_2 = L_{21} i_1 + L_{22} i_2 \qquad (1.4\text{-}23)$$

where L_{11} and L_{22} are defined by (1.4-16) and (1.4-17), respectively, and L_{12} and L_{21} by (1.4-21). The self-inductances L_{11} and L_{22} are always positive;

however, the mutual inductances L_{12} (L_{21}) may be positive or negative, as previously mentioned.

Although the voltage equations given by (1.4-3) and (1.4-4) may be used for purposes of analysis, it is customary to perform a change of variables which yields the well-known equivalent T circuit of two windings coupled by a linear magnetic circuit. To set the stage for this derivation, let us express the flux linkages from (1.4-22) and (1.4-23) as

$$\lambda_1 = L_{l1}i_1 + L_{m1}\left(i_1 + \frac{N_2}{N_1}\,i_2\right) \tag{1.4-24}$$

$$\lambda_2 = L_{l2}i_2 + L_{m2}\left(\frac{N_1}{N_2}\,i_1 + i_2\right) \tag{1.4-25}$$

With λ_1 in terms of L_{m1} and λ_2 in terms of L_{m2}, we see two logical candidates for substitute variables, in particular, $(N_2/N_1)i_2$ or $(N_1/N_2)i_1$. If we let

$$i_2' = \frac{N_2}{N_1}\,i_2 \tag{1.4-26}$$

then we are using the substitute variable i_2' which, when flowing through winding 1, produces the same mmf as the actual i_2 flowing through winding 2; $N_1 i_2' = N_2 i_2$. This is said to be referring the current in winding 2 to winding 1 or to a winding with N_1 turns, whereupon winding 1 becomes the reference winding. On the other hand, if we let

$$i_1' = \frac{N_1}{N_2}\,i_1 \tag{1.4-27}$$

then i_1' is the substitute variable which produces the same mmf when flowing through winding 2 as i_1 does when flowing in winding 1; $N_2 i_1' = N_1 i_1$. This change of variables is said to refer the current of winding 1 to winding 2 or to a winding with N_2 turns, whereupon winding 2 becomes the reference winding.

We will demonstrate the derivation of the equivalent T circuit by referring the current of winding 2 to a winding with N_1 turns; thus i_2' is expressed by (1.4-26). We want the instantaneous power to be unchanged by this substitution of variables. Therefore,

$$v_2' i_2' = v_2 i_2 \tag{1.4-28}$$

Hence,
$$v_2' = \frac{N_1}{N_2}\,v_2 \tag{1.4-29}$$

Flux linkages, which have the units of volts per second, are related to the substitute flux linkages in the same way as voltages. In particular,

$$\lambda_2' = \frac{N_1}{N_2}\,\lambda_2 \tag{1.4-30}$$

Now, replace $(N_2/N_1)i_2$ with i_2' in the expression for λ_1 given by (1.4-24). Next, solve (1.4-26) for i_2 and substitute it into λ_2 given by (1.4-25). Now, multiply this result by N_1/N_2 to obtain λ_2' and then substitute $(N_2/N_1)^2 L_{m1}$ for L_{m2} in λ_2'. If we do all this, we will obtain

$$\lambda_1 = L_{l1}i_1 + L_{m1}(i_1 + i_2') \tag{1.4-31}$$

$$\lambda_2' = L_{l2}'i_2' + L_{m1}(i_1 + i_2') \tag{1.4-32}$$

where

$$L_{l2}' = \left(\frac{N_1}{N_2}\right)^2 L_{l2} \tag{1.4-33}$$

The flux linkage equations given by (1.4-31) and (1.4-32) may also be written as

$$\lambda_1 = L_{11}i_1 + L_{m1}i_2' \tag{1.4-34}$$

$$\lambda_2' = L_{m1}i_1 + L_{22}'i_2' \tag{1.4-35}$$

where

$$L_{22}' = \left(\frac{N_1}{N_2}\right)^2 L_{22} = L_{l2}' + L_{m1} \tag{1.4-36}$$

where L_{22} is defined by (1.4-17).

If we multiply (1.4-4) by N_1/N_2 to obtain v_2', the voltage equations become

$$\begin{bmatrix} v_1 \\ v_2' \end{bmatrix} = \begin{bmatrix} r_1 & 0 \\ 0 & r_2' \end{bmatrix} \begin{bmatrix} i_1 \\ i_2' \end{bmatrix} + \frac{d}{dt} \begin{bmatrix} \lambda_1 \\ \lambda_2' \end{bmatrix} \tag{1.4-37}$$

where

$$r_2' = \left(\frac{N_1}{N_2}\right)^2 r_2 \tag{1.4-38}$$

The above voltage equations, (1.4-37), together with the flux linkage equations, (1.4-34) through (1.4-35), suggest the equivalent T circuit shown in Fig. 1.4-2. This method may be extended to include any number of windings wound on the same core.

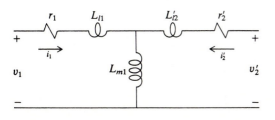

FIGURE 1.4-2
Equivalent T circuit with winding 1 selected as reference winding.

Earlier in this section, we mentioned that in the case of an ideal transformer only the turns-ratio is considered, that is, $v_2 = (N_2/N_1)v_1$ and $i_2 = -(N_1/N_2)i_1$. We can now more fully appreciate the assumptions which are made in this type of analysis. In particular, the resistances r_1 and r_2 and the leakage inductances L_{l1} and L_{l2} are neglected, and it is assumed that the magnetizing inductance is large so that the magnetizing current $i_1 + i_2'$ is negligibly small.

The information presented in this section forms the basis of the equivalent circuits for many types of electric machines. Using a turns-ratio to refer the voltages and currents of rotor circuits of electric machines to a winding with the same number of turns as the stator windings is common practice. In fact, the equivalent circuits for many ac machines are of the same form as shown in Fig. 1.4-2, with the addition of a voltage source referred to as a *speed voltage*. We shall talk much more about this speed voltage later—where it comes from and how it fits into the equivalent circuit.

Example 1C. It is instructive to illustrate the method of deriving an equivalent T circuit from open- and short-circuit measurements. When winding 2 of the 2-winding transformer shown in Fig. 1.4-2 is open-circuited and a voltage of 110 V (rms) at 60 Hz is applied to winding 1, the average power supplied to winding 1 is 6.66 W. The measured current in winding 1 is 1.05 A (rms). Next, with winding 2 short-circuited, the current flowing in winding 1 is 2 A when the applied voltage is 30 V at 60 Hz. The average input power is 44 W. If we assume $L_{l1} = L_{l2}'$, an approximate equivalent T circuit can be determined from these measurements with winding 1 selected as the reference winding.

The average power supplied to winding 1 may be expressed as

$$P_1 = |\tilde{V}_1||\tilde{I}_1| \cos \phi_{pf} \tag{1C-1}$$

where the so-called power factor angle is defined as

$$\phi_{pf} = \theta_{ev}(0) - \theta_{ei}(0) \tag{1C-2}$$

Here, $\tilde{V}_1$ and $\tilde{I}_1$ are phasors with the positive direction of $\tilde{I}_1$ taken in the direction of the voltage drop and $\theta_{ev}(0)$ and $\theta_{ei}(0)$ are the phase angles of $\tilde{V}_1$ and $\tilde{I}_1$, respectively (Appendix B). Solving for ϕ_{pf}, the power factor angle, during the open-circuit test, we have

$$\phi_{pf} = \cos^{-1} \frac{P_1}{|\tilde{V}_1||\tilde{I}_1|} = \cos^{-1} \frac{6.66}{(110)(1.05)} = 86.7° \tag{1C-3}$$

Although $\phi_{pf} = -86.7°$ is also a legitimate solution of (1C-3), the positive solution is taken since $\tilde{V}_1$ leads $\tilde{I}_1$ in an inductive circuit. With winding 2 open-circuited, the input impedance of winding 1 is

$$Z = \frac{\tilde{V}_1}{\tilde{I}_1} = r_1 + j(X_{l1} + X_{m1}) \tag{1C-4}$$

With $\tilde{V}_1$ as the reference phasor, $\tilde{V}_1 = 110\underline{/0°}$, $\tilde{I}_1 = 1.05\underline{/-86.7°}$. Thus,

$$r_1 + j(X_{l1} + X_{m1}) = \frac{110\underline{/0°}}{1.05\underline{/-86.7°}} = 6 + j104.6 \, \Omega \tag{1C-5}$$

In the analysis of sinusoidal steady-state operation, reactances are generally used. That is $X = \omega_e L$, where X is the inductive reactance, L is the inductance, and ω_e ($\omega_e = 2\pi f$) is the electrical angular frequency of the sinusoidal variables (Appendix B). If we neglect core losses, then, from (1C-5), $r_1 = 6\,\Omega$. We also see from (1C-5) that $X_{l1} + X_{m1} = 104.6\,\Omega$,

For the short-circuit test, we will assume that $\tilde{I}_1 = -\tilde{I}_2'$ since transformers are designed so that $X_{m1} \gg |r_2' + jX_{l2}'|$. Hence, using (1C-1) again,

$$\phi_{\text{pf}} = \cos^{-1}\frac{44}{(30)(2)} = 42.8° \tag{1C-6}$$

In this case, the input impedance is $Z = (r_1 + r_2') + j(X_{l1} + X_{l2}')$. This may be determined as

$$Z = \frac{30\underline{/0°}}{2\underline{/-42.8°}} = 11 + j10.2\,\Omega \tag{1C-7}$$

Hence, $r_2' = 11 - r_1 = 5\,\Omega$ and, since it is assumed that $X_{l1} = X_{l2}'$, both are $10.2/2 = 5.1\,\Omega$. Therefore, $X_{m1} = 104.6 - 5.1 = 99.5\,\Omega$. In summary, $r_1 = 6\,\Omega$, $L_{l1} = 13.5\,\text{mH}$, $L_{m1} = 263.9\,\text{mH}$, $r_2' = 5\,\Omega$, $L_{l2}' = 13.5\,\text{mH}$. Make sure we converted from X's to L's correctly.

SP1.4-1. Remove the center leg of the magnetic system shown in Fig. 1B-1. Calculate L_{11}, L_{22}, and L_{12}. Neglect the leakage inductances. [$L_{11} = 299.4\,\text{mH}$, $L_{22} = 107.8\,\text{mH}$. $L_{12} = 179.5\,\text{mH}$]

SP1.4-2. Consider the transformer and parameters calculated in Example 1C. Winding 2 is short-circuited and 12 V (dc) is applied to winding 1. Calculate the steady-state values of i_1 and i_2. Repeat with winding 2 open-circuited. [$I_1 = 2$ A and $I_2 = 0$ in both cases]

SP1.4-3. Calculate the no-load (winding 2 open-circuited) current for the transformer given in Example 1C if $V_1 = \sqrt{2}\,10\cos 100t$. [$\tilde{I}_1 = 0.352\underline{/-77.8°}$ A]

1.5   OPEN- AND SHORT-CIRCUIT CHARACTERISTICS OF STATIONARY MAGNETICALLY COUPLED CIRCUITS

It is instructive to observe the open- and short-circuit characteristics of a transformer with two windings. For this purpose, a transformer with the parameters given in Example 1C was simulated on a computer. The open-circuit characteristics are shown in Figs. 1.5-1 and 1.5-2. The variables plotted are λ, v_1, i_1, v_2', and i_2'. The variable λ is equal to $L_{m1}(i_1 + i_2')$, which is the last term on the right-hand side of (1.4-31) and (1.4-32). This is the flux linkage of winding 1 due to the flux in the transformer iron. It is often referred to as the *magnetizing flux linkage(s)* and denoted λ_m, λ_{mag}, or λ_ϕ, while $i_1 + i_2'$ is called the *magnetizing current*.

Initially the windings are unexcited. At time zero ($t = 0$) the voltage applied to winding 1 with winding 2 open-circuited is $v_1 = \sqrt{2}\,110\cos 377t$ in Fig. 1.5-1 and $v_1 = \sqrt{2}\,110\sin 377t$ in Fig. 1.5-2. The waveforms of the steady-state current i_1 are identical in Figs. 1.5-1 and 1.5-2; however, since the inductive reactance is large, applying a sine wave voltage for v_1 at time zero

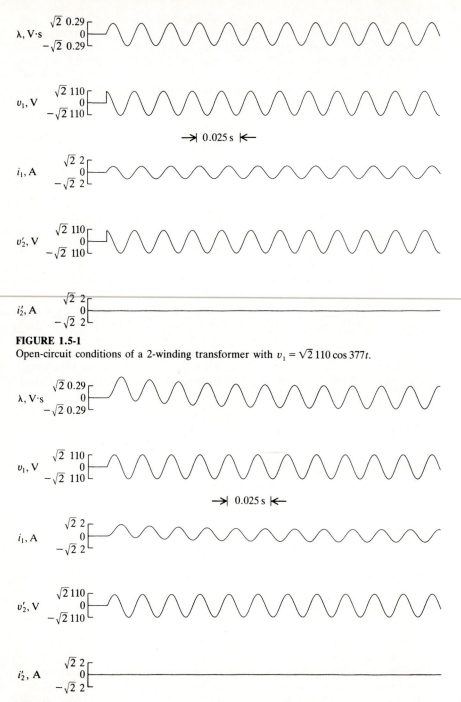

FIGURE 1.5-1
Open-circuit conditions of a 2-winding transformer with $v_1 = \sqrt{2}\,110\cos 377t$.

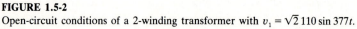

FIGURE 1.5-2
Open-circuit conditions of a 2-winding transformer with $v_1 = \sqrt{2}\,110\sin 377t$.

results in a much larger transient offset in i_1 than when $v_1 = \sqrt{2}\,110 \cos 377t$. (You are asked to show this in a problem at the end of the chapter.) Since $v_1 = \sqrt{2}\,110 \sin 377t$ causes a large transient offset, it makes it easier for us to identify the transient period. Therefore, we shall continue with v_1 as a sine wave. Although it is difficult to determine the time constant for the offset of the current i_1 (or λ) to decay to one-third of its original value, it is on the order of 50 ms. The calculated value of the no-load time constant is $\tau_{nl} = (L_{l1} + L_{m1})/r_1 = 46.2$ ms. Before leaving Figs. 1.5-1 and 1.5-2, note that, during steady-state conditions, I_1 lags V_1 by something close to 90° (86.7°, from Example 1C).

Let us now go to the short-circuit characteristics. The transient and steady-state response with $v_1 = \sqrt{2}\,110 \sin 377t$ and with $v_2' = 0$ are shown in Fig. 1.5-3. There are several things to note. From Fig. 1.5-3, it appears that the time constant associated with the decay of i_1 is small, less than 5 ms. Now let us look at the magnetizing flux linkage λ. We see that it is smaller in amplitude than in the no-load case. We would expect this since during short-circuit conditions $i_1 \cong -i_2'$; the mmf's of the two windings oppose, and the resulting flux in the transformer iron is less than for the no-load condition where $i_2' = 0$. Looking at this in another way, we realize that i_1 and $-i_2'$ will be essentially equal during short-circuit conditions whenever the impedance of the "mag-

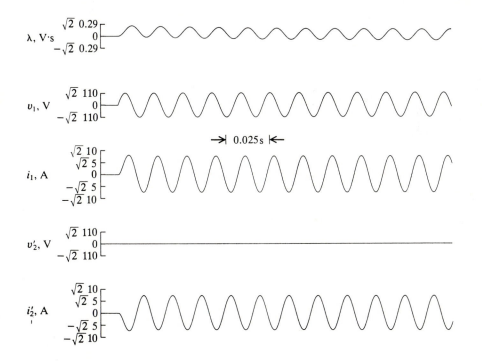

FIGURE 1.5-3
Short-circuit conditions of a 2-winding transformer with $v_1 = \sqrt{2}\,110 \sin 377t$.

netizing branch" ($j\omega_e L_{m1}$) is much larger (say 8 to 10 times larger) than $r_2' + j\omega_e L_{l2}'$. Here, $\omega_e = 377 \, \text{rad/s}$.

It is interesting to note that the decay of the magnetizing flux linkage λ is much slower than the apparent decay of the currents. As we mentioned, the time constant associated with i_1 is small; however, there is indeed a small difference between i_1 and $-i_2'$ and this small current (magnetizing current), which is actually a small part of i_1, must flow in the large inductance L_{m1}. Hence, the magnetizing current is associated with a longer time constant than the much larger component of the current i_1 which circulates through the series r_2' and L_{l2}'.

Let us take a brief look at the effects of saturation of the transformer iron. For this purpose we will assume that the λ versus ($i_1 + i_2'$) plot of the core of the transformer is that shown in Fig. 1.5-4. The slope of the straightline part of this plot is L_{m1}. The saturation characteristics shown in Fig. 1.5-4 were implemented on the computer following the method outlined in [3]. Since λ is small during short-circuit conditions (Fig. 1.5-3), saturation does not occur. However, it is a different situation when we talk about the open-circuit conditions shown in Fig. 1.5-5, which is the same as Fig. 1.5-2 with saturation taken into account. Here, we see that during steady-state open-circuit conditions, the current i_1, which is the total magnetizing current since i_2' is zero, is rich in third harmonic. What is happening? Well, we realize that, if there were

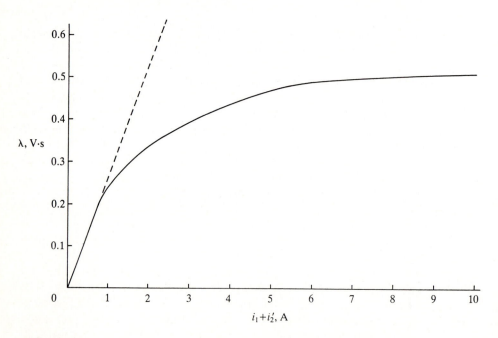

FIGURE 1.5-4
λ versus $i_1 + i_2'$.

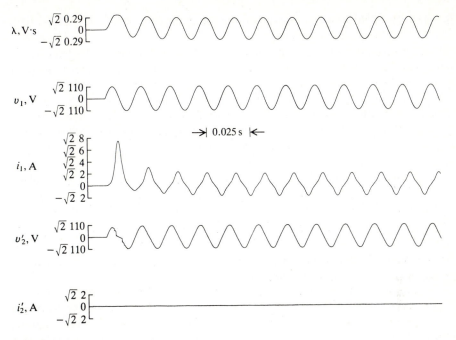

FIGURE 1.5-5
Same as Fig. 1.5-2 with saturation.

no r_1 and L_{l1}, then the time rate of change of λ must equal v_1, the applied voltage. In this case, the peak value of λ would be $\sqrt{2}\,110/377 = \sqrt{2}\,0.29$ V·s. We see from Fig. 1.5-4 that saturation must occur in order for the core to produce this peak value of λ. In the saturated region, a much larger increase in current per unit increase in λ is required than when the transformer is not saturated. Hence, a peaking of the magnetizing current occurs. Now, in real life there is r_1 and L_{l1} and, hence, a voltage drop will occur across each of these components. Thus, the magnitude of $d\lambda/dt$ is somewhat less than that of v_1; nevertheless, saturation must occur to produce the required steady-state peak value of λ, which is approximately $\sqrt{2}\,0.26$ V·s from Fig. 1.5-5.

There is one last item worthy of discussion. Recall that a relatively large transient offset in λ occurs when we apply a sine wave voltage for v_1. This large transient offset drives the core into saturation. Note λ during the first cycle in Fig. 1.5-5. Since the core is highly saturated, the magnetizing current necessary to produce the required λ is very large. In Fig. 1.5-5 we see that this current, which occurs upon "energizing" the transformer, is nearly three times the normal steady-state magnetizing current. In some transformers, this may be as high as 50 to 100 times the normal magnetizing current, and it may take several cycles before reaching steady-state conditions. For this reason, some transformers may "hum" loudly during energization as a result of forces created by the large inrush current. Also, note the waveform of v_2' during the first cycle of energization. The effects of saturation are reflected into the open-circuit voltage of winding 2. Since during saturation the change of the flux linkages is small, the open-circuit voltage will be small, as depicted in Fig. 1.5-5. However, these changes would probably not be as distinct in the open-circuit voltage of an actual transformer.

SP1.5-1. Use the plot of λ in Fig. 1.5-3 to approximate $|\tilde{I}_1 + \tilde{I}_2'|$. $[|\tilde{I}_1 + \tilde{I}_2'| \cong \frac{1}{2}\text{A}]$

SP1.5-2. Calculate, using reasonable approximations, the phase angle between the steady-state current $\tilde{I}_1$ and voltage $\tilde{V}_1$ for the conditions of Fig. 1.5-3. Check your answer from the plots. $[\tilde{V}_1$ leads $\tilde{I}_1$ by 42.8°]

SP1.5-3. Consider the transformer given in Example 1C. Assume $V_1 = \sqrt{2}\,110 \cos 1000t$, and a load is connected across winding 2. The impedance of this load referred to winding 1 is $21 + j5\,\Omega$. Calculate $\tilde{I}_2'$. Make valid approximations to reduce your work. $[\tilde{I}_2' \cong -2.4\underline{/-45°}]$

1.6 THREE-PHASE SYSTEMS AND TRANSFORMER CONNECTIONS

In our study of ac electromechanical motion devices, we will first concentrate on the two-phase versions and then discuss the modifications of this analysis necessary to treat three-phase devices. One wonders if, for this first study, it is really necessary to consider three-phase systems at all since the salient features of each device can be explained from a two-phase treatment. Moreover, three-phase systems have traditionally been associated with power systems since all electric power is transmitted from the point of generation to the cities by three-phase transmission systems. It would appear, from all of this, that we have made a case to let the three-phase system be the concern of the power systems engineer. However, the generator used in your automobile is likely to be a three-phase synchronous generator whose output voltages are rectified. Moreover, development has been underway to convert the small dc motors used in control applications to brushless dc motors. You should not be expected to know this at this time, but brushless dc motors are permanent-magnet motors with the stator windings often connected as a three-phase system. Therefore, it is worthwhile to take at least a brief look at three-phase systems since it is helpful to become aware of the meaning of line-to-neutral and line-to-line quantities. At the same time, consider in passing the connections which are not only common to power system transformers but also to the windings of some low-power ac drive motors used in position control systems, for example.

In a balanced two-phase ac system, the sinusoidal voltage and current of one phase are displaced 90° from the voltage and current of the second phase, and the amplitudes of the sinusoidal voltages (currents) are equal (Appendix C). In a balanced three-phase ac system, the sinusoidal voltage and current of one phase are displaced 120° from the voltage and current of the other two phases. When transformers are used in a two-phase system, they are arranged as two separate single-phase transformers without interconnection. However, in a three-phase system, the three phases are always interconnected and, although there are numerous variations, there are essentially only three basic transformer configurations. We shall touch on each of these three.

Three-phase transformers may be three separately constructed 2-winding single-phase transformers, as we have considered in the last two sections, or one 6-winding transformer wound on a common iron core. It is sufficient for our purposes to deal only with three separate single-phase transformers.

Wye-Wye Connection

The so-called wye-wye (Y-Y) connection is illustrated in Fig. 1.6-1. The transformers are identical; hence $r_{a1} = r_{b1} = r_{c1}$ and $r_{a2} = r_{b2} = r_{c2}$. Here, we

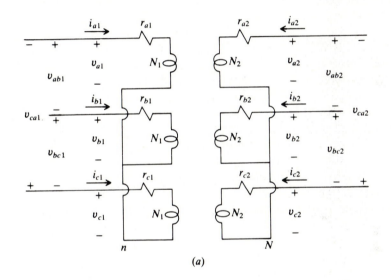

(a)

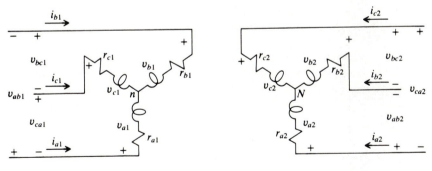

(b)

FIGURE 1.6-1
A Y-Y three-phase transformer connection.

have changed notation to accommodate the three-phase variables. Rather than denoting the windings as winding 1 and winding 2, as in the previous sections, we are using the notation of $a1$ and $a2$ winding, respectively, for the a-phase transformer, $b1$ and $b2$ winding for the b-phase transformer, and $c1$ and $c2$ winding for the c-phase transformer. In Fig. 1.6-1a, the three single-phase transformers are arranged to show clearly the transformer coupling in each phase; however, the circuit diagram shown in Fig. 1.6-1b is generally used since it most readily depicts the type of connection being used. Although the Y-Y connection is not used too often in power systems, it is the connection used in three-phase induction machines, where windings 1 become the windings of the stationary member and windings 2 become the windings of the rotating member. Nevertheless, the Y-Y connection is a good place to start.

The first thing that one tends to notice about Fig. 1.6-1 is that only three conductors enter and three leave the transformers. This does not seem correct. For a single-phase system we would need two conductors and for a two-phase system we would need four. Why are only three required for a three-phase system? We shall talk more about this later but from Appendix C we know that a balanced three-phase set of currents consists of, by definition, equal-amplitude sinusoidal currents displaced by 120°. The sum of these currents is zero; hence, if the system is balanced, $i_{a1} + i_{b1} + i_{c1} = 0$. Thus, current entering one of the 1-windings, for example, returns through the other 1-windings. A fourth wire is not needed for balanced conditions. We are now starting to see the economic advantage of transmitting electric power by a three-phase system. In Chap. 2 we will see another advantage of two- or three-phase electromechanical devices over their single-phase little sisters. The instantaneous power or torque in a single-phase system pulsates about an average value; however, in a multiphase system, the instantaneous power or torque is constant for balanced steady-state conditions. Please do not feel that you should see all these things at this time. We shall go over this constant-power situation in more detail when we discuss two-phase electromechanical motion devices.

In a Y-Y connection, the assigned negative-potential sides of the windings are all connected to form what is called the *neutral point*; n in case of windings 1 in Fig. 1.6-1 and N for windings 2. Either or both neutral points may be grounded or left to "float." The voltages across each transformer winding, v_{a1}, v_{a2}, v_{b1}, etc., are referred to as line-to-neutral voltages while the voltages between two of the three terminals of windings 1 (windings 2) are the line-to-line voltages. The individual transformer currents i_{a1}, i_{a2}, i_{b1}, etc. are the line currents. We can relate the line-to-neutral and line-to-line voltages as

$$v_{ab1} = v_{a1} - v_{b1} \qquad (1.6\text{-}1)$$

$$v_{bc1} = v_{b1} - v_{c1} \qquad (1.6\text{-}2)$$

$$v_{ca1} = v_{c1} - v_{a1} \qquad (1.6\text{-}3)$$

Similar expressions may be written for windings 2.

If the system is balanced, then we can express the steady-state voltages of windings 1 for an *abc* sequence (Appendix C) as

$$V_{a1} = \sqrt{2} V_{s1} \cos \omega_e t \tag{1.6-4}$$

$$V_{b1} = \sqrt{2} V_{s1} \cos \left(\omega_e t - \frac{2\pi}{3} \right) \tag{1.6-5}$$

$$V_{c1} = \sqrt{2} V_{s1} \cos \left(\omega_e t + \frac{2\pi}{3} \right) \tag{1.6-6}$$

where the capital letters are used to denote steady-state conditions. The terminology "*abc* sequence" means that, for balanced operation, phase *a* leads phase *b* by 120° and phase *b* leads phase *c* by 120°. In phasor form,

$$\tilde{V}_{a1} = V_{s1} \underline{/0°} \tag{1.6-7}$$

$$\tilde{V}_{b1} = V_{s1} \underline{/-120°} \tag{1.6-8}$$

$$\tilde{V}_{c1} = V_{s1} \underline{/120°} \tag{1.6-9}$$

The line-to-line voltage may be expressed as

$$\tilde{V}_{ab} = V_{s1} \underline{/0°} - V_{s1} \underline{/-120°}$$
$$= \sqrt{3} V_{s1} \underline{/30°} \tag{1.6-10}$$

$$\tilde{V}_{bc} = V_{s1} \underline{/-120°} - V_{s1} \underline{/120°}$$
$$= \sqrt{3} V_{s1} \underline{/-90°} \tag{1.6-11}$$

$$\tilde{V}_{ca} = V_{s1} \underline{/120°} - V_{s1} \underline{/0°}$$
$$= \sqrt{3} V_{s1} \underline{/150°} \tag{1.6-12}$$

Hence, the line-to-line voltages form a balanced three-phase set which are $\sqrt{3}$ times the magnitude of the line-to-neutral voltages and, for an *abc* sequence, advanced by 30° from them.

For balanced steady-state conditions, the operation of each phase may be considered separately. In fact, we need to consider only one phase since, once we have determined the variables associated with one of the phase transformers, we can express the other phase variables by shifting the phase by either 120° or −120°.

Delta-Delta Connection

The delta-delta (Δ-Δ) connection is illustrated in Fig. 1.6-2. In this type of connection, the line-to-line voltages are the voltages across the individual transformers, that is, $v_{a1} = v_{ab1}$, $v_{a2} = v_{ab2}$, $v_{b1} = v_{bc1}$, etc. There is no neutral connection. The line currents are made up of the currents from two transformers. In particular,

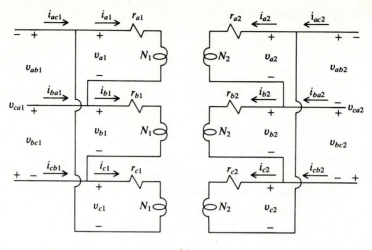

(a)

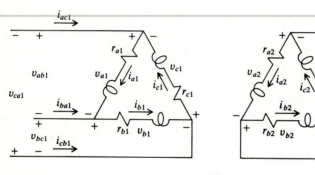

(b)

FIGURE 1.6-2
A Δ-Δ three-phase transformer connection.

$$i_{ac1} = i_{a1} - i_{c1} \qquad (1.6\text{-}13)$$

$$i_{ba1} = i_{b1} - i_{a1} \qquad (1.6\text{-}14)$$

$$i_{cb1} = i_{c1} - i_{b1} \qquad (1.6\text{-}15)$$

If, for example, the transformer currents form a balanced *abc* sequence, then

$$\tilde{I}_{a1} = I_{s1} \underline{/0°} \qquad (1.6\text{-}16)$$

$$\tilde{I}_{b1} = I_{s1} \underline{/-120°} \qquad (1.6\text{-}17)$$

$$\tilde{I}_{c1} = I_{s1} \underline{/120°} \qquad (1.6\text{-}18)$$

The line currents become

$$\tilde{I}_{ac1} = I_{s1}\underline{/0°} - I_{s1}\underline{/120°}$$
$$= \sqrt{3}I_{s1}\underline{/-30°} \qquad (1.6\text{-}19)$$

Thus, $\tilde{I}_{ac1}$ is $\sqrt{3}$ times the amplitude of $\tilde{I}_{a1}$ and shifted 30° back in phase from it. Similarly, $\tilde{I}_{ba1}$ is shifted back 30° from $\tilde{I}_{b1}$ and $\tilde{I}_{cb1}$ back 30° from $\tilde{I}_{c1}$.

Wye-Delta or Delta-Wye Connection

The final three-phase transformer connection which we will consider is the wye-delta (Y-Δ) or delta-wye (Δ-Y) connection. We read transformer connections from the left to right just as we read a book. Hence, it is a Y-Δ or Δ-Y

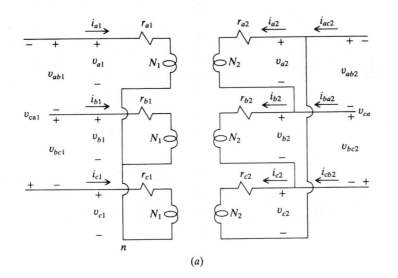

(a)

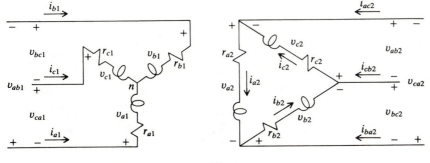

(b)

FIGURE 1.6-3
A Y-Δ three-phase transformer connection.

whichever connection is on the left. Actually, the Δ-Y is generally used to "step up" the voltage whereas the Y-Δ is used for "stepping down" the voltage. The Y-Δ connection is illustrated in Fig. 1.6-3 (p. 29).

Ideal Transformers

Since in electromechanical motion devices the magnetizing flux must traverse the air gap, the leakage inductances and the magnetizing current must be

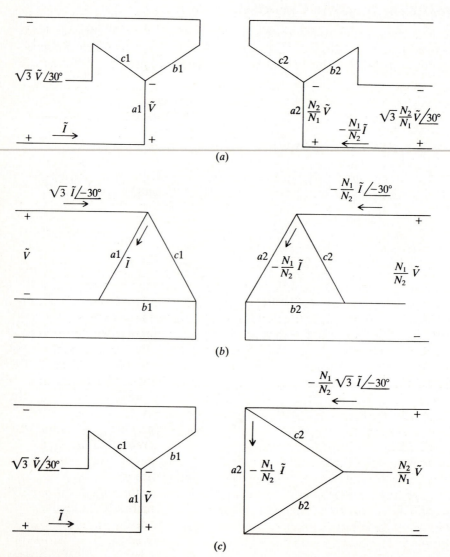

FIGURE 1.6-4
Three-phase ideal transformers for *abc* sequence. (*a*) Y-Y connection; (*b*) Δ-Δ connection; (*c*) Y-Δ connection.

considered when analyzing these devices. In other words, we cannot consider the coupled circuits as an ideal transformer. However, in the analysis of power systems, the system transformers are often assumed to be ideal. In this case, as we have mentioned previously, the winding resistance, leakage inductance, and magnetizing current are all neglected and the voltages (currents) of the transformer windings differ only by the turns ratio. The relations between 1- and 2-winding variables for the Y-Y, Δ-Δ, and Y-Δ connections of ideal transformers are shown in Fig. 1.6-4.

SP1.6-1. Write the expression for the steady-state line-to-line voltage V_{ab} of a Y-Y connected three-phase system for a balanced set of (a) abc-sequence voltages and (b) acb-sequence voltages. Assume that the steady-state line-to-neutral voltage of phase a is $V_a = \sqrt{2}V_s \cos(\omega_e t - 30°)$. [(a) $V_{ab} = \sqrt{6}V_s \cos \omega_e t$; (b) $V_{ab} = \sqrt{6}V_s \cos(\omega_e t - 60°)$]

SP1.6-2. Three single-phase transformers are connected Y-Y. The parameters of each transformer are those given in Example 1C. Assume $V_{a1} = \sqrt{2}\, 110 \cos 1000t$ and a Y-connected load is connected to windings 2. The impedance of each phase of this load referred to windings 1 is $21 + j5\ \Omega$. Calculate $\tilde{I}'_{a2}$. Make valid approximations to reduce your work. [Answer same as SP1.5-3]

SP1.6-3. Refer to Fig. 1.6-2 and (1.6-19). Calculate $\tilde{I}_{ac1}$ for an acb sequence. Assume $\tilde{I}_{a1} = I_{s1}\underline{/0°}$. [$\tilde{I}_{ac1} = \sqrt{3}I_{s1}\underline{/30°}$]

SP1.6-4. In Fig. 1.6-1, $V_{ab1} = \sqrt{6} \cos \omega_e t$ and $I_{a1} = \sqrt{2} \cos(\omega_e t - 30°)$. Assume balanced operation. Calculate the total power flow into the Y-Y transformer. [3 W]

1.7 MAGNETIC SYSTEMS WITH MECHANICAL MOTION

In Chap. 2, relationships are derived for determining the electromagnetic force or torque established in electromechanical systems. Once this development is completed, three examples of elementary electromechanical systems are considered. It is convenient to introduce these three systems here for the purpose of establishing the voltage equations and expressions for the self- and mutual inductances, thereby setting the stage for the analysis to follow in Chap. 2. The first of these electromechanical systems is an elementary version of an electromagnet. It consists of a magnetic core, part of which is movable. The electric system exerts an electromagnetic force upon this movable member, thereby moving it relative to the stationary member. We shall analyze this device, and in Chap. 2 we shall observe its operating characteristics by computer traces. The second system is a rotational device commonly referred to as a reluctance machine. The single-phase reluctance motor, which we shall treat here, is similar in many respects to the electric motors found in clocks and in some types of turntables. Moreover, a large number of stepper motors operate on the reluctance-torque principle. The third device is also a rotational device which has two windings, one on the stationary member and one on the rotational member. This device, although somewhat impracticable, illustrates the concept of windings or magnetic systems in relative motion.

Elementary Electromagnet

An elementary electromagnet which we will consider is shown in Fig. 1.7-1. This system consists of a stationary core with a winding of N turns and a block of magnetic material which is free to slide relative to the stationary member. It is shown in more detail in Chap. 2, wherein a spring, a damper, and an external force are associated with the movable member. We do not need to consider that level of detail here; instead, we will assume that the movable member is at a distance x from the stationary member which may be a function of time, that is $x = x(t)$.

The voltage equation that describes the electric system is

$$v = ri + \frac{d\lambda}{dt} \tag{1.7-1}$$

where the flux linkages are expressed as

$$\lambda = N\Phi \tag{1.7-2}$$

The flux may be written as

$$\Phi = \Phi_l + \Phi_m \tag{1.7-3}$$

where Φ_l is the leakage flux and Φ_m is the magnetizing flux which is common to both the stationary and movable members. If the magnetic system is considered to be linear (saturation neglected), then, as in the case of the stationary coupled circuits, we can express the fluxes in terms of reluctances. That is,

$$\Phi_l = \frac{Ni}{\mathcal{R}_l} \tag{1.7-4}$$

$$\Phi_m = \frac{Ni}{\mathcal{R}_m} \tag{1.7-5}$$

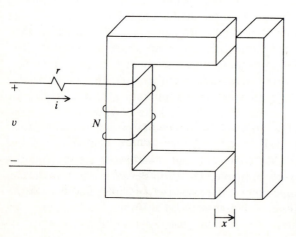

FIGURE 1.7-1
Elementary electromagnet.

where $\mathcal{R}_l$ and $\mathcal{R}_m$ are the reluctances of the leakage and magnetizing paths, respectively.

The flux linkages may now be written as

$$\lambda = \left(\frac{N^2}{\mathcal{R}_l} + \frac{N^2}{\mathcal{R}_m} \right) i \tag{1.7-6}$$

where the leakage inductance is

$$L_l = \frac{N^2}{\mathcal{R}_l} \tag{1.7-7}$$

and the magnetizing inductance is

$$L_m = \frac{N^2}{\mathcal{R}_m} \tag{1.7-8}$$

The reluctance of the magnetizing path is

$$\mathcal{R}_m = \mathcal{R}_i + 2\mathcal{R}_g \tag{1.7-9}$$

where $\mathcal{R}_i$ is the total reluctance of the magnetic material of the stationary and movable members and $\mathcal{R}_g$ is the reluctance of one of the air gaps. If the cross-sectional area of the stationary and movable members is assumed to be equal and of the same material, the reluctances may be expressed as

$$\mathcal{R}_i = \frac{l_i}{\mu_{ri} \mu_0 A_i} \tag{1.7-10}$$

$$\mathcal{R}_g = \frac{x}{\mu_0 A_g} \tag{1.7-11}$$

We will assume that $A_g = A_i$. Even though, as we have mentioned previously, this may be somewhat of an oversimplification, it is sufficient for our purposes. Hence, $\mathcal{R}_m$ may be written as

$$\mathcal{R}_m = \frac{1}{\mu_0 A_i} \left(\frac{l_i}{\mu_{ri}} + 2x \right) \tag{1.7-12}$$

The magnetizing inductance now becomes

$$L_m = \frac{N^2}{(1/\mu_0 A_i)(l_i/\mu_{ri} + 2x)} \tag{1.7-13}$$

In this analysis, the leakage inductance is assumed to be constant. The magnetizing inductance is clearly a function of displacement. That is, $x = x(t)$ and $L_m = L_m(x)$. Heretofore, when dealing with linear magnetic circuits wherein mechanical motion is not present as in the case of a transformer, the change of flux linkages with respect to time was simply $L(di/dt)$. This is not the case here. When the inductance is a function of $x(t)$, then

$$\lambda(i, x) = L(x)i = [L_l + L_m(x)]i \tag{1.7-14}$$

and
$$\frac{d\lambda(i, x)}{dt} = \frac{\partial \lambda}{\partial i} \frac{di}{dt} + \frac{\partial \lambda}{\partial x} \frac{dx}{dt} \qquad (1.7\text{-}15)$$

With (1.7-15) in mind, we see that the voltage equation (1.7-1), becomes

$$v = ri + [L_l + L_m(x)] \frac{di}{dt} + i \frac{dL_m(x)}{dx} \frac{dx}{dt} \qquad (1.7\text{-}16)$$

Equation (1.7-16) is a nonlinear differential equation owing to the last two terms on the right-hand side.

Let us go back to the magnetizing inductance L_m, as given by (1.7-13), for just a moment. In preparation for our work in Chap. 2, let us write (1.7-13) as

$$L_m(x) = \frac{k}{k_0 + x} \qquad (1.7\text{-}17)$$

where
$$k = \frac{N^2 \mu_0 A_i}{2} \qquad (1.7\text{-}18)$$

$$k_0 = \frac{l_i}{2\mu_{ri}} \qquad (1.7\text{-}19)$$

When $x = 0$, $L_m(x)$ is determined by the reluctance of the magnetic material. That is, for $x = 0$,

$$L_m(0) = \frac{k}{k_0} = \frac{N^2 \mu_0 \mu_{ri} A_i}{l_i} \qquad (1.7\text{-}20)$$

Depending upon the parameters of the magnetic material, $L_m(x)$ may be adequately predicted by

$$L_m(x) = \frac{k}{x} \qquad \text{for } x > 0 \qquad (1.7\text{-}21)$$

We will use this approximation in Chap. 2.

Elementary Reluctance Machine

An elementary reluctance machine is shown in Fig. 1.7-2. It consists of a stationary core with a winding of N turns and a movable member which rotates at an angular displacement and angular velocity of θ_r and ω_r, respectively. The displacement is defined as

$$\theta_r = \int_0^t \omega_r(\xi) \, d\xi + \theta_r(0) \qquad (1.7\text{-}22)$$

where ξ is a dummy variable of integration.

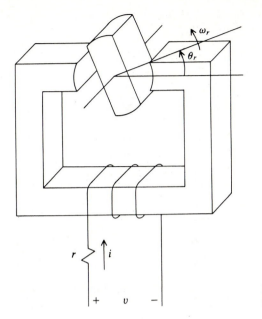

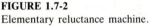

FIGURE 1.7-2
Elementary reluctance machine.

The voltage equation is of the form given by (1.7-1). Similarly, the flux may be divided into a leakage and magnetizing flux, as given by (1.7-3). It is convenient to express the flux linkages as

$$\lambda = (L_l + L_m)i \qquad (1.7\text{-}23)$$

where L_l is the leakage inductance and L_m is the magnetizing inductance. The leakage inductance is essentially constant, independent of θ_r; however, the magnetizing inductance is a periodic function of θ_r. That is, $L_m = L_m(\theta_r)$. In particular, with $\theta_r = 0$,

$$L_m(0) = \frac{N^2}{\mathcal{R}_m(0)} \qquad (1.7\text{-}24)$$

With $\theta_r = 0$, the reluctance of the magnetizing path $\mathcal{R}_m$ is maximum due to the large air gap when the rotor is in the vertical (unaligned) position. Hence, L_m is a minimum in this position. Note that this same situation occurs not only at $\theta_r = 0$ but also when $\theta_r = \pi, 2\pi$, etc.

Now, with $\theta_r = \pi/2$,

$$L_m(\tfrac{1}{2}\pi) = \frac{N^2}{\mathcal{R}_m(\tfrac{1}{2}\pi)} \qquad (1.7\text{-}25)$$

Here, $\mathcal{R}_m$ is a minimum and thus L_m is a maximum. This same situation occurs at $\theta_r = \tfrac{3}{2}\pi, \tfrac{5}{2}\pi$, etc. Hence, the magnetizing inductance varies between maximum and minimum positive values twice per revolution of the rotating member (rotor). Let us make it easy for ourselves and assume that this variation may be adequately approximated by a sinusoidal function. In particu-

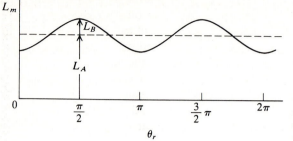

lar, let $L_m(\theta_r)$ be expressed as

$$L_m(\theta_r) = L_A - L_B \cos 2\theta_r \tag{1.7-26}$$

whereupon

$$L_m(0) = L_A - L_B \tag{1.7-27}$$

$$L_m(\tfrac{1}{2}\pi) = L_A + L_B \tag{1.7-28}$$

and $L_A > L_B$. The average value is L_A, as illustrated in Fig. 1.7-3. The self-inductance may now be expressed as

$$\boxed{\begin{aligned} L(\theta_r) &= L_l + L_m(\theta_r) \\ &= L_l + L_A - L_B \cos 2\theta_r \end{aligned}} \tag{1.7-29}$$

The voltage equation is of the form given by (1.7-16) with x replaced by θ_r.

Windings in Relative Motion

The rotational device shown in Fig. 1.7-4 will be used to illustrate windings in relative motion. This device consists of two windings each containing several

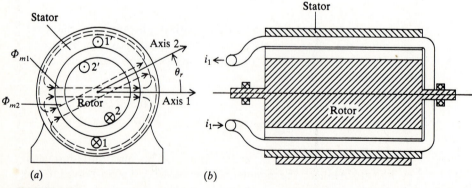

(a)   (b)

FIGURE 1.7-4
Elementary rotational electromechanical device. (a) End view; (b) cross-sectional view.

turns of a conductor. Winding 1 has N_1 turns and it is on the stationary member (stator); winding 2 has N_2 turns and it is on the rotating member (rotor). The $\otimes$ indicates that the assumed direction of positive current flow in the conductors is into the paper while $\odot$ indicates positive current flow in the conductors is out of the paper. In a practical device, the turns of a winding are distributed over an arc (often 30 to 60°) of the stator and rotor. However, in this introductory consideration, it is sufficient to assume that the turns are concentrated in one position, as shown in Fig. 1.7-4. Also, the length of the air gap between the stator and rotor is shown exaggerated relative to the inside diameter of the stator.

The voltage equations may be written as

$$v_1 = r_1 i_1 + \frac{d\lambda_1}{dt} \tag{1.7-30}$$

$$v_2 = r_2 i_2 + \frac{d\lambda_2}{dt} \tag{1.7-31}$$

where r_1 and r_2 are the resistances of winding 1 and winding 2, respectively. The magnetic system is assumed linear; therefore the flux linkages may be expressed as

$$\lambda_1 = L_{11} i_1 + L_{12} i_2 \tag{1.7-32}$$

$$\lambda_2 = L_{21} i_1 + L_{22} i_2 \tag{1.7-33}$$

The self-inductances L_{11} and L_{22} are constants and may be expressed in terms of leakage and magnetizing inductances as

$$L_{11} = L_{l1} + L_{m1}$$
$$= \frac{N_1^2}{\mathcal{R}_{l1}} + \frac{N_1^2}{\mathcal{R}_m} \tag{1.7-34}$$

$$L_{22} = L_{l2} + L_{m2}$$
$$= \frac{N_2^2}{\mathcal{R}_{l2}} + \frac{N_2^2}{\mathcal{R}_m} \tag{1.7-35}$$

where $\mathcal{R}_m$ is the reluctance of the complete magnetic path of Φ_{m1} and Φ_{m2}, which is through the rotor and stator iron and twice across the air gap. Clearly, it is the same for the magnetic system established by either winding 1 or winding 2.

Take a moment to note the designation of axis 1 and axis 2 in Fig. 1.7-4. These axes denote the positive direction of the respective magnetic systems with the assumed positive direction of current flow in the windings (right-hand rule). Now let us consider L_{12}. (Is it clear that $L_{12} = L_{21}$?) When θ_r, which is defined by (1.7-22), is zero, then the coupling between windings 1 and 2 is maximum. In particular, with $\theta_r = 0$, the magnetic system of winding 1 aids that of winding 2 with positive currents assumed. Hence the mutual inductance is positive and

$$L_{12}(0) = \frac{N_1 N_2}{\mathcal{R}_m} \qquad (1.7\text{-}36)$$

When $\theta_r = \pi/2$, the windings are orthogonal. The mutual coupling is zero. Hence,

$$L_{12}(\tfrac{1}{2}\pi) = 0 \qquad (1.7\text{-}37)$$

Again let us make it as simple as possible by assuming that the mutual inductance may be adequately predicted by

$$\boxed{L_{12}(\theta_r) = L_{sr} \cos \theta_r} \qquad (1.7\text{-}38)$$

where L_{sr} is the amplitude of the sinusoidal mutual inductance between the stator and rotor windings as given by (1.7-36).

In writing the voltage equations from (1.7-30) and (1.7-31) the total derivative of the flux linkages is required. This is accomplished by taking the partial derivative of both λ_1 and λ_2 with respect to i_1, i_2, and θ_r. Writing these voltage equations is a problem at the end of the chapter.

SP1.7-1. Let $L_m(x) = k/x$, $i = t$, and $x = t$. Express $d[L_m(x)i]/dt$. [Zero]

SP1.7-2. Express $L(\theta_r)$ of the elementary reluctance machine if minimum reluctance occurs at $\theta_r = 0$. $[L(\theta_r) = L_l + L_A + L_B \cos 2\theta_r]$

SP1.7-3. Express L_{11}, L_{22}, and L_{12} if positive i_2 is reversed from that shown in Fig. 1.7-4. $[L_{11}$ and L_{22} are unchanged; $L_{12} = -L_{sr} \cos \theta_r]$

SP1.7-4. Consider Fig. 1.7-4. $I_1 = 1$ A, $L_{sr} = 0.1$ H, $\omega_r = 100$ rad/s, $\theta_r(0) = 0$, and winding 2 is open-circuited. Express V_2. $[V_2 = -10 \sin 100t]$

1.8   RECAPPING

We will analyze electromechanical motion devices from the coupled-circuits viewpoint. Although the coupled windings of many electromechanical devices are in relative motion, the equivalent circuit of stationary coupled windings (the transformer) is the beginning of the equivalent circuits which we will develop for these devices in later chapters. We will find the concept of referring variables from one winding to the other very useful as we proceed.

The first step in the analysis of electromechanical motion devices of the electromagnetic type is to express the voltage and flux linkage equations in terms of self- and mutual inductances. We will not consider saturation in our analysis; instead we will restrict our work to linear magnetic systems and leave the analytical treatment of saturation to a more advanced study of these devices. In this chapter, we learned that electromagnetic, electromechanical motion devices are characterized by self- or mutual inductances which vary with displacement of the movable member.

In the next chapter, we will first develop an analytical means of determining the electromagnetic force or torque in electromechanical motion devices. Once we have accomplished this, we will be able to express the electromagnetic force in the elementary electromagnet and the electromagnetic torque in the elementary rotational devices which we have just considered.

1.9 REFERENCES

1. D. M. Triezenberg, *Electric Power Systems*, Classnotes, Purdue University, 1978.
2. G. R. Slemon and A. Straughen, *Electric Machines*, Addison-Wesley Publishing Company, Reading, Mass., 1980.
3. P. C. Krause, *Analysis of Electric Machinery*, McGraw-Hill Book Company, New York, 1986.

1.10 PROBLEMS

In all Problems sections, the more lengthy or involved problems are denoted by an asterisk.

1. Consider the magnetic system shown in Fig. 1.2-1. Let $\mu_r = 1500$, $N = 100$ turns, and $i = 2$ A. The cross section of the iron is square, each side 4 cm in length. The air gap is 4 mm in length. The mean length of the iron is 200 times the air gap length. Neglect leakage flux and assume $A_i = A_g$. Calculate the flux.
2. Repeat Example 1A with a second air gap of 2 mm in length cut midway between c and d. Neglect leakage flux and assume $A_i = A_g$.
3. An iron-core transformer which has two windings is shown in Fig. 1.10-1. $N_1 = 50$ turns, $N_2 = 100$ turns, and $\mu_r = 4000$. Calculate L_{12}, L_{m1}, and L_{m2}.

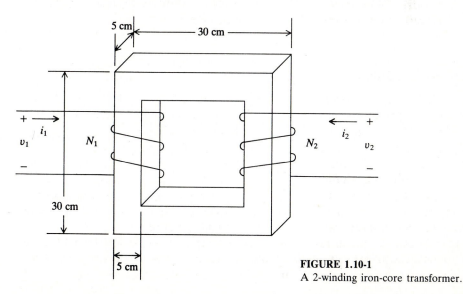

FIGURE 1.10-1
A 2-winding iron-core transformer.

4. An iron "doughnut" (toroid) with two coils is shown in Fig. 1.10-2. $N_1 = 100$ turns and $N_2 = 200$ turns, $\mu_r = 10^4/4\pi$ H/m. Calculate L_{12}.

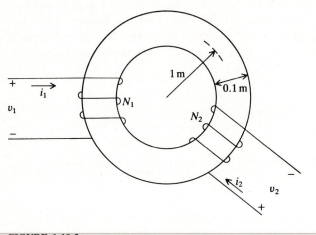

FIGURE 1.10-2
A 2-winding iron-core toroid (not to scale).

5. An air gap is cut through the left leg of the magnetic system shown in Fig. 1B-1 so that the associated reluctance is $10\,\mathcal{R}_y$ rather than $\mathcal{R}_y$. Express L_{12} and L_{21} in terms of N_1, N_2, $\mathcal{R}_x$, and $\mathcal{R}_y$.

6. Two coupled coils have the following parameters: $r_1 = 10\,\Omega$, $L_{l1} = 0.1\,L_{11}$, $r_2 = 2.5\,\Omega$, $L_{l2} = 0.1\,L_{22}$, $L_{11} = 100$ mH, $N_1 = 100$ turns, $L_{22} = 25$ mH, $N_2 = 50$ turns. Develop an equivalent T circuit with (a) winding 1 as the reference winding and (b) winding 2 as reference winding.

7. Assume that the direction of positive current is reversed in winding 2 of Fig. 1.4-1. Express (a) L_{12} in terms of N_1, N_2, and $\mathcal{R}_m$; (b) λ_1 and λ_2 in the form of (1.4-22) and (1.4-23); (c) λ_1 and λ_2' in the form of (1.4-31) and (1.4-32); and (d) v_1 and v_2' in the form of (1.4-37).

8. The parameters of a transformer are: $r_1 = r_2' = 10\,\Omega$, $L_{m1} = 300$ mH, $L_{l1} = L_{l2}' = 30$ mH. A 10-V peak-to-peak 30-Hz sinusoidal voltage is applied to winding 1. Winding 2 is short-circuited. Assume $i_1 = -i_2'$. Calculate the phasor $\tilde{I}_1$ with $\tilde{V}_1$ at zero degrees.

9. A transformer with two windings has the following parameters: $r_1 = r_2 = 1\,\Omega$, $L_{m1} = 1$ H, $L_{l1} = L_{l2} = 0.01$ H, $N_1 = N_2$. A 2-Ω load resistance R_L is connected across winding 2. $V_1 = 2\cos 400t$. (a) Calculate $\tilde{I}_1$. (b) Express I_1.

10. A transformer with two windings has the following parameters: $r_1 = 1\,\Omega$, $L_{l1} = 0.01$ H, $L_{m1} = 0.2$ H, $N_2 = 2N_1$, $r_2 = 2\,\Omega$, $L_{l2} = 0.04$ H, $L_{m2} = 0.8$ H. A 4-Ω resistance R_L is connected across the terminals of winding 2 and a voltage $V_1 = \sqrt{2}\,2\cos 400t$ is applied to winding 1. Calculate and draw the phasor diagram showing $\tilde{V}_1$, $\tilde{I}_1$, $\tilde{V}_2'$, and $\tilde{I}_2'$. Neglect the magnetizing current.

*11. Consider the parameters of the transformer given in Example *1C. Calculate the input impedance measured from winding 1 with winding 2 short-circuited for (a) a dc source, (*b*) a 10-Hz source, and (*c*) a 400-Hz source. In each case, determine the input impedance first with the magnetizing current included and then with it neglected. The magnetizing current cannot be neglected as the frequency approaches zero. Why?

*12. Analytically obtain the expression for i_1 in Figs. 1.5-1 and 1.5-2.

*13. If, in Figs. 1.5-1 and 1.5-2, the resistance r_1 is zero, express i_1 for $t \geq 0$.

*14. Determine the phase of v_1 in Fig. 1.5-3 in order to obtain the maximum offset in i_1. Neglect the magnetizing current.

15. The impedance of each phase of a Δ-connected load is Z_Δ. Show that the equivalent phase impedance of a Y-connected load is $\frac{1}{3} Z_\Delta$.

16. For the elementary electromagnet shown in Fig. 1.7-1, assume that the cross-sectional area of the stationary and movable member is the same and $A_i = A_g = 4 \, cm^2$. Assume $l_i = 20 \, cm$, $N = 500$, and $\mu_{ri} = 1000$. Express $L_m(x)$ given by (1.7-17) and the approximation for $x > 0$ given by (1.7-21). Determine the minimum value of x when this approximate expression for $L_m(x)$ is less than 1.1 the value given by (1.7-17).

*17. Express the voltage v of the elementary electromagnet given by (1.7-16) for $L_m(x)$ given by (1.7-17), $i = \sqrt{2} I_s \cos \omega_e t$ and $x = t$. Approximate v when t is large.

18. Express the voltage equation for the elementary reluctance machine shown in Fig. 1.7-2. Use (1.7-29) for $L(\theta_r)$.

19. Write the voltage equations for the coils in relative motion shown in Fig. 1.7-4. Use L_{11}, L_{22}, and L_{12} as expressed by (1.7-38).

CHAPTER

2

PRINCIPLES OF ELECTROMECHANICAL ENERGY CONVERSION

2.1 INTRODUCTION

The theory of electromechanical energy conversion is the cornerstone for the analysis of electromechanical motion devices. This theory allows us to express the electromagnetic force or torque in terms of device variables such as the currents and the displacement of the mechanical system. Since numerous types of electromechanical devices are used in motion systems, it is desirable to establish methods of analysis which may be applied to a variety of electromechanical devices rather than just to electric machines. Therefore, the theory of electromechanical energy conversion is set forth in considerable detail for the purpose of providing a background sufficient to analyze electromechanical systems other than just those treated in this text. The first part of this chapter is devoted to establishing analytically the relationships which can be used to express the electromagnetic force or torque. Although one may prefer to perform a separate derivation for each device because it is instructive to do so, a general set of formulas are given in tabular form which are applicable to all electromechanical systems with a single mechanical input.

Once the theory of electromechanical energy conversion is established, a detailed analysis of the elementary electromagnet, which was introduced in Chap. 1, is performed with computer traces included to demonstrate its dynamic performance to changes in the applied voltage and the external mechanical force. In the final sections, the expressions for the electromagnetic torque are developed for the elementary single-phase reluctance machine and for windings in relative motion. Brief discussions are given of several steady-state modes of operation of these devices, which help to illustrate, in an elementary form, the positioning of stepper motors and the operation of a clock motor.

The material presented in this chapter is sufficient preparation to study dc machines, covered in Chap. 3. However, Chap. 3 is not necessary for the analysis of electromechanical motion devices of the induction and synchronous types which begins with Chap. 4.

2.2 ENERGY BALANCE RELATIONSHIPS

Electromechanical systems comprise an electric system, a mechanical system, and a means whereby the electric and mechanical systems can interact. Interaction can take place through any and all electromagnetic and electrostatic fields which are common to both systems, and energy is transferred from one system to the other as a result of this interaction. Both electrostatic and electromagnetic coupling fields may exist simultaneously and the electromechanical system may have any number of electric and mechanical subsystems. However, before considering an involved system, it is helpful to analyze the electromechanical system in a simplified form. An electromechanical system with one electric subsystem, one mechanical subsystem, and with one coupling field is depicted in Fig. 2.2-1. Electromagnetic radiation is neglected, and it is assumed that the electric system operates at a frequency sufficiently low so that the electric system may be considered as a lumped-parameter system.

Heat loss will occur in the mechanical system due to friction, and the electric system will dissipate heat due to the resistance of the current-carrying conductors. Eddy current and hysteresis losses occur in the ferromagnetic materials whereas dielectric losses occur in all electric fields. If W_E is the total energy supplied by the electric source and W_M the total energy supplied by the mechanical source, then the energy distribution could be expressed as

$$W_E = W_e + W_{eL} + W_{eS} \qquad (2.2\text{-}1)$$

$$W_M = W_m + W_{mL} + W_{mS} \qquad (2.2\text{-}2)$$

In (2.2-1), W_{eS} is the energy stored in the electric or magnetic fields which are

FIGURE 2.2-1
Block diagram of elementary electromechanical system.

not coupled with the mechanical system. The energy W_{eL} is the heat loss associated with the electric system excluding the coupling field losses. This loss occurs due to the resistance of the current-carrying conductors as well as the energy dissipated in the form of heat owing to hysteresis, eddy currents, and dielectric losses external to the coupling field. The energy W_e is the energy transferred to the coupling field by the electric system. The energies common to the mechanical system may be defined in a similar manner. In (2.2-2), W_{mS} is the energy stored in the moving member and compliances of the mechanical system, W_{mL} is the energy loss of the mechanical system in the form of heat, and W_m is the energy transferred to the coupling field. It is important to note that, with the convention adopted, the energy supplied by either source is considered positive. Therefore, W_E (W_M) is negative when energy is supplied to the electric source (mechanical source).

If W_F is defined as the total energy transferred to the coupling field, then

$$W_F = W_f + W_{fL} \qquad (2.2\text{-}3)$$

where W_f is energy stored in the coupling field and W_{fL} is the energy dissipated in the form of heat due to losses within the coupling field (eddy current, hysteresis, or dielectric losses). The electromechanical system must obey the law of conservation of energy, thus,

$$W_f + W_{fL} = (W_E - W_{eL} - W_{eS}) + (W_M - W_{mL} - W_{mS}) \qquad (2.2\text{-}4)$$

which may be written as

$$W_f + W_{fL} = W_e + W_m \qquad (2.2\text{-}5)$$

This energy balance is shown schematically in Fig. 2.2-2.

The actual process of converting electric energy to mechanical energy (or vice versa) is independent of (1) the loss of energy in either the electric or the mechanical systems (W_{eL} and W_{mL}), (2) the energies stored in the electric or magnetic fields which are not common to both systems (W_{eS}), or (3) the energies stored in the mechanical system (W_{mS}). If the losses of the coupling

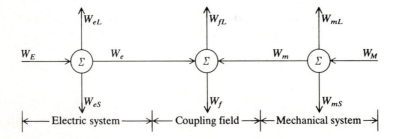

FIGURE 2.2-2
Energy balance.

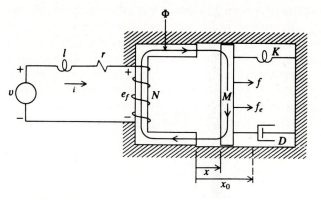

FIGURE 2.2-3
Electromechanical system with magnetic field.

field are neglected, then the field is conservative and (2.2-5) becomes

$$W_f = W_e + W_m \qquad (2.2\text{-}6)$$

Examples of elementary electromechanical systems are shown in Figs. 2.2-3 and 2.2-4. The system shown in Fig. 2.2-3 has a magnetic coupling field whereas the electromechanical system shown in Fig. 2.2-4 employs an electric field as a means of transferring energy between the electric and mechanical systems. In both systems, the space between the movable and stationary members is exaggerated for clarity. In these systems, v is the voltage of the electric source and f is an externally applied mechanical force. The electromagnetic or electrostatic force is denoted f_e. The resistance of the current-

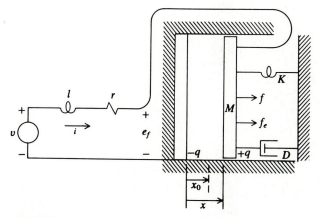

FIGURE 2.2-4
Electromechanical system with electric field.

carrying conductor is denoted by r with l denoting the inductance of a linear (conservative) electromagnetic system which does not couple the mechanical system. In the mechanical system, M is the mass of the movable member, and the linear compliance and damper are represented by a spring constant K and a damping coefficient D, respectively. The displacement x_0 is the zero force or equilibrium position of the mechanical system, which is the steady-state position of the mass with f_e and f equal to zero.

The voltage equation, that describes the electric systems shown in Figs. 2.2-3 and 2.2-4, may be written as

$$v = ri + l\frac{di}{dt} + e_f \tag{2.2-7}$$

where e_f is the voltage drop due to the coupling field. The dynamic behavior of the translational mechanical systems may be expressed by employing Newton's law of motion. Thus,

$$f = M\frac{d^2x}{dt^2} + D\frac{dx}{dt} + K(x - x_0) - f_e \tag{2.2-8}$$

Since power is the time rate of energy transfer, the total energy supplied by the electric source is

$$W_E = \int vi\,dt \tag{2.2-9}$$

The total energy supplied by the mechanical source is

$$W_M = \int f\,dx \tag{2.2-10}$$

which may also be expressed as

$$W_M = \int f\frac{dx}{dt}\,dt \tag{2.2-11}$$

Substituting (2.2-7) into (2.2-9) yields

$$W_E = r\int i^2\,dt + l\int i\frac{di}{dt}\,dt + \int e_f i\,dt \tag{2.2-12}$$

The first term on the right-hand side of (2.2-12) represents the energy loss due to the resistance of the conductors (W_{eL}). The second term represents the energy stored in the linear electromagnetic field external to the coupling field (W_{eS}). Therefore, the total energy transferred to the coupling field from the electric system is

$$W_e = \int e_f i \, dt$$

(2.2-13)

Similarly, for the mechanical system

$$W_M = M \int \frac{d^2 x}{dt^2} \, dx + D \int \left(\frac{dx}{dt}\right)^2 dt + K \int (x - x_0) \, dx - \int f_e \, dx$$

(2.2-14)

Here, the first and third terms on the right-hand side of (2.2-14) represent the kinetic energy stored in the mass and the potential energy stored in the spring, respectively. The sum of these two stored energies is W_{mS}. You may wish to take a moment to show that the first term on the right-hand side of (2.2-14) can be written $\frac{1}{2} M (dx/dt)^2$. The second term is the heat loss due to friction (W_{mL}). Thus, the total energy transferred to the coupling field from the mechanical system is

$$W_m = -\int f_e \, dx$$

(2.2-15)

It is important to note that a positive force f_e is assumed to be in the same direction as a positive displacement dx. Substituting (2.2-13) and (2.2-15) into the energy balance relation, (2.2-6), yields

$$W_f = \int e_f i \, dt - \int f_e \, dx$$

(2.2-16)

The equations set forth may be readily extended to include an electromechanical system with any number of electric and mechanical inputs and any number of coupling fields. Considering the system shown in Fig. 2.2-5, the

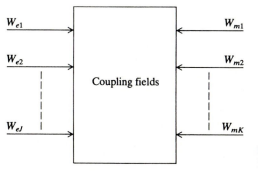

FIGURE 2.2-5
Multiple electric and mechanical inputs.

energy supplied to the coupling field may be expressed as

$$W_f = \sum_{j=1}^{J} W_{ej} + \sum_{k=1}^{K} W_{mk} \qquad (2.2\text{-}17)$$

wherein J electric and K mechanical inputs exist. The total energy supplied to the coupling field from the electric inputs is

$$\sum_{j=1}^{J} W_{ej} = \int \sum_{j=1}^{J} e_{fj} i_j \, dt \qquad (2.2\text{-}18)$$

The total energy supplied to the coupling field from the mechanical inputs is

$$\sum_{k=1}^{K} W_{mk} = -\int \sum_{k=1}^{K} f_{ek} \, dx_k \qquad (2.2\text{-}19)$$

In our analysis of electromechanical systems, we will consider devices with only one mechanical input, for example, the shaft of the electric machine or the moving arm of a magnetic relay. On the other hand, it is necessary to consider systems with multiple electric inputs, for example, the two- or three-phase ac machine or the dc machine with several independent control fields. In all cases, however, the multiple electric inputs have a common coupling field. Therefore, we need not become too ambitious in the following derivations. More specifically, hereafter we will restrict our analysis to electromechanical devices with only one mechanical input. Thus, the k subscript will be dropped from f_e, x, and W_m. This reduces our work considerably without restricting the practical application of our results. With one mechanical input, the energy balance equation becomes

$$W_f = \int \sum_{j=1}^{J} e_{fj} i_j \, dt - \int f_e \, dx \qquad (2.2\text{-}20)$$

In differential form, which will be the form we will use extensively,

$$dW_f = \sum_{j=1}^{J} e_{fj} i_j \, dt - f_e \, dx \qquad (2.2\text{-}21)$$

SP2.2-1. The current flowing in a 1-H inductor which is external to the coupling field is $i = kt + k_0$. Calculate W_{eS}, the energy stored in the inductor at $t = 1$ s. [$W_{eS} = \frac{1}{2}(k + k_0)^2$]

SP2.2-2. Express the undamped natural frequency of the mechanical system described by (2.2-8). [$\omega_n = (K/M)^{1/2}$]

SP2.2-3. Express the instantaneous power delivered to the inductor in SP2.2-1. [$p = k^2 t + k k_0$]

SP2.2-4. For two electric inputs, $e_{f1} = k_1 t$, $i_1 = k_0$, $e_{f2} = k_2 t^2$, and $i_2 = k_3 t$. Express the total energy supplied to the coupling field (W_e) in 2 s. [$W_e = 2k_0 k_1 + 4k_2 k_3$]

2.3 ENERGY IN COUPLING FIELD

Before using (2.2-21) to obtain an expression for the electromagnetic force f_e, it is necessary to derive an expression for the energy stored in the coupling field. Once we have an expression for W_f, we can take the total derivative to obtain dW_f which can then be substituted into (2.2-21). When expressing the energy in the coupling field, it is convenient to neglect all losses associated with the electric or magnetic coupling field, whereupon the field is assumed to be conservative and the energy stored therein is a function of the state of the electrical and mechanical variables. Although the effects of the core losses of the coupling field may be functionally accounted for by appropriately introducing a resistance in the electric circuit, this refinement is generally not necessary since the ferromagnetic material is selected and arranged in laminations so as to minimize the hysteresis and eddy current losses. Moreover, nearly all of the energy stored in the coupling field is stored in the air gap of the electromechanical device. Since air is a conservative medium, all of the energy stored therein can be returned to the electric or mechanical systems. Therefore, the assumption of a lossless coupling field is not as restrictive as it might first appear.

The energy stored in a conservative field is a function of the state of the system variables and not the manner in which the variables reached that state. It is convenient to take advantage of this feature when developing a mathematical expression for the field energy. In particular, it is convenient to fix mathematically the position of the mechanical system associated with the coupling field and then excite the electric system with the displacement of the mechanical system held fixed. During the excitation of the electric inputs, $dx = 0$, hence, W_m is zero even though electromagnetic or electrostatic forces occur. Therefore, with the displacement held fixed, the energy stored in the coupling field during the excitation of the electric inputs is equal to the energy supplied to the coupling field by the electric inputs. Thus, with $dx = 0$, the energy supplied from the electric system may be expressed from (2.2-20) as

$$W_f = \int \sum_{j=1}^{J} e_{fj} i_j \, dt \qquad \text{with } dx = 0 \qquad (2.3\text{-}1)$$

In our discussion thus far, we have considered both the electric and electromagnetic coupling fields. However, our primary interest is the electromagnetic system and, hereafter, we will direct our attention accordingly. Let us consider a singly excited electromagnetic system similar to that shown in Fig. 2.2-3. In this case $e_f = d\lambda/dt$, whereupon (2.3-1) becomes

$$W_f = \int i \, d\lambda \qquad \text{with } dx = 0 \qquad (2.3\text{-}2)$$

Here $j = 1$, however, the subscript is omitted for the sake of brevity. The area to the left of the λi relationship, shown in Fig. 2.3-1, for a singly excited electromagnetic system is the area described by (2.3-2). In Fig. 2.3-1, this area represents the energy stored in the field at the instant when $\lambda = \lambda_a$ and $i = i_a$. The λi relationship need not be linear, it need only be single-valued, a property which is characteristic to a conservative or lossless field. Moreover, since the coupling field is conservative, the energy stored in the field with $\lambda = \lambda_a$ and $i = i_a$ is independent of the excursion of the electrical and mechanical variables before reaching this state.

The area to the right of the λi curve is called the *coenergy* and can be expressed as

$$W_c = \int \lambda \, di \qquad \text{with } dx = 0 \qquad (2.3\text{-}3)$$

Although the coenergy has little or no physical significance, we will find it a convenient quantity for expressing the electromagnetic force. From Fig. 2.3-1, we see that the sum of W_f and W_c is λ times i, that is,

$$\lambda i = W_c + W_f \qquad (2.3\text{-}4)$$

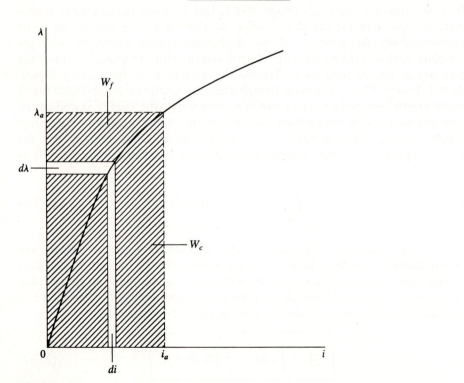

FIGURE 2.3-1
Stored energy and coenergy in a magnetic field of a singly excited electromagnetic device.

which is also valid for multiple electric inputs, where λi in (2.3-4) is replaced by $\sum_{j=1}^{J} \lambda_j i_j$. It should be clear that, for a linear magnetic system, where the λi plots are straightline relationships, $W_f = W_c = \frac{1}{2}\lambda i$.

The displacement x defines completely the influence of the mechanical system upon the coupling field; however, since λ and i are related, only one is needed in addition to x in order to describe the state of the electromechanical system. Therefore, we can select either λ and x as independent variables or i and x. If i and x are selected as the independent variables, it is convenient to express the field energy and the flux linkages as

$$W_f = W_f(i, x) \tag{2.3-5}$$

$$\lambda = \lambda(i, x) \tag{2.3-6}$$

With i and x as independent variables, we must express $d\lambda$ in terms of di before substituting into (2.3-2). Thus from (2.3-6),

$$d\lambda = \frac{\partial \lambda(i, x)}{\partial i} di + \frac{\partial \lambda(i, x)}{\partial x} dx \tag{2.3-7}$$

In the derivation of an expression for the energy stored in the field, dx is set equal to zero. Hence, in the evaluation of field energy, $d\lambda$ is equal to the first term on the right-hand side of (2.3-7). Substituting into (2.3-2) yields

$$W_f(i, x) = \int i \frac{\partial \lambda(i, x)}{\partial i} di$$

$$= \int_0^i \xi \frac{\partial \lambda(\xi, x)}{\partial \xi} d\xi \tag{2.3-8}$$

where ξ is a dummy variable of integration. Evaluation of (2.3-8) gives the energy stored in the field of the singly excited system. The coenergy in terms of i and x may be evaluated from (2.3-3) as

$$W_c(i, x) = \int \lambda(i, x) di$$

$$= \int_0^i \lambda(\xi, x) d\xi \tag{2.3-9}$$

With λ and x as independent variables

$$W_f = W_f(\lambda, x) \tag{2.3-10}$$

$$i = i(\lambda, x) \tag{2.3-11}$$

The field energy may be evaluated from (2.3-2) as

$$W_f(\lambda, x) = \int i(\lambda, x) d\lambda$$

$$= \int_0^\lambda i(\xi, x) d\xi \tag{2.3-12}$$

To evaluate the coenergy with λ and x as independent variables, we need to express di in terms of $d\lambda$. Thus from (2.3-11),

$$di = \frac{\partial i(\lambda, x)}{\partial \lambda} d\lambda + \frac{\partial i(\lambda, x)}{\partial x} dx \qquad (2.3\text{-}13)$$

Since $dx = 0$ in this evaluation, (2.3-3) becomes

$$W_c(\lambda, x) = \int \lambda \frac{\partial i(\lambda, x)}{\partial \lambda} d\lambda$$

$$= \int_0^\lambda \xi \frac{\partial i(\xi, x)}{\partial \xi} d\xi \qquad (2.3\text{-}14)$$

For a linear electromagnetic system, the λi plots are straightline relationships. Thus, for the singly excited magnetically linear system,

$$\lambda(i, x) = L(x)i \qquad (2.3\text{-}15)$$

or

$$i(\lambda, x) = \frac{\lambda}{L(x)} \qquad (2.3\text{-}16)$$

where $L(x)$ is the inductance. Let us evaluate $W_f(i, x)$. With $dx = 0$ and, since $\partial\lambda(i, x)/\partial i = L(x)$, (2.3-7) becomes

$$d\lambda = L(x)\, di \qquad (2.3\text{-}17)$$

Hence, from (2.3-8),

$$W_f(i, x) = \int_0^i \xi L(x)\, d\xi = \tfrac{1}{2} L(x)i^2 \qquad (2.3\text{-}18)$$

It is left to the reader to show that by a similar procedure $W_f(\lambda, x)$, $W_c(i, x)$, and $W_c(\lambda, x)$ are equal to (2.3-18) for this magnetically linear system.

The field energy is a state function and the expression describing the field energy in terms of system variables is valid regardless of the variations in the system variables. For example, (2.3-18) expresses the field energy regardless of the variations in $L(x)$ and i. The fixing of the mechanical system so as to obtain an expression for the field energy is a mathematical convenience and not a restriction upon the result.

In the case of a multiexcited electromagnetic system, an expression for the field energy may be obtained by evaluating the following relation with $dx = 0$:

$$\boxed{W_f = \int \sum_{j=1}^{J} i_j\, d\lambda_j \qquad \text{with } dx = 0} \qquad (2.3\text{-}19)$$

Since the coupling field is considered conservative, (2.3-19) may be evaluated independent of the order in which the flux linkages or currents are brought to their final values. To illustrate evaluation of (2.3-19) for a multiexcited system, we will allow the currents to establish their final states one at a time while all other currents are mathematically fixed either in their unexcited or final states. This procedure may be illustrated by considering a doubly excited electric system with one mechanical input. An electromechanical system of this type could be constructed by placing a second winding, supplied from a second electric system, on either the stationary or movable member of the system shown in Fig. 2.2-3. In this evaluation, it is convenient to use currents and displacement as the independent variables. Hence, for a doubly excited electric system,

$$W_f(i_1, i_2, x) = \int [i_1 \, d\lambda_1(i_1, i_2, x) + i_2 \, d\lambda_2(i_1, i_2, x)] \qquad \text{with } dx = 0$$

$$(2.3\text{-}20)$$

In this determination of an expression for W_f, the mechanical displacement is held constant $(dx = 0)$; thus (2.3-20) becomes

$$W_f(i_1, i_2, x) = \int i_1 \left[\frac{\partial \lambda_1(i_1, i_2, x)}{\partial i_1} \, di_1 + \frac{\partial \lambda_1(i_1, i_2, x)}{\partial i_2} \, di_2 \right]$$

$$+ i_2 \left[\frac{\partial \lambda_2(i_1, i_2, x)}{\partial i_1} \, di_1 + \frac{\partial \lambda_2(i_1, i_2, x)}{\partial i_2} \, di_2 \right] \qquad (2.3\text{-}21)$$

We will evaluate the energy stored in the field by employing (2.3-21) twice. First, we will mathematically increase the current i_1 from zero to its desired final value while holding i_2 at zero. Thus, i_1 is the variable of integration and $di_2 = 0$. Energy is supplied to the coupling field from the source connected to winding 1. As the second evaluation of (2.3-21), i_2 is increased from zero to its desired final value while maintaining i_1 at the value attained in the preceding step. Hence, i_2 is the variable of integration and $di_1 = 0$. During this time, energy is supplied from both sources to the coupling field since $i_1 \, d\lambda_1$ is, in general, nonzero. The total energy stored in the coupling field is the sum of the two evaluations. Following this two-step procedure, the evaluation of (2.3-21) for the total field energy becomes

$$W_f(i_1, i_2, x) = \int i_1 \frac{\partial \lambda_1(i_1, 0, x)}{\partial i_1} \, di_1$$

$$+ \int \left[i_1 \frac{\partial \lambda_1(i_1, i_2, x)}{\partial i_2} \, di_2 + i_2 \frac{\partial \lambda_2(i_1, i_2, x)}{\partial i_2} \, di_2 \right] \qquad (2.3\text{-}22)$$

which should be written

$$W_f(i_1, i_2, x) = \int_0^{i_1} \xi \frac{\partial \lambda_1(\xi, 0, x)}{\partial \xi} \, d\xi$$

$$+ \int_0^{i_2} \left[i_1 \frac{\partial \lambda_1(i_1, \xi, x)}{\partial \xi} + \xi \frac{\partial \lambda_2(i_1, \xi, x)}{\partial \xi} \right] d\xi \qquad (2.3\text{-}23)$$

The first integral on the right-hand side of (2.3-22) or (2.3-23) results from the first step of the evaluation with i_1 as the variable of integration and with $i_2 = 0$

and $di_2 = 0$. The second integral comes from the second step of the evaluation with i_1 equal to its final value ($di_1 = 0$) and i_2 as the variable of integration. The order of allowing the currents to reach their final state is irrelevant; that is, as our first step, we could have made i_2 the variable of integration while holding i_1 at zero ($di_1 = 0$) and then let i_1 become the variable of integration while holding i_2 at its final value. The results would be the same. For three electric inputs, the evaluation procedure would require three steps, one for each current to be brought mathematically to its final state.

Let us now evaluate the energy stored in a magnetically linear electromechanical system with two electric inputs and one mechanical input. For this, let

$$\lambda_1(i_1, i_2, x) = L_{11}(x)i_1 + L_{12}(x)i_2 \qquad (2.3\text{-}24)$$

$$\lambda_2(i_1, i_2, x) = L_{21}(x)i_1 + L_{22}(x)i_2 \qquad (2.3\text{-}25)$$

where the self-inductances $L_{11}(x)$ and $L_{22}(x)$ include the leakage inductances. With the mechanical displacement held constant ($dx = 0$),

$$d\lambda_1(i_1, i_2, x) = L_{11}(x)\,di_1 + L_{12}(x)\,di_2 \qquad (2.3\text{-}26)$$

$$d\lambda_2(i_1, i_2, x) = L_{12}(x)\,di_1 + L_{22}(x)\,di_2 \qquad (2.3\text{-}27)$$

The coefficients on the right-hand side of (2.3-26) and (2.3-27) are the partial derivatives. For example, $L_{11}(x)$ is the partial derivative of $\lambda_1(i_1, i_2, x)$ with respect to i_1. Appropriate substitution into (2.3-23) gives

$$W_f(i_1, i_2, x) = \int_0^{i_1} \xi L_{11}(x)\,d\xi + \int_0^{i_2} [i_1 L_{12}(x) + \xi L_{22}(x)]\,d\xi \qquad (2.3\text{-}28)$$

which yields

$$W_f(i_1, i_2, x) = \tfrac{1}{2}L_{11}(x)i_1^2 + L_{12}(x)i_1 i_2 + \tfrac{1}{2}L_{22}(x)i_2^2 \qquad (2.3\text{-}29)$$

It follows that the total field energy of a linear electromagnetic system with J electric inputs may be expressed as

$$\boxed{\; W_f(i_1, \ldots, i_J, x) = \frac{1}{2} \sum_{p=1}^{J} \sum_{q=1}^{J} L_{pq} i_p i_q \;} \qquad (2.3\text{-}30)$$

Example 2A. Consider the magnetic system described by

$$\lambda(i, x) = (0.1 + kx^{-1})i \qquad (2\text{A-}1)$$

Calculate W_f and W_c. Here, it is convenient to work with i and x as independent variables. Thus, from (2.3-8),

$$W_f = \int_0^i \xi(0.1 + kx^{-1})\,d\xi$$

$$= \tfrac{1}{2}(0.1 + kx^{-1})i^2 \qquad (2\text{A-}2)$$

From (2.3-9),
$$W_c = \int_0^i (0.1 + kx^{-1})\xi \, d\xi$$

$$= \tfrac{1}{2}(0.1 + kx^{-1})i^2 \tag{2A-3}$$

We see that $W_f = W_c$. Did you recognize the fact that this is a linear magnetic system with

$$L(x) = 0.1 + kx^{-1} \tag{2A-4}$$

where the first term on the right-hand side is analogous to the leakage inductance and kx^{-1} is the magnetizing inductance of the electromagnet treated in Chap. 1? Let us make it a nonlinear system; in particular, let

$$\lambda(i, x) = (0.1 + kx^{-1})i^2 \tag{2A-5}$$

From (2.3-8),
$$W_f = \int_0^i \xi 2(0.1 + kx^{-1})\xi \, d\xi$$

$$= \tfrac{2}{3}(0.1 + kx^{-1})i^3 \tag{2A-6}$$

From (2.3-9),
$$W_c = \int_0^i (0.1 + kx^{-1})\xi^2 \, d\xi$$

$$= \tfrac{1}{3}(0.1 + kx^{-1})i^3 \tag{2A-7}$$

We see that W_f and W_c are not equal; however, according to (2.3-4),

$$\lambda i = W_f + W_c \tag{2A-8}$$

Let us show that this is true.

$$[(0.1 + kx^{-1})i^2]i = \tfrac{2}{3}(0.1 + kx^{-1})i^3 + \tfrac{1}{3}(0.1 + kx^{-1})i^3$$

$$(0.1 + kx^{-1})i^3 = (0.1 + kx^{-1})i^3 \tag{2A-9}$$

SP2.3-1. $\lambda = kx^2 i$. Calculate W_f and W_c when $kx^2 = 1$ V·s/A and $i = 2$ A. [$W_f = W_c = 2$ J]

SP2.3-2. $\lambda = kxi^2$. Calculate W_f and W_c when $kx = 1$ V·s/A^2 and $i = 2$ A. [$W_f = \tfrac{16}{3}$ J; $W_c = \tfrac{8}{3}$ J]

SP2.3-3. The current is increased from 2 to 3 A in SP2.3-2. Calculate the change in W_f and W_c. [$\Delta W_f = \tfrac{38}{3}$ J; $\Delta W_c = \tfrac{19}{3}$ J]

SP2.3-4. $i = b(x)\lambda^2$. Express $W_f(\lambda, x)$ and $W_c(\lambda, x)$. [$W_f(\lambda, x) = \tfrac{1}{3}b(x)\lambda^3$; $W_c(\lambda, x) = \tfrac{2}{3}b(x)\lambda^3$]

2.4 GRAPHICAL INTERPRETATION OF ENERGY CONVERSION

Before proceeding to the derivation of expressions for the electromagnetic force, it is instructive to consider briefly a graphical interpretation of the energy conversion process. For this purpose, let us again refer to the elementary system shown in Fig. 2.2-3 and assume that as the movable member moves from $x = x_a$ to $x = x_b$, where $x_b < x_a$, the λi characteristics are given by Fig. 2.4-1 [1]. Let us furthermore assume that, as the member moves from x_a to x_b,

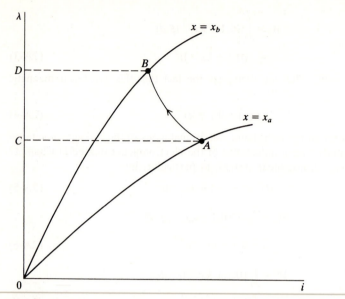

FIGURE 2.4-1
Graphical representation of electromechanical energy conversion for λi path A to B.

the λi trajectory moves from point A to point B. The exact trajectory from A to B is determined by the combined dynamics of the electric and mechanical systems, and any variation in v and f which may occur. Now, the area $0AC0$ represents the original energy stored in the coupling field; area $0BD0$ represents the final energy stored in the field. Therefore, the change in field energy is

$$\Delta W_f = \text{area } 0BD0 - \text{area } 0AC0 \tag{2.4-1}$$

The change in W_e, denoted as ΔW_e, is

$$\Delta W_e = \int_{\lambda_A}^{\lambda_B} i \, d\lambda = \text{area } CABDC \tag{2.4-2}$$

From (2.2-6),

$$\boxed{\Delta W_m = \Delta W_f - \Delta W_e} \tag{2.4-3}$$

Hence, by adding and subtracting the appropriate areas in Fig. 2.4-1, we can obtain ΔW_m. In particular,

$$\Delta W_m = \text{area } 0BD0 - \text{area } 0AC0 - \text{area } CABDC$$
$$= -\text{area } 0AB0 \tag{2.4-4}$$

The energy contributed to the coupling field from the mechanical system ΔW_m

is negative. Energy has been supplied to the mechanical system from the coupling field, part of which came from the energy stored in the field, and part of which came from the electric system. If the member is now moved back to x_a, the λi trajectory may be as shown in Fig. 2.4-2. Here, ΔW_m is still area $0AB0$, but it is positive, which means that energy was supplied from the mechanical system to the coupling field, part of which is stored in the field and part of which is transferred to the electric system.

The energy supplied by the mechanical system during the motion from B to A (area $0AB0$ in Fig. 2.4-2) is larger than the energy supplied to the mechanical system during the original motion from A to B (area $0AB0$ in Fig. 2.4-1). Therefore, the net energy supplied by the mechanical system for the complete cycle is positive. The net ΔW_m for the cycle from A to B back to A is the shaded area shown in Fig. 2.4-3. Since the coupling field energy at point A is uniquely determined from the mechanical displacement and current at point A, the net change in field energy is zero as we move from A to B and back to A. Since ΔW_f is zero for this cycle,

$$\Delta W_m = -\Delta W_e \tag{2.4-5}$$

For the cycle shown, the net ΔW_e is negative since the magnitude of the change in W_e is larger when we went from B to A than from A to B and ΔW_e is negative from B to A. If the trajectory had been in the counterclockwise direction, the net ΔW_e would have been positive and the net ΔW_m negative.

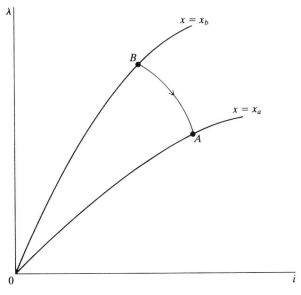

FIGURE 2.4-2
Graphical representation of electromechanical energy conversion for λi path B to A.

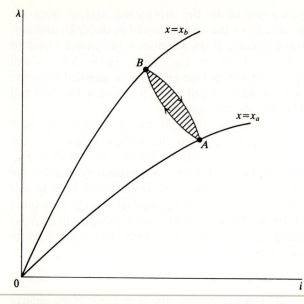

FIGURE 2.4-3
Graphical representation of electromechanical energy conversion for λi path A to B to A.

Example 2B. Let us calculate the change in W_m for the two λi paths between A and B shown in Fig. 2B-1. We will assume that the λi characteristics are for the electromagnet shown in Fig. 2.2-3 as the movable member moves from $x = x_1$ to $x = x_2$, where $x_2 < x_1$. In Fig. 2B-1a, motion from x_1 to x_2 occurs very slowly, whereupon the current remains essentially constant. In Fig. 2B-1b, the motion from x_1 to x_2 is very rapid—so rapid that the flux linkages remain essentially

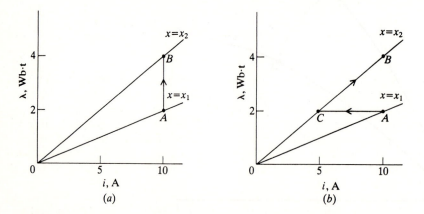

FIGURE 2B-1
Motion from x_1 to x_2. (a) Constant current; (b) constant flux linkage.

constant. Here, the current decreases rapidly to point C, then it increases back to its original value of 10 A. The λi paths are idealized and impracticable; however, they are of academic interest since they form the boundaries for some types of motion.

Before proceeding, let us assume that the voltage applied to the winding of the electromagnet (Fig. 2.2-3) is 10 V and that $r = 1\,\Omega$. We shall also assume that the external inductance l is zero. Since, in Fig. 2B-1, the magnetic system is assumed to be linear, $L_l + L_m(x_1) = 0.2$ H and $L_l + L_m(x_2) = 0.4$ H.

Let us now do what we originally set out to do. First, we will consider Fig. 2B-1a. The change in field energy as we move from A to B is

$$\Delta W_f = W_{fB} - W_{fA} = \text{area } 0B40 - \text{area } 0A20$$
$$= \tfrac{1}{2}(4)(10) - \tfrac{1}{2}(2)(10) = 10\text{ J} \qquad (2B-1)$$

where W_{fA} is the field energy at A and W_{fB} at B. For Fig. 2B-1a, the change in the energy coming from the electric system is

$$\Delta W_e = \int_{\lambda_A}^{\lambda_B} i\, d\lambda$$
$$= \int_2^4 10\, d\lambda = 10(4-2) = 20\text{ J} \qquad (2B-2)$$

Now, $$\Delta W_m = \Delta W_f - \Delta W_e$$
$$= 10 - 20 = -10\text{ J} \qquad (2B-3)$$

Work has been done on the mechanical system to move the movable member from x_1 to x_2. According to (2.4-4), the area $0AB0$ should be 10 J. Check to see if this is true.

Consider Fig. 2B-1b; here the change of field energy ΔW_f is the same as for Fig. 2B-1a. However, ΔW_e is not; in particular,

$$\Delta W_e = \int_{\lambda_A}^{\lambda_B} i\, d\lambda = \text{area } 2CB42$$
$$= (2)(5) + \tfrac{1}{2}(2)(5) = 15\text{ J} \qquad (2B-4)$$

Thus, $$\Delta W_m = 10 - 15 = -5\text{ J} \qquad (2B-5)$$

Since the energy transferred between the coupling field and the mechanical system is different for the two cases, it is logical to assume that the two mechanical systems are different. In particular, as the motion in Fig. 2B-1a is very slow, this could occur in several ways; perhaps the damper is very stiff or we have control over the external force f, and we caused it to change very slowly. In Fig. 2B-1b, the damper may be less stiff and, perhaps, the force was changed instantaneously (stepped). Please recall that these paths are impracticable but of academic interest.

SP2.4-1. Calculate ΔW_m in Fig. 2B-1 for the cycle A to B to C to A. $[\Delta W_m = -5\text{ J}]$
SP2.4-2. $\lambda = a(x)i^2$. The current is held constant at 2 A while $a(x)$ increases from 1 to 2. Calculate ΔW_m. $[\Delta W_m = -\tfrac{8}{3}\text{ J}]$

2.5   ELECTROMAGNETIC AND ELECTROSTATIC FORCES

The energy balance relationships given by (2.2-21) may be arranged as

$$f_e \, dx = \sum_{j=1}^{J} e_{fj} i_j \, dt - dW_f \tag{2.5-1}$$

To obtain an expression for f_e, it is necessary first to express W_f and then take its total derivative. One is tempted to substitute the integrand of (2.3-1) into (2.5-1) for the infinitesimal change of field energy. This procedure is incorrect since the integrand of (2.3-1) was obtained with the mechanical displacement held fixed ($dx = 0$). The total differential of the field energy is required in (2.5-1).

The force or torque in any electromechanical system may be evaluated by employing (2.5-1), that is, $dW_f = dW_e + dW_m$. In many respects, one gains a much better understanding of the energy conversion process of a particular system by starting the derivation of the force or torque expressions with $dW_f = dW_e + dW_m$, as in Example 2C at the end of this section, rather than by selecting a relationship from a table. However, for the sake of completeness, derivation of the force equations will be set forth and tabulated for electromechanical systems with one mechanical input and J electric inputs [2].

For an electromagnetic system, (2.5-1) may be written as

$$f_e \, dx = \sum_{j=1}^{J} i_j \, d\lambda_j - dW_f \tag{2.5-2}$$

With i_j and x selected as independent variables,

$$W_f = W_f(i_1, \ldots, i_J; x) = W_f(\mathbf{i}, x) \tag{2.5-3}$$

$$\lambda_j = \lambda_j(i_1, \ldots, i_J; x) = \lambda_j(\mathbf{i}, x) \tag{2.5-4}$$

In (2.5-3) and (2.5-4) and hereafter, the functional notation of $(i_1, \ldots, i_J; x)$ is abbreviated as $(\mathbf{i}, x)$, where $\mathbf{i}$ is used to denote the complete set of currents associated with the J windings. From (2.5-3) and (2.5-4),

$$dW_f = \sum_{j=1}^{J} \left[\frac{\partial W_f(\mathbf{i}, x)}{\partial i_j} \, di_j \right] + \frac{\partial W_f(\mathbf{i}, x)}{\partial x} \, dx \tag{2.5-5}$$

$$d\lambda_j = \sum_{n=1}^{J} \left[\frac{\partial \lambda_j(\mathbf{i}, x)}{\partial i_n} \, di_n \right] + \frac{\partial \lambda_j(\mathbf{i}, x)}{\partial x} \, dx \tag{2.5-6}$$

The summation index n is used in (2.5-6) so as to avoid confusion with the subscript j since each $d\lambda_j$ must be evaluated for changes in all currents to account for mutual coupling between electric systems. [Recall that we did this in (2.3-21) for $J = 2$.] Substituting (2.5-5) and (2.5-6) into (2.5-2) yields

$$f_e(\mathbf{i}, x)\, dx = \sum_{j=1}^{J} i_j \left\{ \sum_{n=1}^{J} \left[\frac{\partial \lambda_j(\mathbf{i}, x)}{\partial i_n} \, di_n \right] + \frac{\partial \lambda_j(\mathbf{i}, x)}{\partial x} \, dx \right\}$$

$$- \sum_{j=1}^{J} \left[\frac{\partial W_f(\mathbf{i}, x)}{\partial i_j} \, di_j \right] - \frac{\partial W_f(\mathbf{i}, x)}{\partial x} \, dx \qquad (2.5\text{-}7)$$

Gathering terms,

$$f_e(\mathbf{i}, x)\, dx = \left\{ \sum_{j=1}^{J} \left[i_j \frac{\partial \lambda_j(\mathbf{i}, x)}{\partial x} \right] - \frac{\partial W_f(\mathbf{i}, x)}{\partial x} \right\} dx$$

$$+ \sum_{j=1}^{J} \left\{ i_j \sum_{n=1}^{J} \left[\frac{\partial \lambda_j(\mathbf{i}, x)}{\partial i_n} \, di_n \right] - \frac{\partial W_f(\mathbf{i}, x)}{\partial i_j} \, di_j \right\} \qquad (2.5\text{-}8)$$

The previous equation is satisfied provided that

$$f_e(\mathbf{i}, x) = \sum_{j=1}^{J} \left[i_j \frac{\partial \lambda_j(\mathbf{i}, x)}{\partial x} \right] - \frac{\partial W_f(\mathbf{i}, x)}{\partial x} \qquad (2.5\text{-}9)$$

and

$$0 = \sum_{j=1}^{J} \left\{ i_j \sum_{n=1}^{J} \left[\frac{\partial \lambda_j(\mathbf{i}, x)}{\partial i_n} \, di_n \right] - \frac{\partial W_f(\mathbf{i}, x)}{\partial i_j} \, di_j \right\} \qquad (2.5\text{-}10)$$

Although (2.5-10) is of little practical importance, (2.5-9) can be used to evaluate the force on the mechanical system with $\mathbf{i}$ and x selected as independent variables. A second force equation with $\mathbf{i}$ and x as independent variables may be obtained from (2.5-9) by incorporating the expression for coenergy. For a multiexcited system, the coenergy may be expressed as

$$W_c = \sum_{j=1}^{J} i_j \lambda_j - W_f \qquad (2.5\text{-}11)$$

Since $\mathbf{i}$ and x are independent variables, the partial derivative with respect to x is

$$\frac{\partial W_c(\mathbf{i}, x)}{\partial x} = \sum_{j=1}^{J} \left[i_j \frac{\partial \lambda_j(\mathbf{i}, x)}{\partial x} \right] - \frac{\partial W_f(\mathbf{i}, x)}{\partial x} \qquad (2.5\text{-}12)$$

Hence, substituting (2.5-12) into (2.5-9) yields

$$f_e(\mathbf{i}, x) = \frac{\partial W_c(\mathbf{i}, x)}{\partial x} \qquad (2.5\text{-}13)$$

It should be recalled that positive f_e and positive dx are in the same direction. Also, if the magnetic system is linear, $W_c = W_f$.

By a procedure similar to that used above, force equations may be derived with λ and x as independent variables where λ denotes the flux linkages $\lambda_1, \ldots, \lambda_J$ of the J windings. These relations are given in Table 2.5-1 without proof.

TABLE 2.5-1
Electromagnetic force at mechanical input. (For rotational systems replace f_e with T_e and x with θ.)

$$f_e(\mathbf{i}, x) = \sum_{j=1}^{J} \left[i_j \frac{\partial \lambda_j(\mathbf{i}, x)}{\partial x} \right] - \frac{\partial W_f(\mathbf{i}, x)}{\partial x}$$

$$f_e(\mathbf{i}, x) = \frac{\partial W_c(\mathbf{i}, x)}{\partial x}$$

$$f_e(\boldsymbol{\lambda}, x) = -\frac{\partial W_f(\boldsymbol{\lambda}, x)}{\partial x}$$

$$f_e(\boldsymbol{\lambda}, x) = -\sum_{j=1}^{J} \left[\lambda_j \frac{\partial i_j(\boldsymbol{\lambda}, x)}{\partial x} \right] + \frac{\partial W_c(\boldsymbol{\lambda}, x)}{\partial x}$$

In Table 2.5-1, the independent variables to be used are designated in each equation by the abbreviated functional notation. Although only translational mechanical systems have been considered, all force relationships developed herein may be modified for the purpose of evaluating the torque in rotational systems. In particular, when considering a rotational system, f_e is replaced with the electromagnetic torque T_e, and x with the angular displacement θ. These substitutions are justified since the change of mechanical energy in a rotational system is expressed

$$dW_m = -T_e\, d\theta \qquad (2.5\text{-}14)$$

The force equation for an electromechanical system with electric coupling fields may be derived by following a procedure similar to that used in the case of magnetic coupling fields. These relationships are given in Table 2.5-2 without explanation.

TABLE 2.5-2
Electrostatic force at mechanical input. (For rotational systems replace f_e with T_e and x with θ.)

$$f_e(\mathbf{e}_f, x) = \sum_{j=1}^{J} \left[e_{fj} \frac{\partial q_j(\mathbf{e}_f, x)}{\partial x} \right] - \frac{\partial W_f(\mathbf{e}_f, x)}{\partial x}$$

$$f_e(\mathbf{e}_f, x) = \frac{\partial W_c(\mathbf{e}_f, x)}{\partial x}$$

$$f_e(\mathbf{q}, x) = -\frac{\partial W_f(\mathbf{q}, x)}{\partial x}$$

$$f_e(\mathbf{q}, x) = -\sum_{j=1}^{J} \left[q_j \frac{\partial e_{fj}(\mathbf{q}, x)}{\partial x} \right] + \frac{\partial W_c(\mathbf{q}, x)}{\partial x}$$

Example 2C. We have mentioned that one may prefer to determine the electromagnetic force or torque by starting with the relationship $dW_f = dW_e + dW_m$ rather than by selecting a formula from a table. To illustrate this procedure, let

$$\lambda = [1 + a(x)]i^2 \qquad (2C-1)$$

First, we must evaluate the field energy. Since losses in the coupling field are neglected, W_f is a function of state. Hence, W_f may be evaluated by fixing the mechanical displacement. This is done by setting $dx = 0$, whereupon

$$dW_f = dW_e = i \, d\lambda \qquad \text{with } dx = 0 \qquad (2C-2)$$

where dW_e is obtained from (2.2-13) with $e_f = d\lambda/dt$. From (2C-1) with $dx = 0$,

$$d\lambda = 2[1 + a(x)]i \, di \qquad (2C-3)$$

Substituting (2C-3) into (2C-2) and solving for W_f yields

$$W_f = \int_0^i 2[1 + a(x)]\xi^2 \, d\xi = \tfrac{2}{3}[1 + a(x)]i^3 \qquad (2C-4)$$

To obtain an expression for f_e, we go back to the basic relationship that $dW_f = dW_e + dW_m$; however, now $dx \neq 0$. Thus, from (2C-4),

$$dW_f = \frac{2}{3}i^3 \frac{\partial a(x)}{\partial x} \, dx + 2[1 + a(x)]i^2 \, di \qquad (2C-5)$$

Now,

$$dW_e = i \, d\lambda = i\left[i^2 \frac{\partial a(x)}{\partial x} \, dx + 2[1 + a(x)]i \, di \right] \qquad (2C-6)$$

and from (2.2-15),

$$dW_m = -f_e \, dx \qquad (2C-7)$$

Substituting into $dW_f = dW_e + dW_m$ yields

$$\frac{2}{3}i^3 \frac{\partial a(x)}{\partial x} \, dx + 2[1 + a(x)]i^2 \, di = i^3 \frac{\partial a(x)}{\partial x} \, dx + 2[1 + a(x)]i^2 \, di - f_e \, dx \qquad (2C-8)$$

Note the di terms cancel as (2.5-10) tells us they should, and by equating the coefficients of dx,

$$f_e = \frac{1}{3}i^3 \frac{\partial a(x)}{\partial x} \qquad (2C-9)$$

Let us check our work by using information from Table 2.5-1. Since the system is magnetically nonlinear, $W_f \neq W_c$. Thus, knowing W_f, we can use the first entry in Table 2.5-1 or evaluate W_c and use the second entry. We shall do both. From Table 2.5-1 with one electric input ($J = 1$),

$$f_e(i, x) = i \frac{\partial \lambda(i, x)}{\partial x} - \frac{\partial W_f(i, x)}{\partial x} \qquad (2C-10)$$

Now, $\lambda(i, x)$ is expressed by (2C-1) and $W_f(i, x)$ by (2C-4), thus

$$f_e(i, x) = i\left[i^2 \frac{\partial a(x)}{\partial x} \right] - \frac{2}{3}i^3 \frac{\partial a(x)}{\partial x}$$

$$= \frac{1}{3}i^3 \frac{\partial a(x)}{\partial x} \qquad (2C-11)$$

which agrees with our previous result.

Alternatively, from the second entry of Table 2.5-1,

$$f_e(i, x) = \frac{\partial W_c(i, x)}{\partial x} \tag{2C-12}$$

Now, from (2.3-4),

$$W_c(i, x) = \lambda(i, x)i - W_f(i, x)$$
$$= \tfrac{1}{3}[1 + a(x)]i^3 \tag{2C-13}$$

[Obtain this same expression by using (2.3-3) to evaluate W_c.] Substituting (2C-13) into (2C-12) yields

$$f_e(i, x) = \frac{1}{3} i^3 \frac{\partial a(x)}{\partial x} \tag{2C-14}$$

SP2.5-1. $\lambda = kx^2i^2$. Express f_e when $i = 2$ A and $x = 1$ m. $[f_e = (16k/3) \text{ N}]$
SP2.5-2. $\lambda = ki/x$. Express f_e if $i = \sqrt{2}I_s \cos \omega_e t$. $[f_e = -(kI_s^2/2x^2)(1 + \cos 2\omega_e t)]$
SP2.5-3. $i = a(x)\lambda^3$. Express f_e by using the third entry in Table 2.5-1. $[f_e = -\tfrac{1}{4}\lambda^4(\partial a(x)/\partial x)]$
SP2.5-4. In a rotational system, $\lambda = ki^2 \sin \theta$. Express the torque T_e. $[T_e = \tfrac{1}{3}ki^3 \cos \theta]$

2.6   OPERATING CHARACTERISTICS OF AN ELEMENTARY ELECTROMAGNET

From our work in Sec. 1.7 we established that for $x > 0$ the inductance of the electromagnet shown in Fig. 2.2-3 may be adequately approximated by (1.7-21). That is,

$$L(x) = L_l + L_m(x) = L_l + \frac{k}{x} \tag{2.6-1}$$

Now,

$$\lambda(i, x) = L(x)i \tag{2.6-2}$$

and, since the system is magnetically linear,

$$W_f(i, x) = W_c(i, x) = \tfrac{1}{2}L(x)i^2 \tag{2.6-3}$$

From the second entry of Table 2.5-1,

$$f_e(i, x) = \frac{\partial W_c(i, x)}{\partial x} = \frac{1}{2} i^2 \frac{\partial L(x)}{\partial x} \tag{2.6-4}$$

Substituting (2.6-1) into (2.6-4) yields

$$f_e(i, x) = -\frac{ki^2}{2x^2} \tag{2.6-5}$$

The force f_e is always negative in the system shown in Fig. 2.2-3; the electromagnetic force pulls the moving member to the stationary member. In other words, an electromagnetic force is set up so as to minimize the reluctance (maximize the inductance) of the magnetic system.

The differential equations which describe the electromagnet are given by (2.2-7) for the electric system and (2.2-8) for the mechanical system. If the applied voltage v and the applied mechanical force f are both constant, all derivatives with respect to time in (2.2-7) and (2.2-8) are zero during steady-state operation. Thus,

$$v = ri \qquad (2.6\text{-}6)$$

$$f = K(x - x_0) - f_e \qquad (2.6\text{-}7)$$

Equation (2.6-7) may be written as

$$-f_e = f - K(x - x_0) \qquad (2.6\text{-}8)$$

A plot of (2.6-8), with f_e replaced by (2.6-5), is shown in Fig. 2.6-1 for the following system parameters: $r = 10 \, \Omega$, $K = 2667 \, \text{N/m}$, $x_0 = 3 \, \text{mm}$, $k = 6.283 \times 10^{-5} \, \text{H} \cdot \text{m}$.

In Fig. 2.6-1 the plot of the negative of the electromagnetic force is for an applied voltage of $5 \, \text{V}$ whereupon the steady-state current is $0.5 \, \text{A}$. The straight lines represent the right-hand side of (2.6-8) with $f = 0$ (lower straight

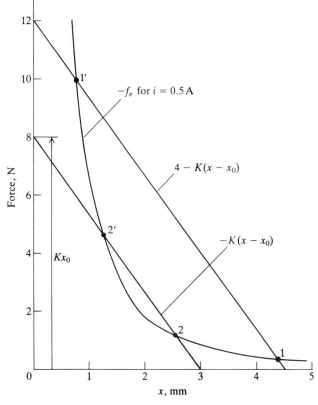

FIGURE 2.6-1
Steady-state operation of electromechanical system shown in Fig. 2.2-3.

line) and $f = 4$ N (upper straight line). Both lines intersect the $-f_e$ curve at two points. In particular, the upper line intersects the $-f_e$ curve at 1 and 1'; the lower line intersects at 2 and 2'. Stable operation occurs at only points 1 and 2. The system will not operate stably at points 1' and 2'. This can be explained by assuming the system is operating at one of these points (1' and 2') and then show that any system disturbance whatsoever will cause the system to move

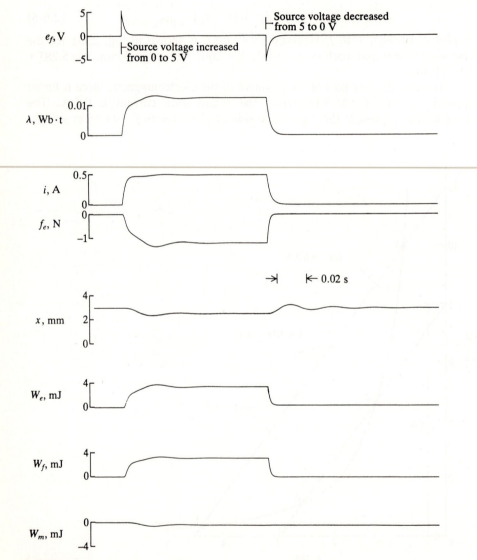

FIGURE 2.6-2
Dynamic performance of the electromechanical system shown in Fig. 2.2-3 during step changes in the source voltage.

away from these points. If, for example, x increases slightly from its value corresponding to point $1'$, the restraining force, $f - K(x - x_0)$, which is a force to pull the movable member to the right, is larger than the electromagnetic force to pull the movable member to the left. Hence, x will continue to increase until the system reaches operating point 1. If x increases beyond its value corresponding to operating point 1, the restraining force, $f - K(x - x_0)$, is less than the electromagnetic force. Therefore, the system will establish steady-state operation at 1. If, on the other hand, x decreases from point $1'$, the electromagnetic force is larger than the restraining force. Therefore, the

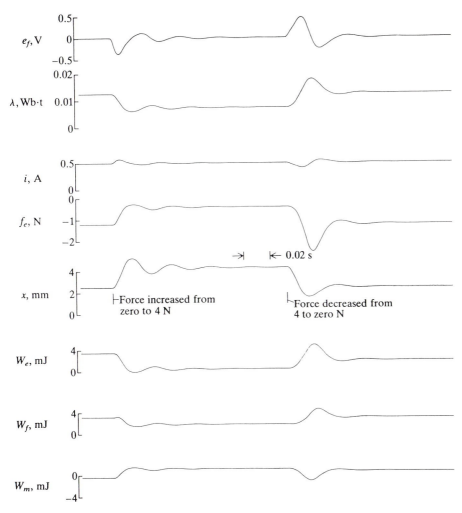

FIGURE 2.6-3
Dynamic performance of the electromechanical system shown in Fig. 2.2-3 during step changes in the applied force.

movable member will move until it comes in contact with the stationary member ($x = 0$). The restraining force which yields a straight line below the $-f_e$ curve will not permit stable operation with $x > 0$. Note also that, at point 2, x is less than x_0 and the spring is extended, exerting a force to the right on the movable member. At point 1, x is greater than x_0 and the spring is in compression, exerting a force to the left on the movable member.

The dynamic behavior of the system during step changes in the source voltage v is shown in Fig. 2.6-2 and Figs. 2.6-3 and 2.6-4 for step changes in the applied force f. The following system parameters were used in addition to those given previously: $L_l = 0$, $l = 0$, $M = 0.055$ kg, $D = 4$ N·s/m. The computer traces shown in Fig. 2.6-2 depict the dynamic performance of the example system when the applied voltage is stepped from zero to 5 V and then back to zero, with the applied mechanical force f held equal to zero. The following system variables are plotted: e_f, λ, i, f_e, x, W_e, W_f, and W_m. The energies are plotted in millijoules (mJ). Initially, the mechanical system is at rest with $x = x_0$

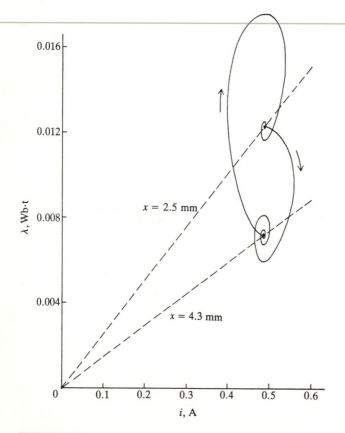

FIGURE 2.6-4
λ versus i plot of the system response shown in Fig. 2.6-3.

(3 mm). When the source voltage is applied, x decreases and, when steady-state operation is reestablished, x is approximately 2.5 mm, which is operating point 2 in Fig. 2.6-1. During the transient period, energy enters the coupling field via W_e. The bulk of this energy is stored in the field (W_f) with a smaller amount transferred to the mechanical system, some of which is dissipated in the damper during the transient period and the remainder is stored in the spring. When the applied voltage is removed, the electric and mechanical systems return to their original states. The change in W_m is small, increasing only slightly. Hence, during the transient period, there is an interchange of energy between the spring and mass which is finally dissipated in the damper. The net change in W_f during the application and removal of the applied voltage is zero, hence the net change in W_e is positive and equal to the negative of the net change in W_m. The energy transferred to the mechanical system during this cycle is dissipated in the damper since f is fixed at zero and the mechanical system returns to its initial rest position with zero energy stored in the spring.

In Fig. 2.6-3, the initial state is that shown in Fig. 2.6-2 with 5 V applied to the electric system. The mechanical force f is increased from zero to 4 N, whereupon energy enters the coupling field from the mechanical system. Energy is transferred from the coupling field to the electric system and dissipated in the resistor, some coming from the mechanical system and some from the energy originally stored in the magnetic field. We have moved from point 2 to point 1 in Fig. 2.6-1. Next, the force is stepped back to zero from 4 N. The electric and mechanical systems return to their original states. During the cycle a net energy has been transferred from the mechanical system to the electric system which is dissipated in the resistance. This cycle is depicted on the λi plot shown in Fig. 2.6-4.

SP2.6-1. In Fig. 2.6-2, e_f jumps to 5 V when the source voltage is stepped from zero to 5 V and jumps from zero to -5 V when the source voltage is stepped from 5 V to zero. Why? [At first $t = 0^+$, $v = L(x)(di/dt)$; at second $t = 0^+$, $L(x)(di/dt) = -ir$]

SP2.6-2. Consider Fig. 2.6-1 with initial operation at point 2. Determine the final operating value of x (a) if f is stepped from zero to -1 N; (b) if f is zero but v is stepped from 5 V to 10 V. [(a) and (b) $x = 0$]

SP2.6-3. Assume that the elementary electromagnet shown in Fig. 2.2-3 portrays the λi characteristics shown in Fig 2.4-1. As the system moves from x_a to x_b, the λi trajectory moves from A to B, as shown in Fig. 2.4-1. Assume steady-state operation exists at A and B. (a) Does the voltage v increase or decrease? (b) Does the applied force f increase or decrease? [(a) and (b) decrease]

SP2.6-4. Why does not $i(t)$ portray an exponential increase when the source voltage is increased from zero to 5 V in Fig. 2.6-2? [$L(x)$]

2.7 SINGLE-PHASE RELUCTANCE MACHINE

An elementary two-pole single-phase reluctance machine, which was first shown in Fig. 1.7-2, is shown in a slightly different form in Fig. 2.7-1. In particular, the notation has been changed, that is, winding 1 is now winding *as*.

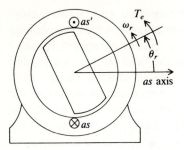

FIGURE 2.7-1
Elementary two-pole single-phase reluctance machine.

We shall see the convenience of this notation as we go along. Next, the stator or stationary member has been changed to depict more accurately the configuration common for this device. The voltage equation may be written as

$$v_{as} = r_s i_{as} + \frac{d\lambda_{as}}{dt} \qquad (2.7\text{-}1)$$

where r_s is the resistance of the *as* winding and

$$\lambda_{as} = L_{asas} i_{as} \qquad (2.7\text{-}2)$$

The self-inductance of the *as* winding, which is denoted L_{asas}, can be approximated by an expression of the same form as that for the device shown in Fig. 1.7-2. In particular, from (1.7-29),

$$L_{asas} = L_{ls} + L_A - L_B \cos 2\theta_r \qquad (2.7\text{-}3)$$

where

$$\theta_r = \int_0^t \omega_r(\xi)\, d\xi + \theta_r(0) \qquad (2.7\text{-}4)$$

and L_{ls} is the leakage inductance.

An expression for the electromagnetic torque may be obtained by using the information given in Table 2.5-1. The magnetic system is linear. Hence $W_f = W_c$, and we can use the second entry to evaluate T_e. In particular,

$$T_e(i_{as}, \theta_r) = \frac{\partial W_c(i_{as}, \theta_r)}{\partial \theta_r} \qquad (2.7\text{-}5)$$

where

$$W_c(i_{as}, \theta_r) = \tfrac{1}{2}(L_{ls} + L_A - L_B \cos 2\theta_r)i_{as}^2 \qquad (2.7\text{-}6)$$

From which

$$T_e(i_{as}, \theta_r) = L_B i_{as}^2 \sin 2\theta_r \qquad (2.7\text{-}7)$$

Although the expression for torque given by (2.7-7) is valid for transient and steady-state conditions, we shall consider only elementary modes of steady-state operation in this section.

As a first example, let i_{as} be a constant current. In this case, the torque may be expressed as

$$T_e = K \sin 2\theta_r \qquad (2.7\text{-}8)$$

where K is a constant equal to $L_B I_{as}^2$. Equation (2.7-8) is plotted in Fig. 2.7-2 with the position of the rotor shown for $\theta_r = 0, \frac{1}{4}\pi, \frac{1}{2}\pi, \ldots, 2\pi$. Let us assume that there is no external torque on the shaft; that is, there is no torque to twist the shaft one way or the other. We learned from the study of the electromagnet in the previous section that there is a force created to minimize the reluctance of the magnetic system. This would suggest that the rest position of the rotor with constant I_{as} would be at the minimum reluctance positions; either $\theta_r = \frac{1}{2}\pi$ or $\frac{3}{2}\pi$, depending upon the initial rotor position. For example, if $0 < \theta_r < \frac{1}{2}\pi$ at the time i_{as} was increased from zero to a constant value (transients neglected), the torque T_e would immediately become $K \sin 2\theta_r$, which is a positive value. Recall that the positive assumed direction of the electromagnetic torque is in the positive direction of θ_r (Fig 2.7-1). Thus, the electromagnetic torque would cause the rotor to rotate to $\theta_r = \frac{1}{2}\pi$. Is this a stable point of operation? Well, at $\theta_r = \frac{1}{2}\pi$, $T_e = 0$. This tells us that an electromagnetic torque is not created to move the rotor if θ_r is exactly at $\frac{1}{2}\pi$. However, let us try the stability test that we used on the electromagnet. If we displace the rotor ever so slightly from this operating point, will it return to $\theta_r = \frac{1}{2}\pi$? If so, then $\theta_r = \frac{1}{2}\pi$ is a stable operating point for a constant stator current. Let θ_r decrease very slightly, whereupon T_e becomes positive; hence, an electromagnetic torque is developed to increase θ_r back to $\frac{1}{2}\pi$. Now let θ_r increase very slightly beyond $\frac{1}{2}\pi$; T_e now becomes negative which forces θ_r back to $\frac{1}{2}\pi$. Certainly, $\theta_r = \frac{1}{2}\pi$ is a

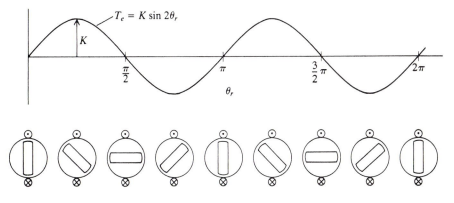

FIGURE 2.7-2
Electromagnetic torque versus angular displacement of a single-phase reluctance machine with constant stator current.

stable operating point and we can see that $\theta_r = -\frac{1}{2}\pi$ would also be a stable operating point.

We still have some questions. What would happen if the rotor were initially positioned at $\theta_r = 0$ when we increased i_{as} from zero to a constant value? Well, from our previous discussion we know that the rotor would come to rest at $\theta_r = \frac{1}{2}\pi$ or $-\frac{1}{2}\pi$. However, since there is no external torque on the rotor and since (2.7-8) tells us that $T_e = 0$ at $\theta_r = 0$, why does it not remain at $\theta_r = 0$? The fact that the external torque is satisfied is certainly one condition for stable operation; however, let us try our stability test again. When we increase θ_r slightly from zero, T_e becomes positive, which would cause θ_r to increase even further. The rotor will move away from $\theta_r = 0$ to $\frac{1}{2}\pi$, a stable operating point. If we decrease θ_r slightly from zero, T_e becomes negative, decreasing θ_r even further and the rotor will come to rest at $-\frac{1}{2}\pi$. Therefore, let us assume that, before increasing the current from zero to a constant value, we took great pains to set θ_r exactly at zero. If then we increased the current from zero to a constant value, the rotor would theoretically remain at $\theta_r = 0$. However, any disturbance, ever so slight, will cause the rotor to rotate away from $\theta_r = 0$ and there is a 50-50 chance as to the direction—counterclockwise or clockwise.

We have established that, with zero external torque on the shaft, $\theta_r = \frac{1}{2}\pi$ and $\frac{3}{2}\pi$ are stable points of operation for a constant stator current, and $\theta_r = 0$ and π are unstable. Let us see what happens when we apply an external torque. For this purpose, let us assume that initially the rotor is positioned at $\theta_r = \frac{1}{2}\pi$ with i_{as} a constant. Assume that we apply a torque to twist the shaft in the counterclockwise direction, increasing θ_r. An electromagnetic torque T_e will be developed to oppose the external torque in an attempt to maintain alignment with the as axis. When the T_e produced, in the attempt to align, is equal and opposite to the external torque, the rotor will stop advancing and come to rest with $\frac{1}{2}\pi < \theta_r < \frac{3}{4}\pi$. Now, the external torque must be less than the peak absolute value of T_e at $\theta_r = \frac{3}{4}\pi$ (K in Fig. 2.7-2). Otherwise, the torque to align cannot satisfy the external torque whereupon the rotor would continue to advance counterclockwise as long as the external torque was present. If, instead of applying a torque to twist the rotor in the counterclockwise direction, the external torque is applied to twist the rotor in the opposite direction (clockwise), θ_r would decrease. If the electromagnetic torque developed in an attempt to maintain alignment with the as axis is large enough to satisfy the external torque, then the rotor will come to rest with $\frac{1}{4}\pi < \theta_r < \frac{1}{2}\pi$. With a little thought, we can convince ourselves that, if T_e can satisfy the external torque requirements, then stable operation will occur only with $\frac{1}{4}\pi < \theta_r < \frac{3}{4}\pi$ or $\frac{5}{4}\pi < \theta_r < \frac{7}{4}\pi$, which represents the negative slope portions of the T_e versus θ_r characteristics.

Although operation of a single-phase reluctance machine with a constant stator current is impracticable, it provides a basic understanding of reluctance torque, which is the operating principle of variable-reluctance stepper motors used in line printers, for example. We will find that, in its simplest form, a variable-reluctance stepper motor consists of three cascaded, single-phase reluctance motors with the rotors on a common shaft and arranged so that their minimum reluctance paths are displaced from each other. We shall talk more about stepper motors in a later chapter.

Before leaving this brief encounter with the single-phase reluctance machine, it is instructive to look at an approximation of a practicable mode of operation of this device. As we have mentioned, the single-phase reluctance machine is used widely for clock motors. To approximate this application, we will assume that the steady-state stator current is sinusoidal. Thus,

$$I_{as} = \sqrt{2}I_s \cos \theta_{esi} \tag{2.7-9}$$

where, in general,

$$\theta_{esi} = \int_0^t \omega_e(\xi) \, d\xi + \theta_{esi}(0) \tag{2.7-10}$$

In the above equations the electrical angular displacement has the subscript *esi*. The *e* denotes association with an electrical variable, *s* with a stator variable, and *i* with current. For steady-state operation (ω_e constant),

$$\theta_{esi} = \omega_e t + \theta_{esi}(0) \tag{2.7-11}$$

We know that θ_r, which is given by (2.7-4), may also be written in a form similar to (2.7-11) for steady-state operation. Substituting (2.7-9) into (2.7-8) and after using a few trigonometric identities from Appendix A, we have

$$
\begin{aligned}
T_e = {}& L_B I_s^2 \sin 2[\omega_r t + \theta_r(0)] \\
& + \tfrac{1}{2} L_B I_s^2 \{\sin 2[(\omega_r + \omega_e)t + \theta_r(0) + \theta_{esi}(0)] \\
& + \sin 2[(\omega_r - \omega_e)t + \theta_r(0) - \theta_{esi}(0)]\}
\end{aligned}
\tag{2.7-12}
$$

We see from the first term that an average steady-state torque is produced when $\omega_r = 0$. More importantly, we see that either the second or the third term produces an average torque whenever $|\omega_r| = \omega_e$, where ω_e is the angular velocity of the electric system (synchronous speed). For this example, let $\omega_r = \omega_e$, then (2.7-12) becomes

$$
\begin{aligned}
T_e = {}& L_B I_s^2 \sin 2[\omega_e t + \theta_r(0)] \\
& + \tfrac{1}{2} L_B I_s^2 \sin 2[2\omega_e t + \theta_r(0) + \theta_{esi}(0)] \\
& + \tfrac{1}{2} L_B I_s^2 \sin 2[\theta_r(0) - \theta_{esi}(0)]
\end{aligned}
\tag{2.7-13}
$$

Recall that $\theta_r(0)$ and $\theta_{esi}(0)$ are the time zero values of the displacement of the rotor and the current, respectively. Also, we can designate time zero as we please.

It appears from (2.7-13) that we should expect the steady-state torque to pulsate at $4\omega_e$ and $2\omega_e$, and the average torque is a double-angle sinusoidal function of $\theta_r(0) - \theta_{esi}(0)$. Although (2.7-13) illustrates that a clock motor develops an average torque, it is difficult to explain how this all takes place.

Even though you probably have numerous questions regarding the operation of a single-phase clock motor, it is perhaps best to delay a more in-depth discussion until we have studied the material on the rotating magnetic field (rotating air gap mmf) in Chap. 4.

SP2.7-1. Initially, the rotor of a single-phase reluctance motor is positioned with $\theta_r = 0$. Instantaneously i_{as} is stepped to 1 A and an external torque of 0.01 N·m is applied to rotate the rotor in the clockwise direction. $L_B = 0.02$ H. Neglect transients and determine the final value of θ_r. [$\theta_r = -105°$]

SP2.7-2. The device in SP2.7-1 is operating steadily at $\theta_r = -105°$ when i_{as} is reversed instantaneously ($I_{as} = -1$ A). Neglect the transients and determine the final value of θ_r. [$\theta_r = -105°$]

SP2.7-3. The friction and windage losses of a clock motor appear as a retarding torque of 0.01 N·m at $\omega_r = \omega_e$. The steady-state current is $I_{as} = 2 \cos \omega_e t$ and $L_B = 0.02$ H. Calculate $\theta_r(0) - \theta_{esi}(0)$ if the clock is rotating steadily at ω_e. [$\theta_r(0) = -105°$]

2.8 WINDINGS IN RELATIVE MOTION

It is instructive to formulate an expression for the electromagnetic torque of the elementary rotational device shown in Fig. 1.7-4. The voltage equations are given by (1.7-30) and (1.7-31), the self-inductances L_{11} and L_{22} are constant as given by (1.7-34) and (1.7-35), and the mutual inductance is given by (1.7-38). Let us rewrite these equations here for convenience. The voltage equations may be written in matrix form as

$$\begin{bmatrix} v_1 \\ v_2 \end{bmatrix} = \begin{bmatrix} r_1 & 0 \\ 0 & r_2 \end{bmatrix} \begin{bmatrix} i_1 \\ i_2 \end{bmatrix} + \frac{d}{dt} \begin{bmatrix} \lambda_1 \\ \lambda_2 \end{bmatrix} \qquad (2.8\text{-}1)$$

which may also be written as

$$\mathbf{v}_{12} = \mathbf{r}\mathbf{i}_{12} + \frac{d}{dt}\boldsymbol{\lambda}_{12} \qquad (2.8\text{-}2)$$

All terms are defined by comparison with (2.8-1). The flux linkage equations are

$$\begin{bmatrix} \lambda_1 \\ \lambda_2 \end{bmatrix} = \begin{bmatrix} L_{l1} + L_{m1} & L_{sr} \cos \theta_r \\ L_{sr} \cos \theta_r & L_{l2} + L_{m2} \end{bmatrix} \begin{bmatrix} i_{as} \\ i_{bs} \end{bmatrix} \qquad (2.8\text{-}3)$$

From Table 2.5-1,

$$T_e(i_1, i_2, \theta_r) = \frac{\partial W_c(i_1, i_2, \theta_r)}{\partial \theta_r} \qquad (2.8\text{-}4)$$

Since the magnetic system is assumed to be linear,

$$W_c(i_1, i_2, \theta_r) = \tfrac{1}{2}L_{11}i_1^2 + L_{12}i_1i_2 + \tfrac{1}{2}L_{22}i_2^2 \qquad (2.8\text{-}5)$$

The self-inductances are constants and the mutual inductance is represented by the off-diagonal terms in (2.8-3). Thus,

$$\boxed{T_e(i_1, i_2, \theta_r) = -i_1i_2L_{sr}\sin\theta_r} \qquad (2.8\text{-}6)$$

Although (2.8-6) is valid regardless of the form of i_1 and i_2, let us consider for a moment the form of the torque if i_1 and i_2 are both positive and constant. For the positive direction of current shown, the torque would then be of the form

$$T_e = -K\sin\theta_r \qquad (2.8\text{-}7)$$

where K is a positive constant equal to $i_1i_2L_{sr}$.

A plot of (2.8-7) is shown in Fig. 2.8-1 with the position of the windings illustrated for $\theta_r = 0, \tfrac{1}{2}\pi, \ldots, 2\pi$. Also shown are the positions of the poles of the magnetic systems created by constant positive current flowing in the windings. It may at first appear that the north (N^s) and south (S^s) poles produced by positive current flowing in the stator winding are positioned incorrectly. However, recall that flux issues from a north pole of a magnet into the air. Since the stator and rotor windings must each be considered as creating separate magnetic systems, we realize, by the right-hand rule, that flux issues from the north pole of the magnetic system established by the stator winding

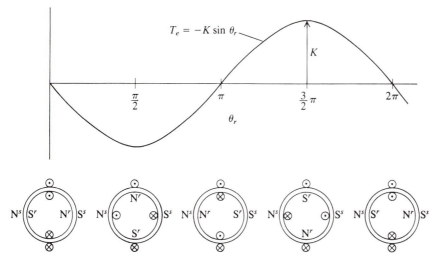

FIGURE 2.8-1
Electromagnetic torque versus angular displacement with constant winding currents.

into the air gap. Similarly, the flux produced by positive current flowing in the rotor winding enters the air gap from the north pole of the magnetic system of the rotor.

Here we see that an electromagnetic torque is produced in an attempt to align the magnetic systems established by currents flowing in the stator and rotor windings; in other words, to align the 1- and 2-axes. We can use our stability test, as we did in the case of the single-phase reluctance motor, to establish the fact that, with no external torque on the rotor, stable positioning occurs at $\theta_r = 0$ while unstable operation occurs at $\theta_r = \pi$. This, of course, assumes that i_1 and i_2 are positive constants. Also, stable operation with an external applied torque occurs over the range of $-\frac{1}{2}\pi < \theta_r < \frac{1}{2}\pi$—the negative slope portion of the T_e versus θ_r characteristic. Although operation with constant winding currents is somewhat impracticable, it does illustrate the principle of positioning of stepper motors with a permanent-magnet rotor which, in many respects, is analogous to holding i_2 constant in the elementary device considered here.

Before leaving this elementary device for good, let us assume that i_2 is constant $(i_2 = I_2)$ and

$$I_1 = \sqrt{2}I \cos \theta_{esi} \tag{2.8-8}$$

where, for steady-state operation,

$$\theta_{esi} = \omega_e t + \theta_{esi}(0) \tag{2.8-9}$$

Also,
$$\theta_r = \omega_r t + \theta_r(0) \tag{2.8-10}$$

Substituting into (2.8-6) yields

$$T_e = -\sqrt{2}II_2 L_{sr} \cos [\omega_e t + \theta_{esi}(0)] \sin [\omega_r t + \theta_r(0)] \tag{2.8-11}$$

which can be expressed as

$$T_e = -\sqrt{2}II_2 L_{sr}\{ \tfrac{1}{2} \sin [(\omega_r + \omega_e)t + \theta_r(0) + \theta_{esi}(0)]$$
$$+ \tfrac{1}{2} \sin [(\omega_r - \omega_e)t + \theta_r(0) - \theta_{esi}(0)]\} \tag{2.8-12}$$

Here we see that an average torque is produced if $|\omega_r| = \omega_e$. In particular, if $\omega_r = \omega_e$, (2.8-12) becomes

$$T_e = -\sqrt{2}II_2 L_{sr}\{ \tfrac{1}{2} \sin [2\omega_e t + \theta_r(0) + \theta_{esi}(0)] + \tfrac{1}{2} \sin [\theta_r(0) - \theta_{esi}(0)]\} \tag{2.8-13}$$

From this brief analysis, we can see that an average torque will be produced with $|\omega_r| = \omega_e$. In fact, with i_2 constant and I_1 given by (2.8-8), this device is an elementary single-phase synchronous machine. However, as in the case of the single-phase reluctance machine, we do not really have all the tools at hand to provide you with an in-depth explanation of its operation. We will be able to do this once we have studied the material in Chap. 4.

SP2.8-1. In the system shown in Fig. 1.7-4, $L_{sr} = 0.1$ H, $i_1 = 2$ A, and $i_2 = 10$ A. (a) A torque of 1 N·m is applied in the clockwise direction. Calculate the steady-state value of θ_r. (b) Repeat with a torque of 2 N·m. [(a) $\theta_r = -30°$; (b) unstable]

SP2.8-2. Reverse the direction of i_1 in Fig. 1.7-4 and determine the range of stable operation for constant, positive currents. [$\frac{1}{2}\pi < \theta_r < \frac{3}{2}\pi$]

SP2.8-3. In Fig. 1.7-4, winding 1 is moved so that $\otimes$ is at three o'clock and $\odot$ at nine. Express T_e. [$T_e = i_1 i_2 L_{sr} \cos \theta_r$]

SP2.8-4. The currents i_1 and i_2 are positive constants in the device shown in Fig. 1.7-4. (a) An external torque is applied to increase θ_r to 45° and then released. Assume damping exists. What is the final position of the rotor? Repeat for θ_r increased to (b) 90°, (c) 120°, (d) 180°, and (e) 210°. [(a) $\theta_r = 0$; (b) 0; (c) 0; (d) 0 or 2π; (e) 2π]

SP2.8-5. In the system shown in Fig. 1.7-4, $L_{sr} = 0.1$ H, $i_1 = 20 \cos \omega_e t$, $i_2 = 2$ A, and $\omega_r = \omega_e$. Calculate $\theta_r(0) - \theta_{esi}(0)$ if an external torque of 1 N·m is applied in the clockwise direction. [$\theta_r(0) = -30°$]

2.9   RECAPPING

The primary purpose of this chapter is to establish a means of expressing the force or torque in electromechanical devices. Therefore, an understanding of all relationships given in Table 2.5-1 is of importance. However, since the magnetic system of the devices which we will consider is assumed to be linear, we will make extensive use of the second entry of Table 2.5-1.

The detailed analysis of the elementary electromagnet and the associated problems at the end of the chapter warrant consideration since this material will not be revisited later in the text. You are now prepared to proceed to the study of dc machines (Chap. 3) or jump to Chap. 4 and begin study of the electromechanical motion devices which operate on the principle of a rotating magnetic field. This includes induction machines, synchronous machines, stepper motors, brushless dc motors, and servomotors.

2.10   REFERENCES

1. D. M. Triezenberg, *Electric Power Systems*, Classnotes, Purdue University, 1978.
2. D. C. White and H. H. Woodson, *Electromechanical Energy Conversion*, John Wiley and Sons, Inc., New York, 1959.

2.11   PROBLEMS

1. A resistor and an inductor are connected as shown in Fig. 2.11-1 with $R = 15\ \Omega$ and $L = 250$ mH. Express the energy stored in the inductor and the energy dissipated by the resistor for $t > 0$ if $i(0) = 10$ A.
2. Consider the spring-mass-damper system shown in Fig. 2.11-2. Let $x_0 = 0$ and assume $f = \cos \omega t$. Express the steady-state response $x(t)$ in the form $x = X_s \cos(\omega t + \phi)$.

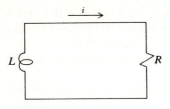

FIGURE 2.11-1
RL circuit.

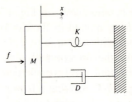

FIGURE 2.11-2
Spring-mass-damper system with applied
force.

*3. Consider the spring-mass-damper system shown in Fig. 2.11-3. At $t = 0$, $x(0) = x_0$ (rest position), and $dx/dt = 1.5\,\text{m/s}$. Also, $M = 0.8\,\text{kg}$, $D = 10\,\text{N}\cdot\text{s/m}$, and $K = 120\,\text{N}\cdot\text{m}$. For $t > 0$, express the energy stored in the spring, W_{mS1}, the kinetic energy of the mass, W_{mS2}, and the energy dissipated by the damper, W_{mL}. You need not evaluate the integral expression for W_{mL}.

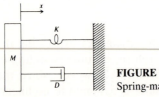

FIGURE 2.11-3
Spring-mass-damper system.

4. Express $W_f(i, x)$ and $W_c(i, x)$ for (a) $\lambda(i, x) = xi^3 + i$; (b) $\lambda(i, x) = -xi^2 + i \sin x$.

*5. The energy stored in the coupling field of a magnetically linear system with two electric inputs may be expressed

$$W_f(\lambda_1, \lambda_2, x) = \tfrac{1}{2}B_{11}\lambda_1^2 + B_{12}\lambda_1\lambda_2 + \tfrac{1}{2}B_{22}\lambda_2^2$$

Express B_{11}, B_{12}, and B_{22} in terms of inductances L_{11}, L_{12}, and L_{22}.

6. An electromechanical system has two electric inputs. The flux linkages may be expressed as

$$\lambda_1(i_1, i_2, x) = x^2 i_1^2 + xi_2 + i_1$$
$$\lambda_2(i_1, i_2, x) = x^2 i_2^2 + xi_1 + i_2$$

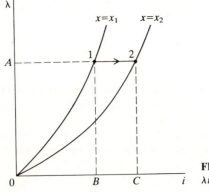

FIGURE 2.11-4
λi characteristics.

Express $W_f(i_1, i_2, x)$ and $W_c(i_1, i_2, x)$ by first making i_2 the variable of integration with $di_1 = 0$ and $i_1 = 0$. Then let i_1 be the variable of integration with $di_2 = 0$ and i_2 at its final value.

7. The mechanical system moves from x_1 to x_2 along the λi path 1 to 2 in Fig. 2.11-4. Identify the following by areas: (a) ΔW_f, (b) ΔW_c, (c) ΔW_e, (d) ΔW_m.

8. Let $i = k\lambda^2 e^{2x}$. Evaluate f_e when $\lambda = 2 \text{ V} \cdot \text{s}$, $x = 1 \text{ m}$, and $k = 1 \text{ A}/(\text{V} \cdot \text{s})^2$.

9. The plunger with mass M shown in Fig. 2.11-5 is free to move within an electromagnet. Although the winding of the electromagnet consists of many turns, only one is shown. The mechanical damping varies directly as the surface area of the plunger within the electromagnet.
   (a) Write the voltage equation for the electric system.
   (b) Write the dynamic equation for the mechanical system. Include the force of gravity.
   (c) Express the mechanical damping D.
   (d) Express the electromagnetic force f_e if $i = a\lambda^2 + b\lambda(x - d)^2$.
   (e) Express the steady-state position x for a constant current flowing in the winding.

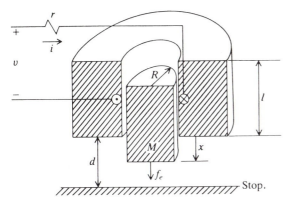
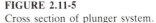

FIGURE 2.11-5
Cross section of plunger system.

10. Express the force of attraction between the iron faces of the air gap in Fig. 1.2-1 in terms of N, the turns of the winding; i, the current flowing in the winding; μ_0; A_g, the cross-sectional area of the air gap; and g, the air gap length. Neglect the reluctance of the iron. (*Hint:* Express the energy in the air gap in terms of air gap length and allow an infinitesimal change in air gap length dg. This procedure is often referred to as "virtual displacement.")

11. Express $f_e(i_1, i_2, x)$ for λ_1 and λ_2 given in Prob. 6.

12. For the electromagnet shown in Fig. 2.2-3, let $L(x) = L_l + L_m(x)$, where L_l is constant and $L_m(x)$ is expressed by (1.7-17). Express f_e for $x = 0$.

13. Following the system transients due to the application of the source voltage in Fig. 2.6-2 ($v = 5 \text{ V}$), the system assumes steady-state operation. Calculate W_{eS}, W_f, and W_{mS} during steady-state operation. [*Hint:* The external inductance l is set equal to zero in this example and $W_{mS} = K \int_{x_0}^{x} (\xi - x_0) \, d\xi$.]

14. Consider Fig. 2.6-3 wherein the force f is changed. The source voltage v is constant and the leakage inductance L_l and the external inductance l are both zero. As f is changed, e_f changes during the transient. Show that, for transient and

steady-state conditions, the energy dissipated in the resistor may be expressed as

$$W_{eL} = \frac{v^2}{r} \int dt - \frac{2v}{r} \int e_f \, dt + \frac{1}{r} \int e_f^2 \, dt$$

Also express W_E and W_e.

15. The dc source for the electromagnet considered in Sec. 2.6 is replaced with a 60-Hz source. With $f = 0$ and $i = \sqrt{2}\,0.5 \cos \omega_e t$, it is found that the system operates with x constant at 2.5 mm.
    (a) Express f_e and justify this observation.
    (b) Calculate the applied voltage.

16. Assume that the steady-state current supplied to a single-phase reluctance motor is sinusoidal. The rotor speed is $\omega_r = \omega_e$. Determine the harmonics which must be present in the applied voltage V_{as} in order for $I_{as} = \sqrt{2}I_s \cos \omega_e t$.

*17. The steady-state currents flowing in the conductors of the device shown in Fig. 1.7-4 are $I_1 = \sqrt{2}I_{s1} \cos \omega_1 t$ and $I_2 = \sqrt{2}I_{s2} \cos (\omega_2 t + \phi_2)$. Assume that during steady-state operation the rotor speed is constant; thus, $\theta_r = \omega_r t + \theta_r(0)$, where $\theta_r(0)$ is the rotor displacement at time zero. Determine the rotor speed(s) at which the device produces a nonzero average torque during steady-state operation if (a) $\omega_1 = \omega_2 = 0$, (b) $\omega_1 = \omega_2 \neq 0$, and (c) $\omega_1 \neq 0$, $\omega_2 = 0$.

18. In Fig. 2.11-6, θ_r and ω_r are positive in the clockwise direction. The peak amplitude of the mutual inductance is L_{sr}. Express (a) the mutual inductance L_{ab} and (b) the electromagnetic torque T_e.

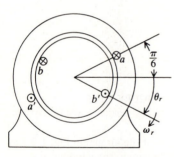

FIGURE 2.11-6
A 2-winding device with clockwise rotation.

19. Consider the electromechanical system shown in Fig. 2.11-7. Assume the peak amplitude of the mutual inductance is L_{sr}.

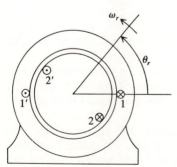

FIGURE 2.11-7
A 2-winding device with counterclockwise rotation.

(*a*) Express the mutual inductance
(*b*) Show the location of the north and south poles of the stator (N^s and S^s) and of the rotor (N^r and S^r) for positive i_1 and negative i_2.
(*c*) Express the electromagnetic torque T_e.

20. The device shown in Fig. 2.11-8 has two stator windings. Winding 1 is fixed in the position shown. However, we are able to wind winding 2 at any angle α relative to winding 1.

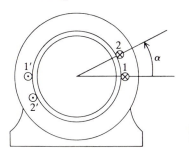

FIGURE 2.11-8
Two stator windings.

(*a*) Express the mutual inductance between windings 1 and 2 as a function of α. Let the peak amplitude be M.
(*b*) Use the concept of virtual displacement to express the torque between the windings.

CHAPTER
3

DIRECT-CURRENT MACHINES

3.1 INTRODUCTION

The direct-current (dc) machine is not as widely used today as it was in the past. For the most part, the dc generator has been replaced by solid-state rectifiers which convert alternating current into direct current with provisions to control the magnitude of the dc voltage. Nevertheless, it is still desirable to devote some time to the dc machine in an introductory course on electromechanical devices since it is still used as a drive motor, especially at the low-power level. This chapter is an attempt to treat dc machines, perhaps a bit more in detail than necessary, but with the flexibility so that one can focus upon the topics which are still of interest to the control or energy systems engineer. In this regard, the dc generator is not considered in detail and, for that matter, neither are the series and compound machines. Today's focus is on the shunt-connected dc motor and the permanent-magnet dc motor which have similar operating characteristics.

A simplified method of analysis is presented rather than an analysis wherein commutation is treated in detail. With this type of analytical approach, the dc machine is considered to be the most straightforward to analyze of all electromechanical devices. Numerous textbooks have been written over the last century on the design, theory, and operation of dc machines. One can add little to the analytical approach which has been used for years. In this chapter, the well-established theory of dc machines is set forth and the dynamic characteristics of the shunt and permanent-magnet machines are illustrated.

The time-domain block diagrams and state equations are then developed for these two types of motors and, in the final section of this chapter, the linear differential equations which describe the dynamic characteristics of these devices are solved by using Laplace transformation techniques. An introductory course in linear differential equations is sufficient background for the material on time-domain block diagrams and state equations. However, those who have not had the opportunity to become familiar with Laplace transformations may omit the final section. It is not a prerequisite for the material in the following chapters.

3.2 ELEMENTARY DIRECT-CURRENT MACHINE

It is instructive to discuss the elementary machine shown in Fig. 3.2-1 prior to a formal analysis of the performance of a practical dc machine. The two-pole elementary machine is equipped with a field winding wound on the stator poles, a rotor coil (a-a'), and a commutator. The commutator is made up of two semicircular copper segments mounted on the shaft at the end of the rotor and insulated from one another as well as from the iron of the rotor. Each terminal of the rotor coil is connected to a copper segment. Stationary carbon brushes ride upon the copper segments whereby the rotor coil is connected to a stationary circuit by a near frictionless contract.

The voltage equations for the field winding and rotor coil are

$$v_f = r_f i_f + \frac{d\lambda_f}{dt} \tag{3.2-1}$$

$$v_{a\text{-}a'} = r_a i_{a-a'} + \frac{d\lambda_{a\text{-}a'}}{dt} \tag{3.2-2}$$

where r_f and r_a are the resistance of the field winding and armature coil, respectively. The rotor of a dc machine is commonly referred to as the *armature*; rotor and armature will be used interchangeably. At this point in the analysis it is sufficient to express the flux linkages as

$$\lambda_f = L_{ff} i_f + L_{fa} i_{a\text{-}a'} \tag{3.2-3}$$

$$\lambda_{a\text{-}a'} = L_{af} i_f + L_{aa} i_{a\text{-}a'} \tag{3.2-4}$$

As a first approximation, the mutual inductance between the field winding and an armature coil may be expressed as a sinusoidal function of θ_r as

$$L_{af} = L_{fa} = -L \cos \theta_r \tag{3.2-5}$$

where L is a constant. As the rotor revolves, the action of the commutator is to switch the stationary terminals from one terminal of the rotor coil to the other. For the configuration shown in Fig. 3.2-1, this switching or commutation occurs at $\theta_r = 0, \pi, 2\pi, \ldots$. At the instant of switching, each brush is in contact with both copper segments whereupon the rotor coil is short-circuited. It is desirable

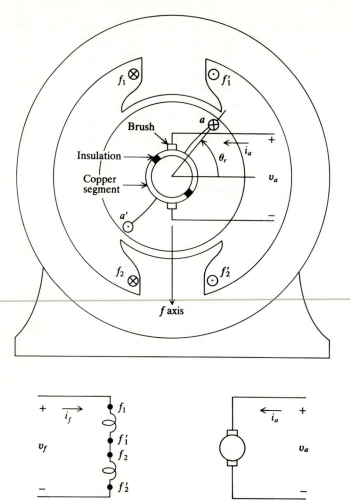

FIGURE 3.2-1
Elementary two-pole dc machine.

to commutate (short-circuit) the rotor coil at the instant the induced voltage is a minimum. The waveform of the voltage induced in the open-circuited armature coil, during constant-speed operation with a constant field winding current, may be determined by setting $i_{a\text{-}a'} = 0$ and i_f equal to a constant. Substituting (3.2-4) and (3.2-5) into (3.2-2) yields the following expression for the open-circuit voltage of coil $a\text{-}a'$ with the field current i_f a constant:

$$v_{a\text{-}a'} = \omega_r L I_f \sin \theta_r \qquad (3.2\text{-}6)$$

where $\omega_r = d\theta_r/dt$ is the rotor speed. The open-circuit coil voltage $v_{a\text{-}a'}$ is zero at $\theta_r = 0, \pi, 2\pi, \ldots$, which is the rotor position during commutation. Commutation is illustrated in Fig. 3.2-2. The open-circuit terminal voltage, v_a corre-

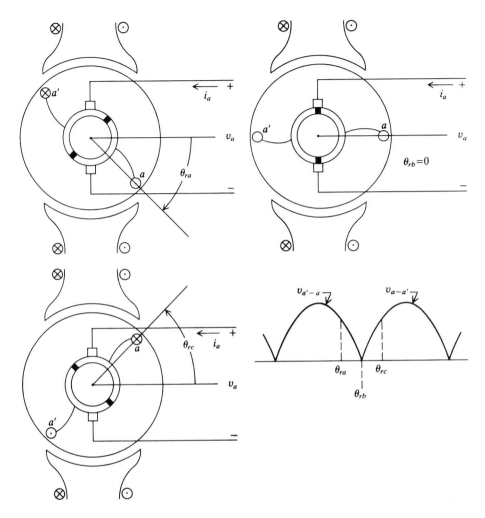

FIGURE 3.2-2
Commutation of the elementary dc machine.

sponding to the rotor positions denoted as θ_{ra}, θ_{rb} $(\theta_{rb} = 0)$, and θ_{rc} are indicated. It is important to note that, during one revolution of the rotor, the assumed positive direction of armature current i_a is down coil side a and out of coil side a' for $0 < \theta_r < \pi$. For $\pi < \theta_r < 2\pi$, positive current is down coil side a' and out of coil side a. In Chaps. 1 and 2 we let positive current flow into the winding denoted without a prime and out the winding denoted with a prime. We will not be able to adhere to this relationship in the case of the armature windings of a dc machine since commutation is involved.

The machine shown in Fig. 3.2-1 is not a practicable machine. Although it could be operated as a generator supplying a resistive load, it could not be operated effectively as a motor supplied from a voltage source owing to the short-circuiting of the armature coil at each commutation. A practicable dc machine, with the rotor equipped with an a winding and an A winding, is shown schematically in Fig. 3.2-3. At the rotor position depicted, coils a_4-a_4' and A_4-A_4' are being commutated. The bottom brush short-circuits the a_4-a_4' coil while the top brush short-circuits the A_4-A_4' coil. Figure 3.2-3 illustrates the instant when the assumed direction of positive current is into the paper in coil sides a_1, A_1; a_2, A_2; . . ., and out in coil sides a_1', A_1'; a_2', A_2'; . . . . It is instructive to follow the path of current through one of the parallel paths from

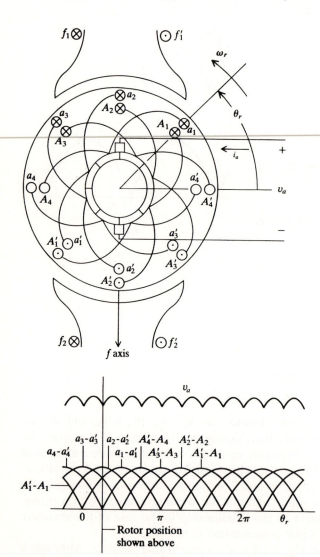

FIGURE 3.2-3
A dc machine with parallel armature windings.

one brush to the other. For the angular position shown in Fig. 3.2-3, positive current enters the top brush and flows down the rotor via a_1 and back through a_1'; down a_2 and back through a_2'; down a_3 and back through a_3' to the bottom brush. A parallel current path exists through A_3-A_3', A_2-A_2', and A_1-A_1'. The open-circuit or induced armature voltage is also shown in Fig. 3.2-3; however, these idealized waveforms require additional explanation. As the rotor advances in the counterclockwise direction, the segment connected to a_1 and A_4 moves from under the top brush, as shown in Fig. 3.2-4. The top brush then

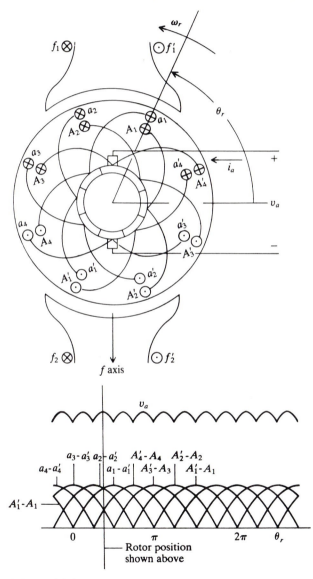

FIGURE 3.2-4

Same as Fig. 3.2-3 with rotor advanced approximately 22.5° counterclockwise.

rides only on the segment connecting A_3 and A_4'. At the same time the bottom brush is riding on the segment connecting a_4 and a_3'. With the rotor so positioned, current flows in A_3 and A_4' and out a_4 and a_3'. In other words, current flows down the coil sides in the upper one half of the rotor and out of the coil sides in the bottom one half. Let us follow the current flow through the parallel paths of the armature windings shown in Fig. 3.2-4. Current now flows through the top brush into A_4' out A_4, into a_1 out a_1', into a_2, out a_2', into a_3 out a_3' to the bottom brush. The parallel path beginning at the top brush is A_3-A_3', A_2-A_2', A_1-A_1', and a_4'-a_4 to the bottom brush. The voltage induced in the coils is shown in Figs. 3.2-3 and 3.2-4 for the first parallel path described. It is noted that the induced voltage is plotted only when the coil is in this parallel path.

In Figs. 3.2-3 and 3.2-4, the parallel windings consist of only four coils. Usually the number of rotor coils is substantially more then four, thereby reducing the harmonic content of the open-circuit armature voltage. In this case, the rotor coils may be approximated as a uniformly distributed winding, as illustrated in Fig. 3.2-5. Therein the rotor winding is considered as current

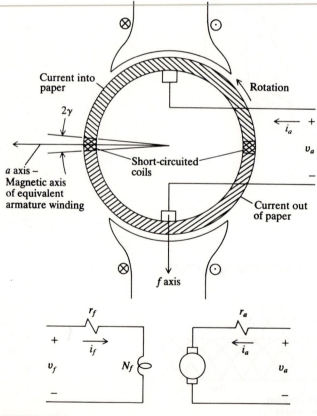

FIGURE 3.2-5
Idealized dc machine with uniformly distributed rotor winding.

sheets which are fixed in space due to the action of the commutator and which establish a magnetic axis positioned orthogonal to the magnetic axis of the field winding. The brushes are shown positioned on the current sheet for the purpose of depicting commutation. The small angular displacement, denoted by 2γ, designates the region of commutation wherein the coils are short-circuited. However, commutation cannot be visualized from Fig. 3.2-5; one must refer to Figs. 3.2-3 and 3.2-4.

In our discussion of commutation, it was assumed that the armature current was zero. With this constraint, the sinusoidal voltage induced in each armature coil crosses through zero when the coil is orthogonal to the field flux. Hence, the commutator was arranged so that commutation would occur when an armature coil was orthogonal to the field flux. When current flows in the armature winding, the flux established therefrom is in an axis orthogonal to the field flux. Thus, a voltage will be induced in the armature coil which is being commutated as a result of "cutting" the flux established by the current flowing in the other armature coils. Arcing at the brushes will occur, and the brushes and copper segments may be damaged with even a relatively small armature current. Although the design of dc machines is not a subject of this text, it is important to mention that brush arcing may be substantially reduced by mechanically shifting the position of the brushes as a function of armature current or by means of interpoles. Interpoles or commutating poles are small stator poles placed over the coil sides of the winding being commutated, midway between the main poles of large horsepower machines. The action of the interpole is to oppose the flux produced by the armature current in the region of the short-circuited coil. Since the flux produced in this region is a function of the armature current, it is desirable to make the flux produced by the interpole a function of the armature current. This is accomplished by winding the interpole with a few turns of the conductor carrying the armature current. Electrically, the interpole winding is between the brush and the terminal. It may be approximated, in the voltage equations, by increasing slightly the armature resistance and inductance (r_a and L_{AA}).

In dc motors, which are subjected to relatively large armature currents or to severe load cycles as in steel mill applications, arcing may occur between commutator segments. This undesirable situation may be minimized by embedding a compensating or pole face winding in the face of each stator pole. It consists of several turns connected in series with the armature circuit. It is wound so that its flux opposes the flux produced by the armature windings under the poles. Electrically the compensating winding may be accounted for, as before, by increasing r_a and L_{AA}.

Another aspect of dc machines, which is important but which cannot be considered in detail in this brief look at dc machines, is the method of winding the armature of multipole machines. Generally, multipole dc machines are used to supply large currents or generate high voltages. To carry large currents, the lap-type of winding is used since this method of winding requires a pair of brushes for each pair of poles. The brushes are connected in parallel and, consequently, large armature currents may be tolerated. The lap winding is similar to the type of winding described in the previous section. For high voltages, a so-called wave winding is used in multipole machines. This method of winding employs only two brushes, but is arranged so that a larger number of coils are connected in series, resulting in a higher armature voltage. Regardless of the type of winding or the number of poles, the equations set forth in the following section for a two-pole machine, wherein ω_r is the rotor angular velocity, may be applied directly to any multipole machine without modification.

Before proceeding to the development of the equations portraying the operating characteristics of these devices, it is instructive to take a brief look at the arrangement of the armature windings and the method of commutation used in many of the low-power permanent-magnet dc motors. Small dc motors used in low-power control systems are often the permanent-magnet type, wherein a constant field flux is established by a permanent magnet rather than by a current flowing in a field winding.

Three rotor positions of a typical low-power permanent-magnet dc motor are shown in Fig. 3.2-6. The rotor is turning in the counterclockwise direction and the rotor position advances from left to right. Physically, these devices may be an inch or less in diameter and in axial length with brushes sometimes as small as a pencil lead. They are mass-produced and relatively inexpensive. Although the brushes actually ride on the outside of the commutator, for convenience they are shown on the inside in Fig. 3.2-6. The armature windings consist of a large number of turns of fine wire, hence, each circle shown in Fig. 3.2-6 represents many conductors. Note that the position of the brushes is shifted approximately 40° relative to a line drawn between the center of the north and south poles. This shift in the brushes was probably determined experimentally by minimizing brush arcing for normal load conditions. Note also that the windings do not span π radians but more like 140°, and there is an odd number of commutator segments.

In Fig. 3.2-6a, only winding 4 is being commutated. As the rotor advances from the position shown in Fig. 3.2-6a, both windings 4 and 1 are being commutated. In Fig. 3.2-6b, winding 1 is being commutated; then windings 1 and 5, and, finally, in Fig. 3.2-6c, only winding 5 is being commutated. The windings are being commutated when the induced voltage is nonzero and one would expect some arcing to occur at the brushes. We must realize, however, that the manufacture and sale of these devices is very competitive, and one often must compromise when striving to produce an acceptable motor at the least cost possible. Although we realize that in some cases it may be a rather crude approximation, we will consider the permanent-magnet dc motor as having current sheets on the armature with orthogonal armature and field magnetic axes as shown in Fig. 3.2-5.

A two-pole 3-hp 180-V 2500-r/min (revolutions per minute) general-purpose shunt-connected dc motor is shown in Fig. 3.2-7 in cutaway. A disassembled two-pole 0.1-hp 6-V 12,000-r/min permanent-magnet dc motor is

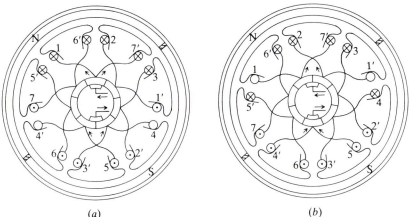

FIGURE 3.2-6
Commutation of a permanent-magnet dc motor.

shown in Fig. 3.2-8. The magnets are samarium cobalt and the device is used to drive hand-held battery-operated surgical instruments. Although some of these terms are new to us, they will be defined as we go along.

SP3.2-1. The peak value of the voltage induced in one coil shown in Fig. 3.2-3 is 1 V. Determine, from Fig. 3.2-3, the maximum and minimum value of v_a. [2.613 V; 2.414 V]
SP3.2-2. Consider Fig. 3.2-3. Indicate the two parallel paths immediately following commutation of a_3-a_3' and A_3-A_3'. [A_3'-A_3, A_4'-A_4, a_1-a_1', and a_2-a_2'; A_2-A_2', A_1-A_1', a_4'-a_4, and a_3'-a_3]

FIGURE 3.2-7
Two-pole 3-hp 180-V 2500-r/min shunt-connected dc motor. (*Courtesy of General Electric.*)

FIGURE 3.2-8
Two-pole 0.1-hp 6-V 12,000-r/min permanent-magnet dc motor. (*Courtesy of Vickers ElectroMech.*)

3.3 VOLTAGE AND TORQUE EQUATIONS

Although rigorous derivation of the voltage and torque equations is possible, it is rather lengthy and little is gained since these relationships may be deduced. The armature coils revolve in a magnetic field established by a current flowing in the field winding. We have established that voltage is induced in these coils by virtue of this rotation. However, the action of the commutator causes the armature coils to appear as a stationary winding with its magnetic axis orthogonal to the magnetic axis of the field winding. Consequently, voltages are not induced in one winding due to the time rate of change of the current flowing in the other (transformer action). Mindful of these conditions, we can write the field and armature voltage equations in matrix form as

$$\begin{bmatrix} v_f \\ v_a \end{bmatrix} = \begin{bmatrix} r_f + pL_{FF} & 0 \\ \omega_r L_{AF} & r_a + pL_{AA} \end{bmatrix} \begin{bmatrix} i_f \\ i_a \end{bmatrix} \tag{3.3-1}$$

where L_{FF} and L_{AA} are the self-inductances of the field and armature windings, respectively, and p is the short-hand notation for the operator d/dt. The rotor speed is denoted as ω_r and L_{AF} is the mutual inductance between the field and the rotating armature coils. The above equation suggests the equivalent circuit shown in Fig. 3.3-1. The voltage induced in the armature circuit, $\omega_r L_{AF} i_f$, is commonly referred to as the counter or back emf. It also represents the open-circuit armature voltage.

There are several other forms in which the field and armature voltage equations are often expressed. For example, L_{AF} may also be written as

$$L_{AF} = \frac{N_a N_f}{\mathcal{R}} \tag{3.3-2}$$

where N_a and N_f are the equivalent turns of the armature and field windings, respectively, and $\mathcal{R}$ is the reluctance. Thus,

$$L_{AF} i_f = N_a \frac{N_f i_f}{\mathcal{R}} \tag{3.3-3}$$

If we now replace $N_f i_f / \mathcal{R}$ with Φ_f, the field flux per pole, then $N_a \Phi_f$ may be substituted for $L_{AF} i_f$ in the armature voltage equation.

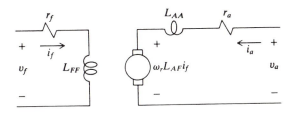

FIGURE 3.3-1
Equivalent circuit of dc machine.

Another substitute variable often used is

$$k_v = L_{AF} i_f \qquad (3.3\text{-}4)$$

We will find that this substitute variable is particularly convenient and frequently used. Even though a permanent-magnet dc machine has no field circuit, the constant field flux produced by the permanent magnet is analogous to a dc machine with a constant k_v.

We can take advantage of previous work to obtain an expression for the electromagnetic torque. In particular, the expression for torque given by (2.8-6) may be used directly to express the torque for the dc machine. If we fix θ_r in Fig. 1.7-4 or Fig. 2.8-1 at $-\frac{1}{2}\pi$, the same relationship exists between the magnetic axes of a dc machine (Fig. 3.2-5) and the magnetic axes of the two-coil machine. Hence, (2.8-6) may be written for the dc machine as

$$T_e = L_{AF} i_f i_a \qquad (3.3\text{-}5)$$

Here again the variable k_v is often substituted for $L_{AF} i_f$. In some instances k_v is multiplied by a factor less than unity when substituted into (3.3-5) so as to approximate the effects of rotational losses. It is interesting that the field winding produces a stationary mmf and, owing to commutation, the armature winding also produces a stationary mmf which is displaced $\frac{1}{2}\pi$ electrical degrees from the mmf produced by the field winding. It follows then that the interaction of these two mmf's produces the electromagnetic torque.

The torque and rotor speed are related by

$$T_e = J \frac{d\omega_r}{dt} + B_m \omega_r + T_L \qquad (3.3\text{-}6)$$

where J is the inertia of the rotor and, in some cases, the connected mechanical load. The units of the inertia are $kg \cdot m^2$ or $J \cdot s^2$. A positive electromagnetic torque T_e acts to turn the rotor in the direction of increasing θ_r. The load torque T_L is positive for a torque, on the shaft of the rotor, which opposes a positive electromagnetic torque T_e. The constant B_m is a damping coefficient associated with the mechanical rotational system of the machine. It has the units of $N \cdot m \cdot s$ and it is generally small and often neglected.

SP3.3-1. When a 12-V permanent-magnet dc motor is driven at $100 \, rad/s$, the open-circuit voltage is 10 V. Calculate k_v. [$k_v = 0.1 \, V \cdot s/rad$]

SP3.3-2. The armature applied voltage of a dc motor is 240 V; the rotor speed is constant at 50 rad/s. The steady-state armature current is 15 A, the armature resistance is $1 \, \Omega$, and $L_{AF} = 1 \, H$. Calculate the steady-state field current. [$I_f = 4.5 \, A$]

SP3.3-3. Calculate the no-load speed ($T_L = 0$) for the permanent-magnet dc motor in SP3.3-1 when rated voltage (12 V) is applied to the armature. [$\omega_r = 120 \, rad/s$]

3.4 BASIC TYPES OF DIRECT-CURRENT MACHINES

The field and armature windings may be excited from separate sources or from the same source with the windings connected differently to form the basic types of dc machines, such as the shunt-connected, the series-connected, and the compound-connected dc machines. The equivalent circuits for each of these machines are given in this section along with an analysis and discussion of their steady-state operating characteristics.

Separate Winding Excitation

When the field and armature windings are supplied from separate voltage sources, the device may operate as either a motor or a generator; it is a motor if it is driving a torque load and a generator if it is being driven by some type of prime mover. The equivalent circuit for this type of machine is shown in Fig. 3.4-1. It differs from that shown in Fig. 3.3-1 in that an external resistance r_{fx} is connected in series with the field winding. This resistance, which is often referred to as a *field rheostat*, is used to adjust the field current if the field voltage is supplied from a constant source.

The voltage equations which describe the steady-state performance of this device may be written directly from (3.3-1) by setting the operator p to zero ($p = d/dt$), whereupon

$$V_f = R_f I_f \qquad (3.4\text{-}1)$$

$$V_a = r_a I_a + \omega_r L_{AF} I_f \qquad (3.4\text{-}2)$$

where $R_f = r_{fx} + r_f$ and capital letters are used to denote steady-state voltages and currents. We know, from the torque relationship given by (3.3-6), that during steady-state operation $T_e = T_L$ if B_m is assumed to be zero. Analysis of steady-state performance is straightforward.

A permanent-magnet dc machine fits into this class of dc machines. As we have mentioned, the field flux is established in these devices by a permanent

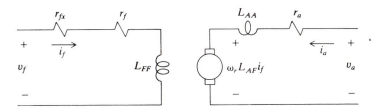

FIGURE 3.4-1
Equivalent circuit for separate field and armature excitation.

magnet. The voltage equation for the field winding is eliminated and $L_{AF}i_f$ is replaced by a constant k_v, which can be measured if it is not given by the manufacturer. Most small, hand-held, fractional-horsepower dc motors are of this type, and speed control is achieved by controlling the amplitude of the applied armature voltage.

Shunt-Connected DC Machine

The field and armature windings may be connected as shown schematically in Fig. 3.4-2. With this connection, the machine may operate either as a motor or a generator. Since the field winding is connected between the armature terminals, this winding arrangement is commonly referred to as a *shunt-connected dc machine* or simply a shunt machine. During steady-state operation, the armature circuit voltage equation is (3.4-2) and, for the field circuit,

$$V_a = I_f R_f \tag{3.4-3}$$

The total current I_t is

$$I_t = I_f + I_a \tag{3.4-4}$$

Solving (3.4-2) for I_a and (3.4-3) for I_f and substituting the results in (3.3-5) yields the following expression for the steady-state electromagnetic torque, positive for motor action, for this type of dc machine:

$$T_e = \frac{L_{AF} V_a^2}{r_a R_f} \left(1 - \frac{L_{AF}}{R_f} \omega_r \right) \tag{3.4-5}$$

The shunt-connected dc machine may operate as either a motor or a generator when connected to a dc source. It may also operate as an isolated self-excited generator, supplying an electric load such as a dc motor or a static load. When the shunt machine is operated from a constant-voltage source, the steady-state operating characteristics are those shown in Fig. 3.4-3. Several

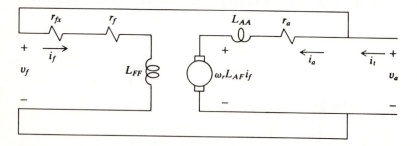

FIGURE 3.4-2
Equivalent circuit of a shunt-connected dc machine.

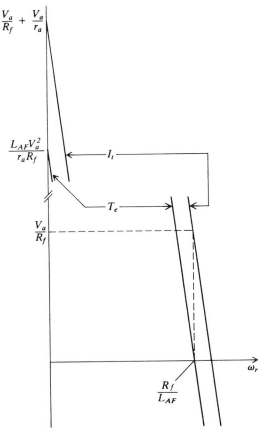

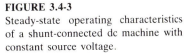

FIGURE 3.4-3
Steady-state operating characteristics of a shunt-connected dc machine with constant source voltage.

features of these characteristics warrant discussion. At stall ($\omega_r = 0$), the steady-state armature current I_a is limited only by the armature resistance. In the case of small, permanent-magnet motors, the armature resistance is quite large and the starting armature current, which results when rated voltage is applied, is generally not damaging. However, larger-horsepower machines are designed with a small armature resistance. Therefore, an excessively high armature current will occur during the starting period if rated voltage is applied to the armature terminals. To prevent high starting current, resistance may be inserted into the armature circuit at stall and decreased either manually or automatically to zero as the machine accelerates to normal operating speed. When silicon-controlled rectifiers (SCR's) or thyristors are used to convert an ac source voltage to dc to supply the dc machine, they may be controlled to provide a reduced voltage during the starting period, thereby preventing a high starting current and eliminating the need to insert resistance into the armature circuit. Other features of the shunt machine with a small armature resistance are the steep torque versus speed characteristics. In other words, the speed of the shunt machine does not change appreciably as the load torque is varied from zero to rated.

When a shunt machine is operated as an isolated generator, it is referred to as a self-excited generator because the field is supplied by the voltage generated by the armature. Although this type of machine is rapidly being replaced by solid-state ac to dc converters, some attention will be given to the self-excited generator in an attempt to satisfy the curiosity of the interested reader. During generator operation, the armature current is negative relative to the positive direction indicated in Fig. 3.4-2. When operated as a generator (T_e and I_a negative), it is assumed that the prime mover maintains the speed of the generator constant regardless of the electric load supplied by the generator.

It would appear from Fig. 3.4-2 that, if the armature terminals of a self-excited generator are short-circuited while the rotor is being driven, the armature current I_a will be zero since the field current I_f and, consequently, the induced voltage $\omega_r L_{AF} I_f$ are all zero during steady-state short-circuit conditions. In practice this is not the case due to the permanent-magnet effect of the iron of the stator poles. In fact, this residual flux may produce a relatively large short-circuit current, depending upon the magnetic characteristics of the iron and the parameters of the machine. Since the existence of residual flux forms the basis of the self-excited mode of operation, it is appropriate to depart from the present analysis to discuss this phenomenon.

A terminal voltage could not be established in the case of a self-excited generator without the residual flux giving rise to an armature voltage once the rotor is driven. This feature may be explained by starting with the magnetization curve of a separately excited dc generator, shown in Fig. 3.4-4. The

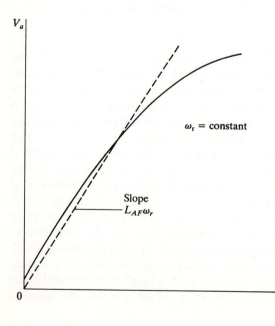

FIGURE 3.4-4
Magnetization curve of separately excited dc generator —open-circuited armature voltage versus field current.

magnetization curve is a plot of the steady-state open-circuit armature voltage versus field current with the rotor driven at a constant speed. The dashed line in Fig. 3.4-4 illustrates the linear relationship between the steady-state open-circuit armature voltage and the field current as predicted by the analysis employed heretofore. The departure of the magnetization curve from the linear relationship at low values of field current is due to the permanent-magnet effect of the stator poles (residual flux). The departure at the higher values of field current is due to the saturation of the iron in the magnetic circuit. The effect of the residual flux upon the self-excited generator wherein the field current is determined by the armature voltage is our immediate interest.

To illustrate this feature, it is convenient to assume that the machine is being driven at a constant speed with the field winding open-circuited. In this mode of operation, the open-circuit terminal voltage is determined by the residual flux and the speed of the rotor. If now the field winding is connected to the armature terminals and if this connection is made so that the field current, resulting from the residual voltage initially appearing at the terminals of the machine, produces a flux which aids the residual flux, the armature voltage will increase. This "building-up" process will continue until the field voltage (terminal voltage) versus field current characteristic intersects the plot of the generated terminal voltage versus field current. In this case, the generated terminal voltage versus field current characteristic differs only slightly from the magnetization curve owing to the voltage drop across the armature resistance. The slope of the straightline relationship between the field voltage (terminal voltage) and field current is the total field resistance R_f.

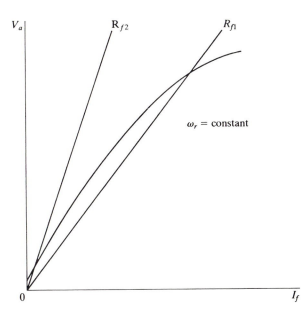

FIGURE 3.4-5
Influence of field resistance upon terminal voltage of self-excited dc generator ($I_t = 0$).

Hence, for a given rotor speed, the building up of the self-excited generator is predicted, first, upon the sense of the flux relative to the residual flux and, second, by the value of field resistance. The relationship between the steady-state terminal voltage with $I_t = 0$ ($I_f = -I_a$) and the field current of a self-excited generator is shown in Fig. 3.4-5 (p. 99). Therein $R_{f2} > R_{f1}$. A total field resistance of R_{f1} will yield a terminal voltage in the normal operating range; however, the terminal voltage will not build up to a normal value if the total field resistance is too large. For example, if the total field resistance is R_{f2}, the terminal voltage will fail to build up beyond a relatively small value.

After consideration of the nonlinear characteristics of the magnetization curve, one wonders if the performance of the dc machine may be accurately predicted with a linear approximation of the magnetization curve. The linear approximation is not valid if the field current is varied over a wide range, causing the machine to operate in the linear region as well as in the saturated region of the magnetization curve. The nonlinear features of the magnetic system may be taken into account either by graphical methods, for analysis of steady-state performance, or a computer may be employed to study either a steady-state or transient performance with the effects of saturation included.

Series-Connected DC Machine

When the field is connected in series with the armature circuit as shown in Fig. 3.4-6, the machine is referred to as a *series-connected dc machine* or a series machine. It is convenient to add the subscript s to denote quantities associated with the series field. It is important to mention the physical difference between the field winding of a shunt machine and that of a series machine. If the field winding is to be a shunt-connected winding, it is wound with a large number of turns of small-diameter wire, making the resistance of the field winding quite large. However, since the series-connected field winding is in series with the armature, it is designed so as to minimize the voltage drop across it. Thus, the winding is wound with a few turns of low-resistance wire.

Although the series machine does not have wide application, a series field

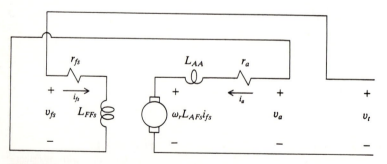

FIGURE 3.4-6
Equivalent circuit for a series-connected dc machine.

is often used in conjunction with a shunt field to form a compound-connected dc machine which is more common. In the case of a series machine (Fig. 3.4-6),

$$v_t = v_{fs} + v_a \qquad (3.4\text{-}6)$$

$$i_a = i_{fs} \qquad (3.4\text{-}7)$$

where v_{fs} and i_{fs} denote the voltage and current associated with the series field. The subscript s is added to avoid confusion with the shunt field when both fields are used in a compound machine.

If the constraints given by (3.4-6) and (3.4-7) are substituted into the armature voltage equation, the steady-state performance of the series-connected dc machine may be described by

$$V_t = (r_a + r_{fs} + L_{AFs}\omega_r)I_a \qquad (3.4\text{-}8)$$

From (3.3-5),

$$
\begin{aligned}
T_e &= L_{AFs}I_a^2 \\
&= \frac{L_{AFs}V_t^2}{(r_a + r_{fs} + L_{AFs}\omega_r)^2}
\end{aligned}
\qquad (3.4\text{-}9)
$$

The steady-state torque-speed characteristic for a typical series machine is shown in Fig. 3.4-7. The stall torque is quite high since it is proportional to the square of the armature current for a linear magnetic system. However, saturation of the magnetic system due to large armature currents will cause the torque to be less than that calculated from (3.4-9). At high rotor speeds, the

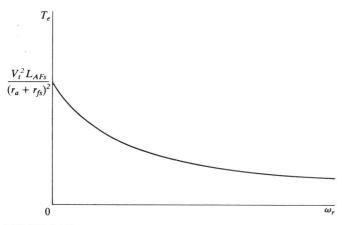

FIGURE 3.4-7
Steady-state torque-speed characteristics of a series-connected dc machine.

torque decreases less rapidly with increasing speed. In fact, if the load torque is small, the series motor may accelerate to speeds large enough to cause damage to the machine. Consequently, the series motor is used in applications such as traction motors for trains and buses or in hoists and cranes where a high starting torque is required and an appreciable load torque exists under normal operation.

Compound-Connected DC Machine

A compound-connected or compound dc machine, which is equipped with both a shunt and a series field winding, is illustrated in Fig. 3.4-8. In most compound machines, the shunt field dominates the operating characteristics while the series field, which consists of a few turns of low-resistance wire, has a secondary influence. It may be connected so as to aid or oppose the flux produced by the shunt field. If the compound machine is to be used as a generator, the series field is connected so as to aid the shunt field (cumulative compounding). Depending upon the strength of the series field, this type of connection can produce a "flat" terminal voltage versus load current charac-teristic, whereupon a near-constant terminal voltage is achieved from no load to full load. In this case, the machine is said to be "flat-compounded." An "overcompounded" machine occurs when the strength of the series field causes the terminal voltage at full load to be larger than at no load. The meaning of an "undercompounded" machine is obvious. In the case of compound dc motors, the series field is often connected to oppose the flux produced by the

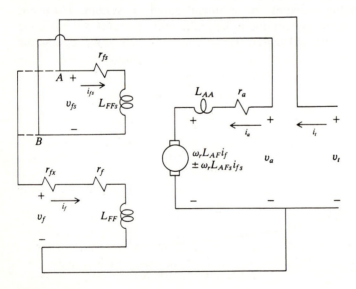

FIGURE 3.4-8
Equivalent circuit of a compound dc machine.

shunt field (differential compounding). If properly designed, this type of connection can provide a near-constant speed from no-load to full-load torque.

The voltage equations for a compound dc machine may be written as

$$\begin{bmatrix} v_f \\ v_t \end{bmatrix} = \begin{bmatrix} R_f + pL_{FF} & \pm pL_{FS} & 0 \\ \omega_r L_{AF} \pm pL_{FS} & \pm \omega_r L_{AFs} + r_{fs} + pL_{FFs} & r_a + pL_{AA} \end{bmatrix} \begin{bmatrix} i_f \\ i_{fs} \\ i_a \end{bmatrix}$$

(3.4-10)

where L_{FS} is the mutual inductance between the shunt and the series fields. The plus and minus signs are used so that either a cumulative or a differential connection may be described.

The shunt field may be connected ahead of the series field (long-shunt connection) or behind the series field (short-shunt connection) as shown by A and B, respectively, in Fig. 3.4-8. The long-shunt connection is commonly used. In this case

$$v_t = v_f = v_{fs} + v_a \qquad (3.4\text{-}11)$$

$$i_t = i_f + i_{fs} \qquad (3.4\text{-}12)$$

where

$$i_{fs} = i_a \qquad (3.4\text{-}13)$$

The steady-state performance of a long-shunt-connected compound machine may be described by the following equations:

$$V_t = \left[\frac{r_a + r_{fs} \pm L_{AFs}\omega_r}{1 - (L_{AF}/R_f)\omega_r} \right] I_a \qquad (3.4\text{-}14)$$

The torque for the long-shunt connection may be obtained by employing (3.3-5) for each field winding. In particular,

$$T_e = L_{AF} I_f I_a \pm L_{AFs} I_{fs} I_a$$
$$= \frac{L_{AF} V_t^2 [1 - (L_{AF}/R_f)\omega_r]}{R_f (r_a + r_{fs} \pm L_{AFs}\omega_r)} \pm \frac{L_{AFs} V_t^2 [1 - (L_{AF}/R_f)\omega_r]^2}{(r_a + r_{fs} \pm L_{AFs}\omega_r)^2} \qquad (3.4\text{-}15)$$

Example 3A. A permanent-magnet dc motor similar to that shown in Fig. 3.2-6 is rated at 6 V with the following parameters: $r_a = 7\,\Omega$, $L_{AA} = 120\,mH$, $k_T = 2\,oz \cdot in/A$, $J = 150\mu oz \cdot in \cdot s^2$. According to the motor information sheet, the no-load speed is approximately 3350 r/min and the no-load armature current is approximately 0.15 A. Let us attempt to interpret this information.

First, let us convert k_T and J to units which we have been using in this text. In this regard, we will convert the inertia to $kg \cdot m^2$ which is the same as $N \cdot m \cdot s^2$. To do this, we must convert ounces to newtons and inches to meters (Appendix A). Thus,

$$J = \frac{150 \times 10^{-6}}{(3.6)(39.37)} = 1.06 \times 10^{-6} \, \text{kg} \cdot \text{m}^2 \tag{3A-1}$$

We have not seen k_T before. It is the torque constant and, if expressed in the appropriate units, it is numerically equal to k_v. When k_v is used in the expression for T_e ($T_e = k_v i_a$), it is often referred to as the *torque constant* and denoted k_T. When used in the voltage equation, it is always denoted k_v. Now we must convert oz·in to N·m, whereupon k_T equals our k_v; hence,

$$k_v = \frac{2}{(16)(0.225)(39.37)} = 1.41 \times 10^{-2} \, \text{N} \cdot \text{m/A} = 1.41 \times 10^{-2} \, \text{V} \cdot \text{s/rad} \tag{3A-2}$$

What do we do about the no-load armature current? What does it represent? Well, probably it is a measure of the friction and windage losses. We could neglect it, but we will not. Instead, let us include it as B_m. First, however, we must calculate the no-load speed. We can solve for the no-load rotor speed from the steady-state armature voltage equation for the shunt machine, (3.4-2), with $L_{AF} i_f$ replaced by k_v:

$$\omega_r = \frac{V_a - r_a I_a}{k_v} = \frac{6 - (7)(0.15)}{1.41 \times 10^{-2}} = 351.1 \, \text{rad/s}$$

$$= \frac{(351.1)(60)}{2\pi} = 3353 \, \text{r/min} \tag{3A-3}$$

Now at this no-load speed,

$$T_e = k_v i_a = (1.41 \times 10^{-2})(0.15) = 2.12 \times 10^{-3} \, \text{N} \cdot \text{m} \tag{3A-4}$$

Since T_L and $J(d\omega_r/dt)$ are zero for this steady-state no-load condition, (3.3-6) tells us that (3A-4) is equal to $B_m \omega_r$; hence,

$$B_m = \frac{2.12 \times 10^{-3}}{\omega_r} = \frac{2.12 \times 10^{-3}}{351.1} = 6.04 \times 10^{-6} \, \text{N} \cdot \text{m} \cdot \text{s} \tag{3A-5}$$

Example 3B. The permanent-magnet dc machine described in Example 3A is operating with rated applied armature voltage and a load torque T_L of 0.5 oz·in. Our task is to determine the efficiency where percent eff = (power output/power input)100.

First let us convert oz·in to N·m:

$$T_L = \frac{0.5}{(16)(0.225)(39.37)} = 3.53 \times 10^{-3} \, \text{N} \cdot \text{m} \tag{3B-1}$$

In Example 3A we determined k_v to be 1.41×10^{-2} V·s/rad and B_m to be 6.04×10^{-6} N·m·s.

During steady-state operation, (3.3-6) becomes

$$T_e = B_m \omega_r + T_L \tag{3B-2}$$

From (3.3-5), with $L_{AF} i_f$ replaced by k_v, the steady-state electromagnetic torque is

$$T_e = k_v I_a \tag{3B-3}$$

Substituting (3B-3) into (3B-2) and solving for ω_r yields

$$\omega_r = \frac{k_v}{B_m} I_a - \frac{1}{B_m} T_L \tag{3B-4}$$

From (3.4-2) with $L_{AF}i_f = k_v$,

$$V_a = r_a I_a + k_v \omega_r \tag{3B-5}$$

Substituting (3B-4) into (3B-5) and solving for I_a yields

$$
\begin{aligned}
I_a &= \frac{V_a + (k_v/B_m) T_L}{r_a + (k_v^2/B_m)} \\
&= \frac{6 + [(1.41 \times 10^{-2})/(6.04 \times 10^{-6})](3.53 \times 10^{-3})}{7 + (1.41 \times 10^{-2})^2/(6.04 \times 10^{-6})} = 0.357 \text{ A}
\end{aligned} \tag{3B-6}
$$

From (3B-4),

$$
\begin{aligned}
\omega_r &= \frac{1.41 \times 10^{-2}}{6.04 \times 10^{-6}} 0.357 - \frac{1}{6.04 \times 10^{-6}} (3.53 \times 10^{-3}) \\
&= 249 \text{ rad/s}
\end{aligned} \tag{3B-7}
$$

The power input is

$$P_{in} = V_a I_a = (6)(0.357) = 2.14 \text{ W} \tag{3B-8}$$

The power output is

$$P_{out} = T_L \omega_r = (3.53 \times 10^{-3})(249) = 0.88 \text{ W} \tag{3B-9}$$

The efficiency is

$$
\begin{aligned}
\text{Percent eff} &= \frac{P_{out}}{P_{in}} 100 \\
&= \frac{0.88}{2.14} 100 = 41.1 \text{ percent}
\end{aligned} \tag{3B-10}
$$

This low efficiency is characteristic of low-power dc motors due to the relatively large armature resistance. In this regard, it is interesting to determine the losses due to $i^2 r$, friction, and windage.

$$P_{i^2 r} = r_a I_a^2 = (7)(0.357)^2 = 0.89 \text{ W} \tag{3B-11}$$

$$P_{fw} = (B_m \omega_r)\omega_r = (6.04 \times 10^{-6})(249)^2 = 0.37 \text{ W} \tag{3B-12}$$

Let us check our work:

$$P_{in} = P_{i^2 r} + P_{fw} + P_{out} = 0.89 + 0.37 + 0.88 = 2.14 \text{ W} \tag{3B-13}$$

which is equal to (3B-8).

SP3.4-1. A 12-V permanent-magnet dc motor has an armature resistance of 12 Ω and $k_v = 0.01$ V·s/rad. Calculate the steady-state stall torque (T_e with $\omega_r = 0$). [$T_e = 0.01$ N·m]

SP3.4-2. Determine T_e in Example 3B. [$T_e = 0.713$ oz·in]

SP3.4-3. The field current of a dc shunt motor is 1 A. $L_{AF} = 1.8$ H, $r_a = 0.6 \Omega$, and $V_a = 240$ V. Calculate the steady-state no-load speed. [$\omega_r = 133.3$ rad/s]

SP3.4-4. The field circuit of the dc shunt motor in SP3.4-3 is accidently open-circuited ($I_f = 0$) while the voltage remains applied to the armature with $T_L = 0$. Assume that the residual field strength due to the permanent-magnet effect of the field iron is approximately 2 percent of the field strength when $I_f = 1$ A. Estimate the no-load speed. [$\omega_r \cong 6667$ rad/s; be careful not to let this happen.]

SP3.4-5. For the dc shunt machine of SP3.4-3, calculate the steady-state rotor speed at which the total current $I_t = -10$ A. This is generator action; the rotor is being driven by a prime mover. [$\omega_r = 137$ rad/s]

SP3.4-6. A dc series motor is connected to a 100-V source. $L_{AFs} = 0.3$ H, $r_a = 2\,\Omega$, and $r_{fs} = 1\,\Omega$. Calculate the stall ($\omega_r = 0$) steady-state torque. [$T_e = 333.3$ N·m]

3.5 DYNAMIC CHARACTERISTICS OF PERMANENT-MAGNET AND SHUNT dc MOTORS

The permanent-magnet and shunt dc motors are widely used and, therefore, are appropriate candidates to illustrate the dynamic performance of typical dc machines. Two modes of dynamic operation are of interest—starting from stall and changes in load torque with the machine supplied from a constant-voltage source.

Dynamic Performance During Starting

In the previous section, it was pointed out that, if the armature resistance is small, damaging armature current could result if rated voltage is applied to the armature terminals when the machine is stalled ($\omega_r = 0$). With the machine at stall, the counter emf is zero and the armature current is opposed only by the voltage drop across the armature resistance and inductance. However, we have noted in Examples 3A and 3B that low-power permanent-magnet dc motors characteristically have a large armature resistance making it possible to "direct-line" start these devices without damaging the stator windings. The no-load starting characteristics ($T_L = 0$) of the permanent-magnet dc motor described in Example 3A are shown in Fig. 3.5-1. The armature voltage v_a, the armature current i_a, and the rotor speed ω_r are plotted. Initially the motor is at stall and, at time zero, 6 V is applied to the armature terminals. The peak transient armature current is limited to approximately 0.55 A by the inductance and resistance of the armature and the fact that the rotor is accelerating from stall, thereby developing the voltage $k_v \omega_r$ which opposes the applied voltage. After about 0.25 s, steady-state operation is achieved with the no-load armature current of 0.15 A. (From Example 3A, $B_m = 6.04 \times 10^{-6}$ N·m·s.) It is noted that the rotor speed is slightly oscillatory (underdamped), as illustrated by the small overshoot of the final value.

When starting larger-horsepower dc machines, it is generally necessary to limit the starting current. As mentioned, this can be accomplished either by phase-controlling an ac-to-dc converter, if it is used to provide the armature

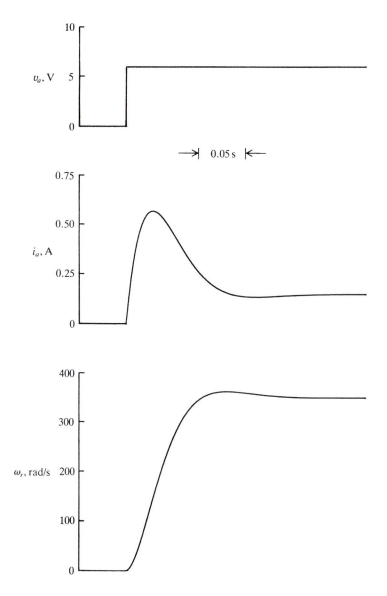

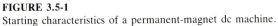

FIGURE 3.5-1
Starting characteristics of a permanent-magnet dc machine.

voltage, or by inserting resistance in series with the armature circuit during starting if the motor is supplied from a constant-voltage source. However, since resistance starting of dc machines is rapidly being replaced by ac-to-dc converters, we will not consider this method further. For those who wish more information regarding this method of starting, it is treated in detail in [1].

Dynamic Performance During Sudden Changes in Load Torque

In Example 3B, we calculated the efficiency of the permanent-magnet dc motor given in Example 3A with a load torque of 0.5 oz · in (3.53×10^{-3} N · m). Let us assume that this load torque was suddenly applied with the motor initially operating at the no-load condition ($I_a = 0.15$ A). The dynamic characteristics following a step change in load torque T_L from zero to 0.5 oz · in are shown in Fig. 3.5-2. The armature current i_a and the rotor speed ω_r are plotted. Since $T_e = k_v i_a$ and since k_v is constant, T_e differs from i_a by a constant multiplier. It is noted that the system is slightly oscillatory. Also, it is noted that the change in the steady-state rotor speed is quite large. From Example 3A or Fig. 3.5-2, we see that no-load speed is 351.1 rad/s. With $T_L = 0.5$ oz · in, the rotor speed is 249 rad/s from Example 3B or Fig. 3.5-2. There has been approximately a 30 percent decrease in speed for this increase in load torque. This is characteristic of a low-power permanent-magnet motor owing to the high armature resistance.

The dynamic performance of a shunt motor during a step decrease in load torque is shown in Fig. 3.5-3. The parameters and rated conditions of this machine are as follows: $R_f = 240 \, \Omega$, $L_{FF} = 120$ H, $L_{AF} = 1.8$ H, $r_a = 0.6 \, \Omega$, $L_{AA} = 0.012$ H. The machine is a 240-V 5-hp dc shunt motor. The inertia of the machine and connected load is 1 kg · m². Also, it is clear that $R_f = r_{fx} + r_f$.

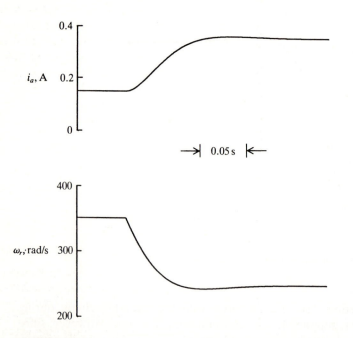

FIGURE 3.5-2
Dynamic performance of a permanent-magnet dc motor following a sudden increase in load torque from zero to 0.5 oz · in.

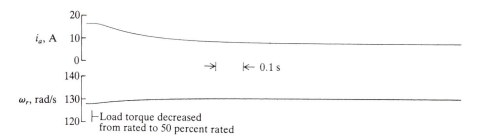

FIGURE 3.5-3
Dynamic performance of a 5-hp shunt motor following a sudden decrease in load torque from rated to 50 percent rated.

Rated armature current is 16.2 A and rated rotor speed is 127.7 rad/s. Here, rated armature current and rated speed are determined by setting v_a and v_f equal to 240 V and the power output equal to 5 hp, where the power output is the electromagnetic torque multiplied by the rotor speed.

Initially, the motor is operating steadily with rated load torque (29.2 N · m) applied to the rotor shaft. The load torque is then stepped to 50 percent of rated value, whereupon the motor speeds up and reestablishes steady-state operation at the reduced load conditions. The armature current i_a and rotor speed ω_r are plotted. Since the field current is constant, the electromagnetic torque T_e is identical to the armature current, differing only by a constant multiplier. It is important to note the small change in rotor speed from full load to 50 percent of full-load torque. At full load, the rotor speed is 127.7 and 130.4 rad/s at 50 percent of full load. This is approximately a 2 percent change in rotor speed for a 50 percent change in load torque. We first noted this "steep" torque-speed characteristic of a shunt machine, which is due to the low value of armature resistance, shown in Fig. 3.4-3. With the field current constant, the voltage and torque equations which describe the shunt motor are linear (neglecting saturation); hence, the response of the machine variables when the load torque is stepped back to rated from 50 percent rated would be the reverse of that shown in Fig. 3.5-3.

The operating characteristics and dynamic response of a larger-horsepower shunt motor shall be illustrated using a 200-hp machine with the following parameters: 200 hp, 250 V, 600 r/min, $R_f = 12\ \Omega$, $L_{FF} = 9$ H, $L_{AF} = 0.18$ H, $r_a = 0.012\ \Omega$, $L_{AA} = 0.00035$ H. The total inertia of rotor and load is 30 kg · m². When rated horsepower, voltage, and speed are all given, as in this case, the values are generally approximate and account for various machine losses. Not all conditions can be met, especially when we consider the machine as an idealized machine. If, for example, we select 600 r/min (62.8 rad/s) as rated speed, rated load torque is then 2375 N · m [(200)(746)/62.8]. However, if this load torque is applied to the rotor and if v_a and v_f are both 250 V, then the calculated steady-state armature current is 633 A and the rotor speed is 617 r/min (64.6 rad/s), with an output N · m as rated load torque. The response

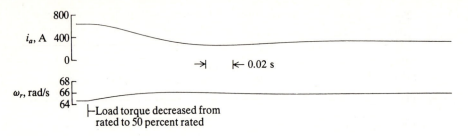

FIGURE 3.5-4
Dynamic performance of a 200-hp shunt motor following a sudden change in load torque from rated to 50 percent rated.

of this 200-hp machine to a decrease (step) in load torque from rated to 50 percent rated is shown in Fig. 3.5-4. Here, again, we see the small change in speed (approximately 1.6 percent) from full load to 50 percent of full load due to the small armature resistance.

SP3.5-1. Plot the T_e versus ω_r characteristic for the permanent-magnet dc motor given in Example 3A. [A straight line between $(12.1 \times 10^{-3}, 0)$ and $(0, 425.5)$]

SP3.5-2. Assume that the peak value of i_a in Fig. 3.5-1 is 0.55 A. Calculate $k_v \omega_r$ at peak i_a. [$k_v \omega_r = 2.15$ V]

SP3.5-3. Determine the steady-state armature current for the 5-hp dc shunt motor considered in this section with $\omega_r = 0$ and rated voltage applied to the armature terminals. [Nearly 25 times rated current; be careful not to do this.]

SP3.5-4. The 5-hp shunt-connected dc motor considered in this section is operating at rated conditions. $I_f = 1$ A, $T_L = 29.2$ N·m, $\omega_r = 127.7$ rad/s, with $V_t = 240$ V. At $t = 0$ the terminal voltage is stepped to $V_t = (1.1)(240)$ V. Neglect L_{AA} and calculate i_a at $t = 0^+$. [At $t = 0^+$, $i_a = 56.9$ A]

SP3.5-5. Calculate the final rotor speed for the condition given in SP3.5-4. [$\omega_r = 128.9$ rad/s]

3.6 TIME-DOMAIN BLOCK DIAGRAMS AND STATE EQUATIONS

Although the analysis of control systems is not our intent, it is worthwhile to set the stage for this type of analysis by means of a "first look" at time-domain block diagrams and state equations. In this section, we will consider only the shunt and permanent-magnet dc machines. The series and compound machines are treated in problems at the end of the chapter.

Shunt-Connected DC Machine

Block diagrams, which portray the interconnection of the system equations, are used extensively in control system analysis and design. Although block diagrams are generally depicted by using the Laplace operator, we shall not do this here since this would require a background in Laplace transformations. Instead, we shall work with the time-domain equations, using the p operator to denote differentiation with respect to time and the operator $1/p$ to denote integration. Those familiar with Laplace transformations will have no trouble converting the time-domain block diagrams to transfer functions by using the Laplace operator.

Arranging the equations of a shunt machine into a block diagram representation is straightforward. The field and armature voltage equations, (3.3-1), and the relationship between torque and rotor speed, (3.3-6), may be written as

$$v_f = R_f(1 + \tau_f p)i_f \tag{3.6-1}$$

$$v_a = r_a(1 + \tau_a p)i_a + \omega_r L_{AF} i_f \tag{3.6-2}$$

$$T_e - T_L = (B_m + Jp)\omega_r \tag{3.6-3}$$

where the field time constant $\tau_f = L_{FF}/R_f$ and the armature time constant $\tau_a = L_{AA}/r_a$. Here, again, p denotes d/dt and $1/p$ will denote integration. Solving (3.6-1) for i_f, (3.6-2) for i_a, and (3.6-3) for ω_r yields

$$i_f = \frac{1/R_f}{\tau_f p + 1} v_f \tag{3.6-4}$$

$$i_a = \frac{1/r_a}{\tau_a p + 1} (v_a - \omega_r L_{AF} i_f) \tag{3.6-5}$$

$$\omega_r = \frac{1}{Jp + B_m} (T_e - T_L) \tag{3.6-6}$$

A few comments are in order regarding these expressions. In (3.6-4), we see that the field voltage v_f is multiplied by the operator $(1/R_f)/(\tau_f p + 1)$ to obtain the field current i_f. The fact that we are multiplying the voltage by an operator to obtain current is in no way indicative of the procedure that we might actually use to calculate the current i_f given the voltage v_f. We are simply expressing the dynamic relationship between the field voltage and current in a form convenient for drawing block diagrams. However, to calculate i_f given v_f, we may prefer to express (3.6-4) in its equivalent form (3.6-1) and use the method of Appendix E to solve the given first-order differential equation.

The operator $(1/R_f)/(\tau_f p + 1)$ in (3.6-4) may also be interpreted as a transfer function relating the field voltage and current. Those of you who are familiar with Laplace transform methods are likely accustomed to seeing transfer functions expressed in terms of the Laplace operator s instead of the differentiation operator p. In fact, the same transfer functions are obtained by using Laplace transform theory with p replaced by s. However, we shall reserve discussion of Laplace transform methods until the final section of this chapter since we do not need this material in our present discussion.

The time-domain block diagram portraying (3.6-4) through (3.6-6) with $T_e = L_{AF}i_f i_a$ is shown in Fig. 3.6-1. This diagram consists of a set of linear blocks, wherein the relationship between the input and corresponding output variable is depicted in transfer function form and a pair of multipliers which represent nonlinear blocks. Since the system is nonlinear, it is not possible to apply previously used techniques (or, for that matter, Laplace transform methods) for solving the differential equations implied by this block diagram. For this, we would use a computer. However, for certain dc machines, for example, permanent-magnet machines or separately excited shunt machines, where the field current i_f is maintained at a constant value, the multipliers in Fig. 3.6-1 are no longer needed and conventional methods of analyzing linear systems may be applied with relative ease.

The so-called state equations of a system represent the formulation of the state variables into a matrix form convenient for computer implementation, particularly for linear systems. The state variables of a system are defined as a minimal set of variables such that knowledge of these variables at any initial time t_0 plus information on the input excitation subsequently applied is sufficient to determine the state of the system at any time $t > t_0$ [2]. In the case of dc machines, the field current i_f, the armature current i_a, the rotor speed ω_r, and the rotor position θ_r are the state variables. However, since θ_r can be established from ω_r by using

$$\theta_r = \int_0^t \omega_r(\xi)\, d\xi + \theta_r(0) \tag{3.6-7}$$

and since θ_r is considered a state variable only when the shaft position is a controlled variable, we will omit θ_r from consideration in this development.

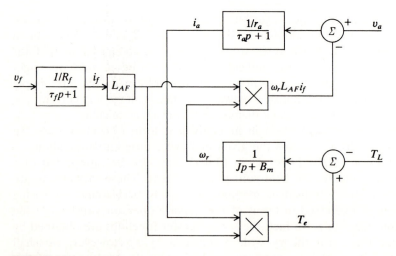

FIGURE 3.6-1
Time-domain block diagram of a shunt-connected dc machine.

The formulation of the state equations for the shunt machine can be readily achieved by straightforward manipulation of the field and armature voltage equations given by (3.3-1) and the equation relating torque and rotor speed given by (3.3-6). In particular, solving the field voltage equation, (3.3-1), for di_f/dt yields

$$\frac{di_f}{dt} = -\frac{R_f}{L_{FF}} i_f + \frac{1}{L_{FF}} v_f.$$ (3.6-8)

Solving the armature voltage equation, (3.3-1), for di_a/dt yields

$$\frac{di_a}{dt} = -\frac{r_a}{L_{AA}} i_a - \frac{L_{AF}}{L_{AA}} i_f \omega_r + \frac{1}{L_{AA}} v_a$$ (3.6-9)

If we wish, we could use k_v for $L_{AF} i_f$; however, we shall not make this substitution. Solving (3.3-6) for $d\omega_r/dt$ with $T_e = L_{AF} i_f i_a$ yields

$$\frac{d\omega_r}{dt} = -\frac{B_m}{J} \omega_r + \frac{L_{AF}}{J} i_f i_a - \frac{1}{J} T_L$$ (3.6-10)

All we have done here is to solve the equations for the highest derivative of the state variables while substituting (3.3-5) for T_e into (3.3-6). Now let us write the state equations in matrix (or vector matrix) form as

$$p \begin{bmatrix} i_f \\ i_a \\ \omega_r \end{bmatrix} = \begin{bmatrix} -\dfrac{R_f}{L_{FF}} & 0 & 0 \\ 0 & -\dfrac{r_a}{L_{AA}} & 0 \\ 0 & 0 & -\dfrac{B_m}{J} \end{bmatrix} \begin{bmatrix} i_f \\ i_a \\ \omega_r \end{bmatrix} + \begin{bmatrix} 0 \\ -\dfrac{L_{AF}}{L_{AA}} i_f \omega_r \\ \dfrac{L_{AF}}{J} i_f i_a \end{bmatrix} + \begin{bmatrix} \dfrac{1}{L_{FF}} & 0 & 0 \\ 0 & \dfrac{1}{L_{AA}} & 0 \\ 0 & 0 & -\dfrac{1}{J} \end{bmatrix} \begin{bmatrix} v_f \\ v_a \\ T_L \end{bmatrix}$$

(3.6-11)

where p is the operator d/dt. Equation (3.6-11) is the state equation(s); however, note that the second term (vector) on the right-hand side contains the products of state variables causing the system to be nonlinear.

Permanent-Magnet DC Machine

As we have mentioned previously, the equations which describe the operation of a permanent-magnet dc machine are identical to those of a shunt-connected dc machine with the field current held constant. Thus, the work in this section applies to both. For the permanent-magnet machine, $L_{AF} i_f$ is replaced by k_v, which is a constant determined by the strength of the magnet, the reluctance of the iron, and the number of turns of the armature winding. The time-domain block diagram may be developed for the permanent-magnet machine by using (3.6-2) and (3.6-3) with k_v substituted for $L_{AF} i_f$. The time-domain block diagram for a permanent-magnet dc machine is shown in Fig. 3.6-2.

Since k_v is constant, the state variables are now i_a and ω_r. From (3.6-9),

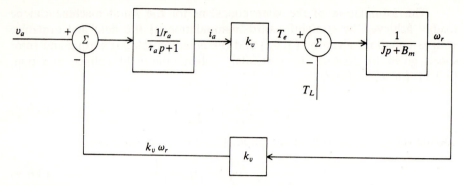

FIGURE 3.6-2
Time-domain block diagram of a permanent-magnet dc machine.

for a permanent-magnet machine,

$$\frac{di_a}{dt} = -\frac{r_a}{L_{AA}} i_a - \frac{k_v}{L_{AA}} \omega_r + \frac{1}{L_{AA}} v_a \tag{3.6-12}$$

From (3.6-10),
$$\frac{d\omega_r}{dt} = -\frac{B_m}{J} \omega_r + \frac{k_v}{J} i_a - \frac{1}{J} T_L \tag{3.6-13}$$

The system is described by a set of linear differential equations. In matrix form, the state equations become

$$p\begin{bmatrix} i_a \\ \omega_r \end{bmatrix} = \begin{bmatrix} -\dfrac{r_a}{L_{AA}} & -\dfrac{k_v}{L_{AA}} \\ \dfrac{k_v}{J} & -\dfrac{B_m}{J} \end{bmatrix} \begin{bmatrix} i_a \\ \omega_r \end{bmatrix} + \begin{bmatrix} \dfrac{1}{L_{AA}} & 0 \\ 0 & -\dfrac{1}{J} \end{bmatrix} \begin{bmatrix} v_a \\ T_L \end{bmatrix} \tag{3.6-14}$$

The form in which the state equations are expressed in (3.6-14) is called the fundamental form. In particular, the previous matrix equation may be expressed symbolically as

$$p\mathbf{x} = \mathbf{Ax} + \mathbf{Bu} \tag{3.6-15}$$

which is called the fundamental form, where p is the operator d/dt, $\mathbf{x}$ is the state vector (column matrix of state variables), and $\mathbf{u}$ is the input vector (column matrix of inputs to the system). We see that (3.6-14) and (3.6-15) are identical in form. Methods of solving equations of the fundamental form given by (3.6-15) are well known. Consequently, it is used extensively in control system analysis [2].

Example 3C. Once the permanent-magnet dc motor is portrayed in block diagram form (Fig. 3.6-2), it is often advantageous, for control design purposes, to express transfer functions between state and input variables. Although transfer functions are generally written with the Laplace operator, it is a simple task to convert the transfer functions given here to transfer functions in terms of the Laplace operator. Our task is to derive transfer functions between the state variables (i_a and ω_r) and the input variables (v_a and T_L) for the permanent-magnet dc machine. From (3.6-10), with $L_{AF}i_f$ replaced by k_v, we have

$$i_a = \frac{1/r_a}{\tau_a p + 1}(v_a - k_v \omega_r) \tag{3C-1}$$

From (3.6-11), with $T_e = k_v i_a$,

$$\omega_r = \frac{1}{Jp + B_m}(k_v i_a - T_L) \tag{3C-2}$$

It is apparent that we could have obtained these same equations from the block diagram given in Fig. 3.6-2. If (3C-1) is substituted into (3C-2), we obtain, after considerable work,

$$\boxed{\omega_r = \frac{(1/k_v \tau_a \tau_m)v_a - (1/J)(p + 1/\tau_a)T_L}{p^2 + (1/\tau_a + B_m/J)p + (1/\tau_a)(1/\tau_m + B_m/J)}} \tag{3C-3}$$

where a new time constant has been introduced. The inertia time constant, which is what τ_m is called, is defined as

$$\boxed{\tau_m = \frac{Jr_a}{k_v^2}} \tag{3C-4}$$

The transfer function between ω_r and v_a may be obtained from (3C-3) by setting T_L equal to zero in (3C-3) and dividing both sides by v_a. Similarly, the transfer function between ω_r and T_L is obtained by setting v_a to zero and dividing by T_L. To calculate ω_r given v_a and T_L, we note that p is d/dt and p^2 is d^2/dt^2 and, if we mulitiply each side of (3C-3) by the denominator of the right-side of the equation, we would have a second-order differential equation in terms of the state variable ω_r. As before, the method of Appendix E may be used to solve this second-order equation.

The characteristic or force-free equation for this linear system is obtained by setting the denominator equal to zero. It is of the general form

$$\boxed{p^2 + 2\alpha p + \omega_n^2 = 0} \tag{3C-5}$$

We are aware that α is the exponential damping coefficient and ω_n is the undamped natural frequency [3]. The damping factor is defined as

$$\boxed{\zeta = \frac{\alpha}{\omega_n}} \tag{3C-6}$$

Let us denote b_1 and b_2 as the negative values of the roots of this second-order equation,

$$b_1, b_2 = \zeta\omega_n \pm \omega_n\sqrt{\zeta^2 - 1} \tag{3C-7}$$

If $\zeta > 1$, the roots are real and the natural response consists of two exponential terms with negative real exponents. When $\zeta < 1$, the roots are a conjugate complex pair and the natural response consists of an exponentially decaying sinusoid.

Now, the transfer function relationship between i_a and the input variables v_a and T_L may be obtained by substituting (3C-2) into (3C-1). After some work, we obtain

$$i_a = \frac{(1/\tau_a r_a)(p + B_m/J)v_a + (1/k_v\tau_a\tau_m)T_L}{p^2 + (1/\tau_a + B_m/J)p + (1/\tau_a)(1/\tau_m + B_m/J)} \tag{3C-8}$$

SP3.6-1. Express the transfer function relationship between θ_r of a permanent-magnet dc machine and the input variables v_a and T_L. [$\theta_r = (1/p)$(3C-3)]

SP3.6-2. A permanent-magnet dc machine is operating without load torque ($T_L = 0$) and $B_m = 0$. Express the transfer function between i_a and v_a. [(3C-8) with T_L and B_m both zero]

3.7 SOLUTION OF DYNAMIC CHARACTERISTICS BY LAPLACE TRANSFORMATION

This section is devoted to examples of Laplace transformation solution of several of the dynamic characteristics of the permanent-magnet and shunt machines illustrated in the previous section. If you are not familiar with the Laplace transformation method of solving linear differential equations, you should omit this section since this method is not in any way prerequisite for the material in subsequent chapters.

Example 3D. The purpose of this example is to illustrate the Laplace transformation solution of the starting characteristics of the permanent-magnet dc motor depicted in Fig. 3.5-1. The rotor is initially at stall with $T_L = 0$ and $B_m = 6.04 \times 10^{-6}$ N·m·s. The armature voltage is stepped from zero to 6 V while the change in T_L is zero. Let us construct the block diagram in terms of the Laplace operator s for this situation. This block diagram is shown in Fig. 3D-1, and the Δ is used to denote changes from the predisturbance, steady-state values. We see that, since the change in T_L is zero, $\Delta T_L(s) = 0$ and, consequently, it does not appear in the block diagram or in the transfer functions.

We can express $\Delta i_a(s)$ and $\Delta\omega_r(s)$ from (3C-8) and (3C-3), respectively, by replacing the operator p with the Laplace operator s as well as all time-domain state variables (i_a and ω_r) and input variables (v_a and T_L) with the corresponding Δ variables as functions of the Laplace operator. In particular, i_a becomes $\Delta i_a(s)$, ω_r becomes $\Delta\omega_r(s)$, v_a becomes $\Delta v_a(s)$, and, as we mentioned previously, the change in T_L is zero; thus $\Delta T_L(s) = 0$. Hence, from (3C-8),

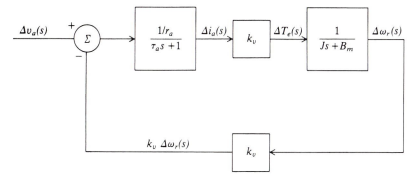

FIGURE 3D-1
Block diagram of permanent-magnet dc motor for a step change in applied voltage.

$$\Delta i_a(s) = \frac{(1/\tau_a r_a)(s + B_m/J)\,\Delta v_a(s)}{s^2 + (1/\tau_a + B_m/J)s + (1/\tau_a)(1/\tau_m + B_m/J)} \qquad (3D\text{-}1)$$

From (3C-3),

$$\Delta\omega_r(s) = \frac{(1/k_v\tau_a\tau_m)\,\Delta v_a(s)}{s^2 + (1/\tau_a + B_m/J)s + (1/\tau_a)(1/\tau_m + B_m/J)} \qquad (3D\text{-}2)$$

Also, $$\Delta T_e(s) = k_v\,\Delta i_a(s) \qquad (3D\text{-}3)$$

We are now ready to solve for the roots of the characteristic equation. For this purpose, it is convenient to make a few preliminary calculations. Recall that the parameters for this motor are given in Example 3A.

$$\tau_a = \frac{L_{AA}}{r_a} = \frac{120 \times 10^{-3}}{7} = 0.017 \text{ sec} \qquad (3D\text{-}4)$$

$$\tau_m = \frac{Jr_a}{k_v^2} = \frac{(1.06 \times 10^{-6})(7)}{(1.41 \times 10^{-2})^2} = 0.037 \text{ sec} \qquad (3D\text{-}5)$$

Here, we used sec rather than s to denote seconds to avoid confusion with the Laplace operator. The characteristic equation is obtained by setting the denominator of (3D-1) or (3D-2) equal to zero. It is of the form of (3C-5), from which

$$2\alpha = \frac{1}{\tau_a} + \frac{B_m}{J} = \frac{1}{0.017} + \frac{6.04 \times 10^{-6}}{1.06 \times 10^{-6}}$$

$$= 58.82 + 5.7 = 64.52 \text{ sec}^{-1} \qquad (3D\text{-}6)$$

$$\omega_n^2 = \frac{1}{\tau_a}\left(\frac{1}{\tau_m} + \frac{B_m}{J}\right) = \frac{1}{0.017}\left(\frac{1}{0.037} + \frac{6.04 \times 10^{-6}}{1.06 \times 10^{-6}}\right)$$

$$= 58.82(27.03 + 5.7) = 1925 \text{ sec}^{-2} \qquad (3D\text{-}7)$$

From (3D-6), $$\alpha = \tfrac{1}{2}(64.52) = 32.26 \text{ sec}^{-1} \qquad (3D\text{-}8)$$

From (3D-7), $$\omega_n = (1925)^{1/2} = 43.9 \text{ rad/sec} \qquad (3D\text{-}9)$$

The damping factor is

$$\zeta = \frac{\alpha}{\omega_n} = \frac{32.26}{43.9} = 0.735 \tag{3D-10}$$

Since $\zeta < 1$, we know that the roots of the characteristic equation will be a conjugate complex pair and the transient response will contain damped sinusoids. From (3C-7),

$$b_1, b_2 = \zeta\omega_n \pm \omega_n\sqrt{\zeta^2 - 1}$$

$$= 0.735 \times 43.9 \pm 43.9\sqrt{(0.735)^2 - 1}$$

$$= 32.3 \pm j29.8 \text{ sec}^- \tag{3D-11}$$

which is of the form

$$b_1, b_2 = \gamma \pm j\beta \tag{3D-12}$$

Now for a step in applied voltage,

$$\Delta v_a(s) = \frac{K}{s} \tag{3D-13}$$

where $K = 6$ V. Hence (3D-1) becomes

$$\Delta i_a(s) = \frac{(K/\tau_a r_a)(s + a_1)}{s(s + b_1)(s + b_2)} \tag{3D-14}$$

where, depending upon the damping factor, b_1 and b_2 may be real or a conjugate complex pair and

$$a_1 = \frac{B_m}{J} = 5.7 \text{ sec}^{-1} \tag{3D-15}$$

Therefore, (3D-2) becomes

$$\Delta\omega_r(s) = \frac{K/k_v \tau_a \tau_m}{s(s + b_1)(s + b_2)} \tag{3D-16}$$

We must now obtain the inverse Laplace transform of (3D-14) and (3D-16). For this, we will use the partial-fraction expansion. In particular, (3D-14) and (3D-16) may be written as

$$\Delta i_a(s) = \frac{K}{\tau_a r_a}\left(\frac{A}{s} + \frac{B}{s + b_1} + \frac{C}{s + b_2}\right) \tag{3D-17}$$

$$\Delta\omega_r(s) = \frac{K}{k_v \tau_a \tau_m}\left(\frac{D}{s} + \frac{E}{s + b_1} + \frac{F}{s + b_2}\right) \tag{3D-18}$$

Equating (3D-14) and (3D-17) yields

$$A(s + b_1)(s + b_2) + Bs(s + b_2) + Cs(s + b_1) = s + a_1 \tag{3D-19}$$

Similarly, from (3D-16) and (3D-18),

$$D(s + b_1)(s + b_2) + Es(s + b_2) + Fs(s + b_1) = 1 \tag{3D-20}$$

By equating coefficients of (3D-19), we obtain

$$A = \frac{a_1}{b_1 b_2} \tag{3D-21}$$

$$B = \frac{a_1 - b_1}{b_1(b_1 - b_2)} \tag{3D-22}$$

$$C = \frac{a_1 - b_2}{b_2(b_2 - b_1)} \tag{3D-23}$$

Similarly, from (3D-20),

$$D = \frac{1}{b_1 b_2} \tag{3D-24}$$

$$E = \frac{1}{b_1(b_1 - b_2)} \tag{3D-25}$$

$$F = \frac{1}{b_2(b_2 - b_1)} \tag{3D-26}$$

Taking the inverse Laplace transform of (3D-17) and (3D-18) yields (Appendix A)

$$i_a = i_a(0) + \frac{K}{\tau_a r_a} \left[\frac{a_1}{b_1 b_2} + \frac{a_1 - b_1}{b_1(b_1 - b_2)} e^{-b_1 t} + \frac{a_1 - b_2}{b_2(b_2 - b_1)} e^{-b_2 t} \right] \tag{3D-27}$$

$$\omega_r = \omega_r(0) + \frac{K}{k_v \tau_a \tau_m} \left[\frac{1}{b_1 b_2} + \frac{1}{b_1(b_1 - b_2)} e^{-b_1 t} + \frac{1}{b_2(b_2 - b_1)} e^{-b_2 t} \right] \tag{3D-28}$$

where $i_a(0)$ and $\omega_r(0)$ are the predisturbance steady-state values which, in this case, are both zero.

If b_1 and b_2 are real, we would be essentially done; however, $\zeta < 1$ for this machine, hence b_1 and b_2 form a conjugate complex pair as expressed by (3D-12). Substituting (3D-12) for b_1 and b_2 into (3D-27) and (3D-28), and after a little exercise in complex algebra and making use of Euler's identity a couple of times, we obtain

$$i_a = i_a(0) + \frac{K}{\tau_a r_a} \frac{a_1}{\gamma^2 + \beta^2} \left[-e^{-\gamma t} \left(\cos \beta t - \frac{\gamma^2 + \beta^2 - a_1 \gamma}{a_1 \beta} \sin \beta t \right) \right] \tag{3D-29}$$

$$\omega_r = \omega_r(0) + \frac{K}{k_v \tau_a \tau_m} \frac{1}{\gamma^2 + \beta^2} \left[1 - e^{-\gamma t} \left(\cos \beta t + \frac{\gamma}{\beta} \sin \beta t \right) \right] \tag{3D-30}$$

Substituting in the appropriate values yields

$$i_a = 0.149[1 - e^{-32.3t}(\cos 29.8t - 10.3 \sin 29.8t)] \tag{3D-31}$$

$$\omega_r = 350.3[1 - e^{-32.3t}(\cos 29.8t + 1.08 \sin 29.8t)] \tag{3D-32}$$

Example 3E. Figures 3.5-3 and 3.5-4 illustrate the dynamic response to a load torque change for a 5- and 200-hp dc shunt motor, respectively. Let us use the material presented in Examples 3C and 3D to obtain these responses analytically. Since the terminal voltage is constant, equal to its predisturbance value $\Delta v_a(s) = 0$, and since $B_m = 0$, (3C-3) may be written as

$$\Delta \omega_r(s) = \frac{-(1/J)(s + 1/\tau_a) \Delta T_L(s)}{s^2 + (1/\tau_a)s + 1/\tau_a \tau_m} \tag{3E-1}$$

where Δ is used to denote change from predisturbance steady-state values and the operator p in (3C-3) has been replaced by the Laplace operator s. With $\Delta v_a(s) = 0$, (3C-8) yields

$$\Delta i_a(s) = \frac{(1/k_v \tau_a \tau_m) \Delta T_L(s)}{s^2 + (1/\tau_a)s + 1/\tau_a \tau_m} \tag{3E-2}$$

A step change in load torque is expressed as

$$\Delta T_L(s) = \frac{K}{s} \tag{3E-3}$$

where, for the moment, we will let K be a constant with magnitude and sign; however, it is not the same K used in Example 3D. In Fig. 3.5-3, for the 5-hp motor, $K = -\frac{1}{2}(29.2) \, \text{N} \cdot \text{m}$ and in Fig. 3.5-4, for the 200-hp motor $K = -\frac{1}{2}(2375) \, \text{N} \cdot \text{m}$. Hence, (3E-1) becomes

$$\Delta \omega_r(s) = \frac{-(K/J)(s + 1/\tau_a)}{s[s^2 + (1/\tau_a)s + 1/\tau_a \tau_m]} \tag{3E-4}$$

and (3E-2) becomes

$$\Delta i_a(s) = \frac{K/k_v \tau_a \tau_m}{s[s^2 + (1/\tau_a)s + 1/\tau_a \tau_m]} \tag{3E-5}$$

The characteristic equation is obtained by setting the denominator of (3E-1) or (3E-2) equal to zero. Comparing this result with (3C-5) we can write

$$\omega_n = \sqrt{\frac{1}{\tau_a \tau_m}} \tag{3E-6}$$

$$\alpha = \frac{1}{2\tau_a} \tag{3E-7}$$

$$\zeta = \frac{\sqrt{\tau_a \tau_m}}{2\tau_a} \tag{3E-8}$$

The negative values of the roots of the characteristic equation may be written in the form given by (3C-7) or as

$$b_1, b_2 = \frac{1}{2\tau_a} \pm \sqrt{\frac{1}{4\tau_a^2} - \frac{1}{\tau_a \tau_m}} \tag{3E-9}$$

Equations (3E-4) and (3E-5) may now be written as

$$\Delta \omega_r(s) = \frac{-(K/J)(s + a_1)}{s(s + b_1)(s + b_2)} \tag{3E-10}$$

$$\Delta i_a(s) = \frac{K/k_v \tau_a \tau_m}{s(s + b_1)(s + b_2)} \tag{3E-11}$$

where

$$a_1 = \frac{1}{\tau_a} \tag{3E-12}$$

It is interesting to note that (3E-10) and (3E-11) are similar in form to (3D-14) and (3D-16), respectively. Since we have taken the inverse Laplace transform of these expressions in Example 3D, we shall make use of that work. First, however, let us do some preliminary calculations. For the 5-hp machine,

$$\tau_a = \frac{L_{AA}}{r_a} = \frac{0.012}{0.6} = 0.02 \text{ sec} \tag{3E-13}$$

$$\tau_m = \frac{Jr_a}{k_v^2} = \frac{Jr_a}{(L_{AF}i_f)^2}$$

$$= \frac{(1)(0.6)}{[(1.8)(240/240)]^2} = 0.185 \text{ sec} \tag{3E-14}$$

Note that in the above calculation, i_f is obtained by dividing the rated terminal voltage (240 V) by the field resistance (240 Ω). As in Example 3D, sec is used to denote seconds to avoid confusion with the Laplace operator. From (3E-6) through (3E-9),

$$\omega_n = \sqrt{\frac{1}{(0.02)(0.185)}} = 16.44 \text{ rad/sec} \tag{3E-15}$$

$$\alpha = \frac{1}{(2)(0.02)} = 25 \text{ sec}^{-1} \tag{3E-16}$$

$$\zeta = \frac{\sqrt{(0.02)(0.185)}}{(2)(0.02)} = 1.52 \tag{3E-17}$$

$$a_1 = \frac{1}{\tau_a} = \frac{1}{0.02} = 50 \text{ sec}^{-1} \tag{3E-18}$$

$$b_1, b_2 = \frac{1}{(2)(0.02)} \pm \sqrt{\frac{1}{4(0.02)^2} - \frac{1}{(0.02)(0.185)}}$$

$$= 6.17, 43.83 \text{ sec}^{-1} \tag{3E-19}$$

We can use (3D-27) and (3D-28) as guides to express i_a and ω_r, respectively. In particular,

$$\omega_r = \omega_r(0) - \frac{K}{J}\left[\frac{a_1}{b_1 b_2} + \frac{a_1 - b_1}{b_1(b_1 - b_2)}e^{-b_1 t} + \frac{a_1 - b_2}{b_2(b_2 - b_1)}e^{-b_2 t}\right] \tag{3E-20}$$

$$i_a = i_a(0) + \frac{K}{k_v \tau_a \tau_m}\left[\frac{1}{b_1 b_2} + \frac{1}{b_1(b_1 - b_2)}e^{-b_1 t} + \frac{1}{b_2(b_2 - b_1)}e^{-b_2 t}\right] \tag{3E-21}$$

We have previously determined that $\omega_r(0) = 127.7 \text{ rad/sec}$ and $i_a(0) = 16.2 \text{ A}$. Thus, since b_1 and b_2 are real ($\zeta > 1$), (3E-20) becomes

$$\omega_r = 127.7 + \frac{29.2}{(2)(1)}\left[\frac{50}{(6.17)(43.83)} + \frac{50 - 6.17}{6.17(6.17 - 43.83)}e^{-6.17t}\right.$$

$$\left. + \frac{50 - 43.83}{43.83(43.83 - 6.17)}e^{-43.83t}\right]$$

$$= 130.4 - 2.75e^{-t/0.162} + 0.05e^{-t/0.0228} \tag{3E-22}$$

The current, (3E-21), can be written as

$$i_a = 8.1 + 9.43e^{-t/0.162} - 1.33e^{-t/0.0228} \tag{3E-23}$$

Note that the dynamic responses consist of two exponential terms; one time constant is approximately equal to τ_m and the other is approximately equal to τ_a. The response is dominated by the exponential term containing τ_m.

In the case of the 200-hp motor,

$$\tau_a = 0.029 \text{ sec} \tag{3E-24}$$

$$\tau_m = 0.0256 \text{ sec} \tag{3E-25}$$

$$\omega_n = 36.7 \text{ rad/sec} \tag{3E-26}$$

$$\alpha = 17.2 \text{ sec}^{-1} \tag{3E-27}$$

$$\zeta = 0.47 \tag{3E-28}$$

$$a_1 = 34.5 \text{ sec}^{-1} \tag{3E-29}$$

$$b_1, b_2 = 17.2 \pm j32.4 \text{ sec}^{-1} = \gamma \pm j\beta \tag{3E-30}$$

Here, we see that b_1 and b_2 are a conjugate complex pair, hence (3D-29) and (3D-30) may be used as guides for i_a and ω_r, respectively. In particular,

$$i_a = i_a(0) + \frac{K}{k_v \tau_a \tau_m} \frac{1}{\gamma^2 + \beta^2} \left[1 - e^{-\gamma t}\left(\cos \beta t + \frac{\gamma}{\beta} \sin \beta t\right)\right] \tag{3E-31}$$

$$\omega_r = \omega_r(0) - \frac{K}{J} \frac{a_1}{\gamma^2 + \beta^2} \left[1 - e^{-\gamma t}\left(\cos \beta t - \frac{\gamma^2 + \beta^2 - a_1 \gamma}{a_1 \beta} \sin \beta t\right)\right] \tag{3E-32}$$

Substituting in the appropriate values yields

$$i_a = 633 - 317[1 - e^{-17.2t}(\cos 32.4t + 0.531 \sin 32.4t)] \tag{3E-33}$$

$$\omega_r = 64.64 + 1.02[1 - e^{-17.2t}(\cos 32.4t - 0.673 \sin 32.4t)] \tag{3E-34}$$

which can be expressed more compactly as

$$i_a = 316 + 359e^{-t/0.058} \cos(32.4t - 28°) \tag{3E-35}$$

$$\omega_r = 65.65 - 1.22e^{-t/0.058} \cos(32.4t + 33.9°) \tag{3E-36}$$

SP3.7-1. The starting characteristics of a permanent-magnet dc motor are expressed analytically in Example 3D. Once the machine reaches steady-state operation, the armature voltage is stepped from 6 to 12 V. Express i_a and ω_r. [$i_a = 0.149 + $ (3D-31); $\omega_r = 350.3 + $ (3D-32), where $t = 0$ is the instant the voltage is stepped]

SP3.7-2. Derive $\Delta\omega_r(s)/\Delta T_L(s)$ for a shunt machine with L_{AA} and B_m neglected and $\Delta v_a(s) = \Delta i_f(s) = 0$.

$$\left[\frac{\Delta\omega_r(s)}{\Delta T_L(s)} = -\frac{1}{J(s + 1/\tau_m)}\right]$$

SP3.7-3. Derive $\Delta T_e(s)/\Delta T_L(s)$ for a shunt machine with $B_m = 0$ and $\Delta v_a(s) = \Delta i_f(s) = 0$.

$$\left[\frac{\Delta T_e(s)}{\Delta T_L(s)} = \frac{1/\tau_a \tau_m}{s^2 + (1/\tau_a)s + 1/\tau_a \tau_m}\right]$$

3.8 RECAPPING

The dc machine is unique in that it exerts a torque on the rotating member as a result of the interaction of two stationary, orthogonal magnetic systems. One is produced by current flowing in the windings on the stationary member (field) and the other is caused by the current flowing in the windings of the rotating member (armature). Although dc machines are being replaced in many applications by either electronic equipment or other types of electromechanical devices, the permanent-magnet dc motor is still used quite widely in low-power control systems. Since the dynamic characteristics of this device can be described by linear differential equations, we have tried to set the stage for those who are interested in control system analysis by formulating the state equations of a permanent-magnet dc motor in fundamental form. Also, transfer functions have been developed with the control systems student in mind. However, a background in Laplace transformations is not necessary to follow the presentation of this information.

In Chap. 6 we will consider the brushless dc motor which is rapidly replacing the permanent-magnet dc motor. The equations which describe the operation of these two devices are very similar. Hence, much of the material presented here for the permanent-magnet dc motor can be applied directly to the analysis of the brushless dc motor.

3.9 REFERENCES

1. P. C. Krause, *Analysis of Electric Machinery*, McGraw-Hill Book Company, New York, 1986.
2. B. C. Kuo, *Automatic Control Systems*, Prentice-Hall Inc., Englewood Cliffs, N.J., 1987.
3. W. H. Hayt, Jr., and J. E. Kemmerly, *Engineering Circuit Analysis*, McGraw-Hill Book Company, New York, 1978.

3.10 PROBLEMS

1. The parameters of a dc shunt machine are $R_f = 240\,\Omega$, $L_{FF} = 120\,H$, $L_{AF} = 1.8\,H$, $r_a = 0.6\,\Omega$, $L_{AA} = 0$. The load torque is $5\,N \cdot m$ and $V_a = V_f = 240\,V$. Calculate the steady-state rotor speed.
2. The power input to a dc shunt motor during rated-load conditions is $100\,W$. The rotor speed is $2000\,r/min$ and the armature voltage is $100\,V$. The armature resistance is $2\,\Omega$ and $R_f = 200\,\Omega$. Calculate the no-load rotor speed.
3. A permanent-magnet dc motor has the following parameters: $r_a = 8\,\Omega$ and $k_v = 0.01\,V \cdot s/rad$. The shaft load torque is approximated as $T_L = K\omega_r$, where $K = 5 \times 10^{-6}\,N \cdot m \cdot s$. The applied voltage is $6\,V$ and $B_m = 0$. Calculate the steady-state rotor speed ω_r in rad/s.
4. A 250-V 600-r/min 200-hp dc shunt motor is delivering rated horsepower at rated speed. $R_f = 12\,\Omega$, $L_{AF} = 0.18\,H$, and $r_a = 0.012\,\Omega$.
   (a) Calculate the terminal voltage which must be applied to this machine to satisfy this load condition.
   (b) Calculate the full-load ohmic losses and determine the efficiency.
5. Losses in a dc machine include ohmic, no-load rotational, and stray-load losses.

The no-load rotational losses range from approximately 2 to 14 percent of the rated output. These losses include windage, friction, and core losses. Stray-load losses include the increase in core losses due to load and eddy current losses induced by the armature current. These losses are generally taken to be 1 percent of the rating of the machine. Another loss in the machine is due to the brush voltage drop. This is accounted for by increasing the armature resistance or assuming a constant-voltage drop across the brushes regardless of load. Repeat Prob. 4, assuming the machine has 5 percent rotational losses, 1 percent stray-load losses, and a total voltage drop across the brushes of 2 V which should be added to that calculated in Prob. 4a. Assume the rotational and stray-load losses can be represented by a resistance connected across the terminals of the machine.

6. The parameters of a dc shunt machine are $r_a = 10\,\Omega$, $R_f = 50\,\Omega$, and $L_{AF} = 0.5\,H$. Neglect B_m and $V_a = V_f = 25\,V$. Calculate (a) the steady-state stall torque, (b) the no-load speed, and (c) the steady-state rotor speed with $T_L = 3.75 \times 10^{-3}\omega_r$.

7. A permanent-magnet dc motor is driven by a mechanical source at 3820 r/min. The measured open-circuit armature voltage is 7 V. The mechanical source is disconnected, and a 12-V electric source is connected to the armature. With zero-load torque, $I_a = 0.1\,A$ and $\omega_r = 650\,rad/s$. Calculate k_v, B_m, and r_a.

*8. Express the maximum steady-state power output of a dc shunt motor ($P_{out} = T_e\omega_r$) if the field current i_f and armature voltage v_a are held constant. Let $B_m = 0$. (*Hint*: First express the rotor speed for maximum power output.)

9. The parameters of a 5-hp dc shunt machine are $r_a = 0.6\,\Omega$, $L_{AA} = 0.012\,H$, $R_f = 120\,\Omega$, $L_{FF} = 120\,H$, and $L_{AF} = 1.8\,H$. $V_a = V_f = 240\,V$. Calculate the steady-state rotor speed ω_r for $I_t = 0$; generator action.

10. A dc series motor requires 100 W at full load. The full-load speed is 2000 r/min and the terminal voltage is 100 V, $r_a = 2\,\Omega$, and $r_{fs} = 1\,\Omega$. Calculate the stall torque of the motor (T_e with $\omega_r = 0$).

11. The torque load of a dc series motor is $T_L = 100\,N \cdot m$. $L_{AFs} = 0.6\,H$, $r_a = 2\,\Omega$, and $r_{fs} = 3\,\Omega$. The voltage applied to the motor is 200 V. Calculate the steady-state rotor speed. Assume $B_m = 0$.

12. A dc compound motor requires 100 W at full load. The full-load speed is 2000 r/min and the voltage applied to the terminals of the machine is 100 V. $r_a = 2\,\Omega$, $R_f = 200\,\Omega$, and $r_{fs} = 1\,\Omega$. The series field is differentially connected. Calculate the steady-state no-load speed if $L_{AFs} = 0.1L_{AF}$.

13. Modify the state equations given by (3.6-14) for a permanent-magnet dc machine to include θ_r as a state variable.

14. Write the field voltage equation of a shunt dc machine in terms of k_v and express the transfer function between k_v and v_f.

*15. A separately excited dc machine is operating with no load ($T_L = 0$) and fixed field current. The armature resistance and inductance, r_a and L_{AA}, are small and can be neglected. Assume $B_m = 0$. Express the transfer function between i_a and v_a. Show that this motor appears as a capacitor to v_a.

*16. Write the state equations in fundamental form for the coupled circuits considered in Sec. 1.4. Start with (1.4-37) and use λ_1 and λ_2' as state variables. Relate currents and flux linkages by (1.4-34) and (1.4-35).

17. Develop the time-domain block diagram for the coupled circuits considered in Prob. 16.

18. Develop the time-domain block diagram for a series-connected dc machine.

*19. Develop the time-domain block diagram for the compound machine with a long-shunt connection.

20. We see from Figs. 3.5-2 and 3.5-4 that the permanent-magnet dc motor and the 200-hp dc shunt motor both demonstrated a slightly oscillatory speed response to a step charge in load torque. However, from Fig. 3.5-3, we note that the rotor speed response of the 5-hp dc shunt motor appears to be slightly overdamped. Use information in Example 3C to support these observations.

21. Construct the block diagram in terms of the Laplace operator for the permanent-magnet dc motor valid for load torque changes with the armature applied voltage held constant. Write the transfer function $\Delta\omega_r(s)/\Delta T_L(s)$.

22. Formulate the following transfer functions for a permanent-magnet dc machine:

(a) $\dfrac{\Delta i_a(s)}{\Delta T_L(s)}$ with $\Delta v_a(s) = 0$

(b) $\dfrac{\Delta i_a(s)}{\Delta v_a(s)}$ with $\Delta T_L(s) = 0$

(c) $\dfrac{\Delta T_e(s)}{\Delta v_a(s)}$ with $\Delta T_L(s) = 0$

*23. The field current of a dc shunt motor is held constant. Express the transfer function $\Delta i_a(s)/\Delta v_a(s)$ for no-load conditions $[\Delta T_L(s) = 0]$. Show that, if $B_m = 0$, this transfer function is identical in form to that of a series rLC circuit. Express r, L, and C in terms of machine parameters.

*24. Consider the 200-hp dc shunt motor described in Sec. 3.5. Assume that the machine is initially operating at rated conditions. The armature voltage is suddenly stepped from rated to 110 percent rated value. The load torque and field voltage remain constant and $B_m = 0$. Express i_a.

*25. Consider the 200-hp dc shunt motor described in Sec. 3.5. The machine is operating with no-load and rated terminal voltage; however, the field and armature circuits are supplied from separate sources. The armature is disconnected from its source and a 0.5-Ω resistor is immediately connected across the armature. Neglect r_a, L_{AA}, and B_m. Assume $J = 20 \text{ kg} \cdot \text{m}^2$. Express ω_r. This method of slowing a dc machine is referred to as *dynamic braking*.

CHAPTER
4

FEATURES COMMON TO ROTATING MAGNETIC FIELD ELECTROMECHANICAL DEVICES

4.1 INTRODUCTION

In the previous chapter, we found the dc machine to be a bit involved in that it has windings on both the stationary and rotating members, and these circuits are in relative motion whenever the armature (rotor) rotates. However, due to the action of the commutator, the resultant mmf produced by currents flowing in the rotating windings is stationary. In other words, the rotor windings appear to be stationary, magnetically. Therefore, with a constant current in the field (stator) winding, torque is produced and rotation results owing to the force established to align two stationary, orthogonal magnetic fields.

In rotational electromechanical devices other than dc machines, torque is produced as a result of one or more magnetic fields which rotate about the air gap of the device. Reluctance machines, induction machines, synchronous machines, stepper motors, and brushless dc motors, which are actually permanent-magnet synchronous machines, all develop torque in this manner. There

126

are features of these devices which are common to all. In particular, the winding arrangement of the stator and the method of producing a rotating magnetic field due to stator currents are the same for $2, 4, 6, \ldots, P$-pole devices. It seems logical, therefore, to cover these common features once and for all rather than repeat this material for each electromechanical device covered in subsequent chapters.

In our analysis, we will assume that the arrangement of a stator winding may be approximated by a winding distributed sinusoidally in terms of displacement about the air gap. With a sinusoidally distributed winding, the resulting air gap mmf will also be a sinusoidal function of displacement about the air gap. In this chapter, the sinusoidally distributed winding and the waveform of the resulting air gap mmf (magnetic field) are considered from an empirical point of view. Once the expression for the rotating magnetic field (rotating air gap mmf) is established, the concept of magnetic poles is discussed. Until this section only two-pole devices are considered. Fortunately, we are able to show that, with a simple substitution of variables, P-pole devices, where $P = 2, 4, 6, \ldots$, may all be treated as if they had only two poles. In the final section of this chapter, some time is devoted to a nonanalytical discussion of the operation of the reluctance, stepper, induction, and synchronous machines. Although some of this material may seem a bit premature and without theoretical basis, it is worthwhile in that it alerts us to the electromechanical devices which will be covered in later chapters.

Once the material in this chapter has been covered, the reader is prepared for any of the subsequent chapters except the final chapter wherein the servomotor and single-phase induction machines are analyzed. This final chapter has the chapter on induction machines (Chap. 5) as a prerequisite. Otherwise, none of the chapters is a prerequisite for the other. Hopefully, this flexibility will permit the reader to cover the desired topics in an efficient manner.

4.2 WINDINGS

In Chaps. 1 and 2, we considered two windings in relative motion; one on the stator and one on the rotor (Fig. 1.7-4). When introducing this device in Sec. 1.7, we assumed the turns of the windings to be concentrated in one position even though, in practice, they would normally be distributed over a 30 to 60° arc. Let us redraw Fig. 1.7-4 with the following changes. First, we will disregard the rotor winding and focus our attention on the stator winding. We will see the reason for this when we discover that the arrangement of the stator winding(s) is common for the electromechanical motion devices considered in subsequent chapters. Finally, we will call winding 1 of Fig. 1.7-4 winding *as*, and assume that it is distributed in slots over the inner circumference of the stator, as shown in Fig. 4.2-1. This is more characteristic of the stator winding of a single-phase electromechanical motion device than is a concentrated winding.

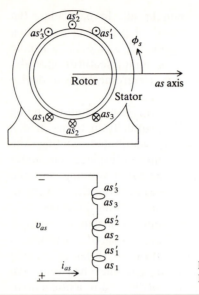

FIGURE 4.2-1
Elementary two-pole single-phase stator winding.

The winding is depicted as a series of individual coils. Each coil is placed in a slot in the stator steel. If we follow the path of positive current i_{as} flowing in the *as* winding, we see that current enters as_1, depicted by $\otimes$, to indicate that the assumed direction of positive current is down the length of the stator in an axial direction (into the paper). Current flows down the length of the stator, loops at the end (end turn, see Fig. 1.7-4), and flows back down the length of the stator and out at as_1', depicted by $\odot$. Note that as_1 and as_1' are placed in stator slots which span π radians. This is characteristic of a two-pole machine; we shall talk more about poles later. Now, as_1 around to as_1' is referred to as a *coil* and as_1 or as_1' is a coil side. In practice, a coil will contain more than one conductor. That is, current will flow into as_1 in a conductor and out as_1' via the same conductor. The conductor, which is, of course, insulated, may then be looped back to as_1 and the winding of the conductor around the as_1-as_1' path repeated, thereby forming a coil with numerous turns. The number of conductors in a coil side tells us the number of turns in this coil. This number will be denoted as nc_s.

Now then, once we have wound nc_s turns in the as_1-as_1' coil, we will take the same conductor and repeat this winding process to form the as_2-as_2' coil. We will assume that the same number of turns (nc_s) make up the as_2-as_2' coil as the as_1-as_1' coil and, similarly, for the as_3-as_3'. We could have wound a different number of turns in each coil but we will assume that this was not done. With the same number of turns (nc_s) in each of the coils, the winding is said to be distributed over a span from as_1 to as_3 or 60°. Once the winding is wound, we can use the right-hand rule to give a meaning to the *as* axis shown in Fig. 4.2-1. It is, by definition, the principal direction of magnetic flux established by positive current flowing in the *as* winding. It is said to denote the positive direction of the magnetic axis of the *as* winding.

In Fig. 4.2-2, we have added a second stator winding—the *bs* winding. Note that its magnetic axis is displaced $\frac{1}{2}\pi$ from that of the *as* winding. Although it is a matter of choice, we will assume that the positive direction of i_{bs} is such that the positive magnetic axis of the *bs* winding is at $\phi_s = \frac{1}{2}\pi$ rather than at $\phi_s = -\frac{1}{2}\pi$. Actually, we have not defined ϕ_s yet, but we can see that it is an angular displacement about the stator, referenced to the *as* axis. Placing the *bs* winding $\frac{1}{2}\pi$ from the *as* winding makes this the stator configuration for a two-pole two-phase electromechanical device. Unfortunately, this fact adds little to our present appreciation of the meaning of two-pole or two-phase; however, we can now establish the meaning of symmetrical as it is used in electromechanical devices. If we wind the orthogonal *bs* winding exactly as we did the *as* winding, that is, the same number of turns per coil (nc_s) with the same size of wire, then the turns and resistance of the *as* and *bs* windings will be identical, whereupon, the stator windings are said to be symmetrical. For a two-pole three-phase symmetrical electromechanical device, there are three identical stator windings, *as*, *bs*, and *cs*, displaced 120° from each other. Essentially all multiphase electromechanical devices are equipped with symmetrical stators.

SP4.2-1. The total number of turns of the *as* winding in Fig. 4.2-2 is 15. The stator windings are symmetrical. For the *as* and *bs* windings, calculate (*a*) the turns per coil, nc_s, and (*b*) the conductors per coil side. [(*a*) 5; (*b*) 5]

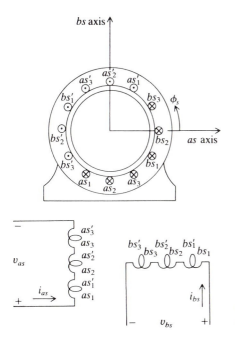

FIGURE 4.2-2
Elementary two-pole two-phase stator windings.

SP4.2-2. A fourth coil, as_4-as'_4, with nc_s turns is placed in the same slot as the as_2-as'_2 coil in Fig. 4.2-2. Is the stator symmetrical? A fourth coil, bs_4-bs'_4, with nc_s turns is placed in the same slot as the bs_2-bs'_2 coil in Fig. 4.2-2. Is the stator now symmetrical? [No; yes]

4.3 AIR GAP mmf—SINUSOIDALLY DISTRIBUTED WINDINGS

In the analysis of electromechanical motion devices it is generally assumed that the stator windings and, in many cases, the rotor windings may be approximated as sinusoidally distributed windings. That is, the distribution of a stator phase winding may be approximated as a sinusoidal function of ϕ_s, and the waveform of the resulting mmf dropped across the air gap (air gap mmf) of the device may also be approximated as a sinusoidal function of ϕ_s. Actually, most electric machines, particularly large machines, are designed so that the windings, especially the stator windings, produce a relatively good approximation of a sinusoidally distributed air gap mmf so as to minimize the voltage and current harmonics. To establish a truly sinusoidal air gap mmf waveform (often referred to as a *space sinusoid*), the winding must also be distributed sinusoidally. Except in cases where the harmonics due to the winding configuration are of importance, it is typically assumed that all windings may be approximated as sinusoidally distributed windings. We will make this assumption in our analysis of the electromechanical motion devices considered in subsequent chapters.

For this discussion, let us add a few coils to the *as* winding shown in Fig. 4.2-1, whereupon we have Fig. 4.3-1. Although we may be going a bit far by adding coils which now span 120°, it helps to illustrate our point. For the purpose of establishing an expression for the air gap mmf, it is convenient to employ the so-called developed diagram of the cross-sectional view of the device shown in Fig. 4.3-1. This developed diagram is obtained by "flattening out" the rotor and stator as depicted in Fig. 4.3-2. To relate the developed

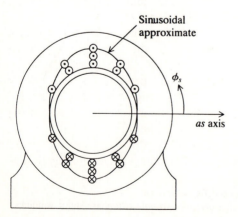

FIGURE 4.3-1
Approximate sinusoidal distribution of the *as* winding.

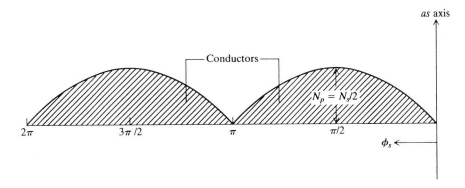

FIGURE 4.3-2
Developed diagram with sinusoidally distributed stator winding.

diagram to the cross-sectional view of the device, it is helpful to define the displacement ϕ_s to the left of the as axis since this allows us to position the stator above the rotor.

From Fig. 4.3-1 or the developed diagram in Fig. 4.3-2, we see that the winding distribution from $0 < \phi_s < \pi$ may be approximated by the expression

$$N_{as} = N_p \sin \phi_s \qquad \text{for } 0 < \phi_s < \pi \qquad (4.3\text{-}1)$$

and from $\pi < \phi_s < 2\pi$ as

$$N_{as} = -N_p \sin \phi_s \qquad \text{for } \pi < \phi_s < 2\pi \qquad (4.3\text{-}2)$$

where N_p is the peak turns density in turns/radian which could be determined from a Fourier analysis of the actual winding distribution. If N_s represents the number of turns of the equivalent sinusoidally distributed winding, then

$$N_s = \int_0^\pi N_p \sin \phi_s \, d\phi_s = 2N_p \qquad (4.3\text{-}3)$$

It is important to note that the number of turns is obtained by integrating the turns density N_{as} from zero to π rather than from zero to 2π. Also, N_s represents not the total turns of the winding, but those of the equivalent sinusoidally distributed winding that corresponds to the fundamental component of the actual winding distribution.

Now, the sinusoidally distributed winding will produce an mmf that is positive in the direction of the as axis, to the right in Fig. 4.3-1, for positive i_{as}. Also, since the reluctance of the steel of this device is much smaller (neglecting saturation) than the reluctance of the air gap, it can be assumed that all of the mmf is dropped across the air gap. It seems logical that, if the windings are sinusoidally distributed in space (ϕ_s), then the mmf dropped across the air gap will also be sinusoidal in space.

For the purposes of establishing an expression for the air gap mmf associated with the as winding, mmf$_{as}$, let us travel a closed path which starts at the center of the rotor as shown in Fig. 4.3-3a. From the center of the rotor, let us proceed to the right along the *as* axis, crossing the air gap at $\phi_s = 0$. Now that we are in the stator, we will turn to the right and proceed clockwise, encircling the lower windings until we get to $\phi_s = \pi$, whereupon we will cross again the air gap and return to the center of the rotor. The total current enclosed is $N_s i_{as}$ which, by Ampere's law, is equal to the mmf drop ($\int \mathbf{H} \cdot d\mathbf{L}$) around the given path. If the reluctance of the rotor and stator steel is small compared with the air gap reluctance, it can be assumed, for practical purposes, that one half of the mmf is dropped across the air gap at $\phi_s = 0$ and one half at $\phi_s = \pi$. By definition, mmf_{as} is positive for an mmf drop across the air gap from the rotor to the stator. Thus, mmf_{as} is positive at $\phi_s = 0$ and negative at $\phi_s = \pi$, assuming positive i_{as}. Moreover, since one half of this mmf ($N_s i_{as}$) is dropped at $\phi_s = 0$ and one half at $\phi_s = \pi$, $\text{mmf}_{as}(0) = (N_s/2)i_{as}$ and $\text{mmf}_{as}(\pi) = -(N_s/2)i_{as}$. This suggests that, for arbitrary ϕ_s, mmf_{as} might be expressed as

$$\boxed{\text{mmf}_{as} = \frac{N_s}{2} i_{as} \cos \phi_s} \tag{4.3-4}$$

Equation (4.3-4) tells us that the air gap mmf is zero at $\phi_s = \pm \frac{1}{2}\pi$. Let us check this by traveling a path starting again at the center of the rotor, but this time we will go straight up crossing the air gap at $\phi_s = \frac{1}{2}\pi$, as shown in Fig. 4.3-3b. Now that we are in the stator, we have two choices—we can go down the right side of the stator steel as in Fig. 4.3-2b or down the left side. In either case, we will cross the air gap into the rotor at $\phi_s = -\frac{1}{2}\pi$ and complete our closed path by ending up at the center of the rotor. Regardless of the path

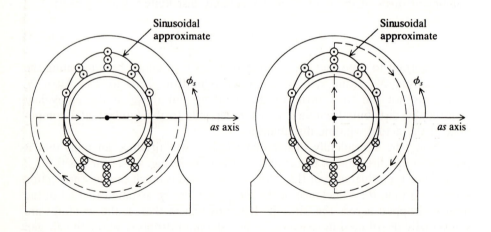

(a)   (b)

FIGURE 4.3-3
Closed paths used to establish mmf_{as}.

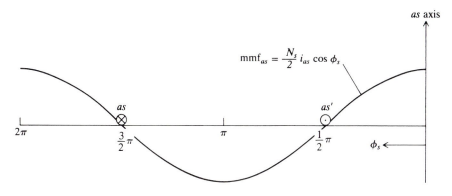

FIGURE 4.3-4
An mmf_{as} due to sinusoidally distributed as winding.

selected, we enclosed as much current going in one direction $\odot$ as in the other $\otimes$. Since the net current enclosed is zero, the mmf drop is zero along the given path, implying that $\text{mmf}_{as} = 0$ at $\phi_s = \pm\frac{1}{2}\pi$. The fact that mmf_{as} is a positive (negative) maximum at $\phi_s = 0$ ($\phi_s = \pi$) and that $\text{mmf}_{as} = 0$ at $\phi_s = \pm\frac{1}{2}\pi$, together with the fact that the winding in Fig. 4.3-1 is sinusoidally distributed, strongly suggests that (4.3-4) is the correct expression of mmf_{as}. The mmf due to the sinusoidally distributed as winding is depicted in Fig. 4.3-4, where the locations of maximum turns density are depicted by $\odot$ and $\otimes$. Hereafter, we will use this to depict a sinusoidally distributed winding.

Let us now consider the bs winding of a two-phase device. Recall that it is an identical winding displaced $\frac{1}{2}\pi$ from the as winding, as shown in Fig. 4.2-2. It follows that the air gap mmf due to a sinusoidally distributed bs winding may be expressed as

$$\boxed{\text{mmf}_{bs} = \frac{N_s}{2} i_{bs} \sin \phi_s} \qquad (4.3\text{-}5)$$

Example 4A. The air gap mmf of the elementary two-pole single-phase stator winding shown in Fig. 4.2-1 is established for those who wish to examine the mmf distribution associated with nonsinusoidally distributed windings. In Fig. 4.2-1, which is shown again in Fig. 4A-1, each of the three coils contains nc_s turns. We will apply Ampere's law to the closed path depicted in Fig. 4A-1. In particular,

$$\oint \mathbf{H} \cdot d\mathbf{L} = i_n \qquad (4\text{A-}1)$$

where i_n is the current enclosed by the closed path. In Fig. 4A-1, the part of the closed path defined by the straight line from o to a is assumed to be at an angle ϕ_s measured from the as axis. If the reluctance of the rotor and stator steel is small, we can neglect the field intensity therein, whereupon (4A-1) becomes

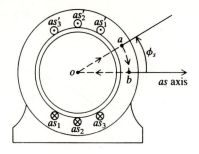

FIGURE 4A-1
Closed path for evaluation of mmf_{as}.

$$H_r(\phi_s)g(\phi_s) - H_r(0)g(0) = i_n \qquad (4A\text{-}2)$$

Here, $H_r(\phi_s)$ represents the radial component of the **H** field in the air gap and $g(\phi_s)$ is the air gap length which is constant in a uniform air gap machine. The stator mmf is defined as the line integral of **H**. Therefore (4A-2) may be written as

$$\text{mmf}_{as}(\phi_s) - \text{mmf}_{as}(0) = i_n \qquad (4A\text{-}3)$$

If, in Fig. 4A-1, $0 < \phi_s < \frac{1}{3}\pi$, then current is not enclosed by the closed path. If we assume that $\text{mmf}_{as}(0) = 0$, then (4A-3) implies that $\text{mmf}_{as}(\phi_s) = 0$ for $0 < \phi_s < \frac{1}{3}\pi$. On the other hand, if $\frac{1}{3}\pi < \phi_s < \frac{1}{2}\pi$, the enclosed current is $-nc_s i_{as}$. The minus sign is needed since, in accordance with the right-hand rule, a positive enclosed current flows into the paper $\otimes$ for the selected clockwise path. Again, if we assume $\text{mmf}_{as}(0) = 0$, then (4A-3) implies that $\text{mmf}_{as} = -nc_s i_{as}$ for $\frac{1}{3}\pi < \phi_s < \frac{1}{2}\pi$. Continuing with this procedure results in the mmf distribution depicted in Fig. 4A-2.

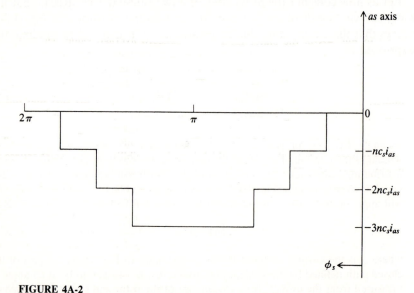

FIGURE 4A-2
Plot of air gap mmf assuming $\text{mmf}_{as}(0) = 0$.

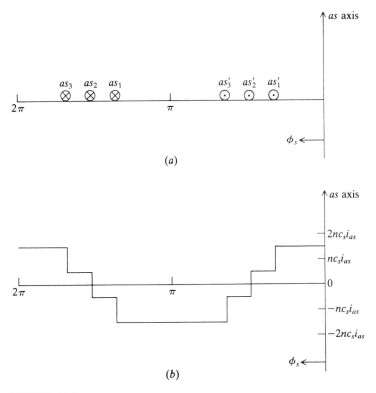

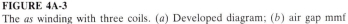

FIGURE 4A-3
The *as* winding with three coils. (*a*) Developed diagram; (*b*) air gap mmf

In Fig. 4A-2, the stator mmf has a nonzero average value. However, in obtaining the distribution of Fig. 4A-2, it was assumed, quite arbitrarily, that $\text{mmf}_{as}(0)$ is zero. If we had assumed a nonzero value, the distribution of Fig. 4A-2 would be shifted up or down. The question arises as to what value of $\text{mmf}_{as}(0)$ should we have assumed? The answer rests upon the fact that mmf_{as} must have zero average value to satisfy Gauss' law. In other words, the flux entering the stator from the air gap must equal the flux entering the air gap from the stator. The air gap mmf depicted in Fig. 4A-2 would imply that all of the flux enters the air gap from the stator, which is an impossibility since a point source of flux does not exist. Hence, mmf_{as} in Fig. 4A-2 must be shifted upward by $\frac{3}{2}nc_s i_{as}$ so that distribution has zero average value. The resulting air gap mmf is depicted in Fig. 4A-3, which may be considered a coarse approximation of a sinusoidally distributed mmf. By Fourier analysis, which is an analytical method of expressing any periodic function in terms of sinusoidal components, it can be shown that the fundamental component of the resulting air gap mmf has a peak amplitude of $1.74nc_s i_{as}$.

SP4.3-1. Express the sinusoidal approximation of mmf_{as} if the assumed positive direction of i_{as} is reversed in Fig. 4.2-1. [$\text{mmf}_{as} = -(4.3\text{-}4)$]

SP4.3-2. If $I_{as} = 1\,\text{A}$, $I_{bs} = -1\,\text{A}$, and $N_s = 10$ in Fig. 4.2-2, express the sum of the sinusoidal approximations of the two air gap mmf's. [$\text{mmf}_{as} + \text{mmf}_{bs} = \sqrt{2}\,5\cos(\phi_s + \pi/4)$]

SP4.3-3. Assume that winding 1 in Fig. 1.7-4 has one turn. Sketch the air gap mmf due to current i_1 flowing in this winding. [$\frac{1}{2}i_1$ for $-\frac{1}{2}\pi < \phi_s < \frac{1}{2}\pi$ and $-\frac{1}{2}i_1$ for $\frac{1}{2}\pi < \phi_s < \frac{3}{2}\pi$]

4.4   ROTATING AIR GAP mmf—TWO-POLE DEVICES

Considerable insight into the operation of electromechanical motion devices can be gained from an analysis of the air gap mmf produced by current flowing in the stator winding(s). We will consider the rotating air gap mmf's produced by currents flowing in the stator windings of single-, two-, and three-phase devices.

Single-Phase Devices

Let us first consider the device shown in Fig. 4.4-1 which illustrates a single-phase stator winding. We will assume that the *as* winding is sinusoidally distributed with *as* and *as'* placed at the point of maximum turns density. As mentioned previously, we will use this means of depicting a sinusoidally distributed winding. Actually, Fig. 4.4-1 is Fig. 4.2-1 with the winding assumed to be sinusoidally distributed.

If we assume that the current flowing in the *as* winding is a constant, then, as in our work in Sec. 2.8, the *as* winding would establish a stationary magnetic system with a north pole from $\frac{1}{2}\pi < \phi_s < \frac{3}{2}\pi$ and a south pole from $-\frac{1}{2}\pi < \phi_s < \frac{1}{2}\pi$. The air gap mmf is directly related to these poles; indeed, the flux flowing from the north pole and into the south pole is caused by the air gap mmf.

Let us see what happens when the current flowing in the *as* winding is a sinusoidal function of time. For this, we will assume steady-state operation with

$$I_{as} = \sqrt{2}I_s\cos[\omega_e t + \theta_{esi}(0)] \qquad (4.4\text{-}1)$$

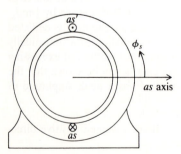

FIGURE 4.4-1
Elementary two-pole single-phase sinusoidally distributed stator winding.

Recall that capital letters without a raised tilde are used to denote steady-state instantaneous variables. As defined in Appendix B, I_s is the rms value of the current, ω_e is the electrical angular velocity, and $\theta_{esi}(0)$ is the angular position corresponding to the time zero value of the instantaneous current. From Appendix B, the phasor representation of I_{as} is $\tilde{I}_{as} = I_s \underline{/\theta_{esi}(0)}$.

The air gap mmf is expressed for the as winding by (4.3-4). Substituting (4.4-1) into (4.3-4) yields

$$\text{mmf}_{as} = \frac{N_s}{2} \sqrt{2} I_s \cos[\omega_e t + \theta_{esi}(0)] \cos \phi_s \tag{4.4-2}$$

Let us consider this expression for a moment. If we stand at $\phi_s = 0$, we would see the air gap mmf vary as a cosinusoidal function of time. In particular, if we selected $\theta_{esi}(0)$ to be zero, mmf_{as} at $\phi_s = 0$ would be a positive maximum (south pole) at $t = 0$ and a negative maximum (north pole) when $\omega_e t = \pi$. Since we are interested in establishing a rotating air gap mmf, this does not look too promising. It would appear that all we have is a stationary, pulsating magnetic field; however, let us use a trigonometric identity from Appendix A to write (4.4-2) as

$$\text{mmf}_{as} = \frac{N_s}{2} \sqrt{2} I_s \{ \tfrac{1}{2} \cos[\omega_e t + \theta_{esi}(0) - \phi_s] + \tfrac{1}{2} \cos[\omega_e t + \theta_{esi}(0) + \phi_s] \} \tag{4.4-3}$$

The arguments of the cosine terms are functions of time and displacement ϕ_s. If we can make an argument constant, then the cosine of this argument would be constant. Let us see if this is possible by setting both arguments equal to a constant. In particular, for the first term on the right-hand side of (4.4-3) let

$$\omega_e t + \theta_{esi}(0) - \phi_s = C_1 \tag{4.4-4}$$

For the second term,

$$\omega_e t + \theta_{esi}(0) + \phi_s = C_2 \tag{4.4-5}$$

Taking the derivative of (4.4-4) and (4.4-5) with respect to time yields, for the first term on the right-hand side,

$$\frac{d\phi_s}{dt} = \omega_e \tag{4.4-6}$$

and for the second term,

$$\frac{d\phi_s}{dt} = -\omega_e \tag{4.4-7}$$

What does this mean? Well, (4.4-6) tells us that, if we run around the air gap of Fig. 4.4-1 in the counterclockwise direction at an angular velocity of ω_e, the first term on the right-hand side of (4.4-3) will appear as a constant mmf, hence, a constant set of north and south poles. On the other hand, (4.4-7) tells us that, if we run clockwise at ω_e, the second term on the right-hand side of

(4.4-3) will appear as a constant mmf. In other words, the pulsating air gap mmf that we noted when standing at $\phi_s = 0$ (or any fixed value of ϕ_s) can be thought of as two, one-half amplitude, oppositely rotating air gap mmf's (magnetic fields), each rotating at the angular velocity of ω_e, which is the electrical angular velocity of the current. Since we have two oppositely rotating sets of north and south poles (magnetic fields), it would seem that the single-phase machine could develop an average torque as a result of interacting with either set. Indeed, we will find that a single-phase electromechanical device with the stator winding as shown in Fig. 4.4-1 can develop an average torque in either direction of rotation. It might appear that this is a four-pole device rather than a two-pole device, since there are two two-pole sets. However, only one set interacts with the rotor to produce a torque with a nonzero average.

Two-Phase Devices

Let us consider the two-pole two-phase stator shown in Fig. 4.4-2. The *as* and *bs* windings are shown with two circles for each winding, which is now our way of depicting sinusoidally distributed windings. For balanced steady-state conditions, the stator currents may be expressed (Appendix C) as

$$I_{as} = \sqrt{2}I_s \cos[\omega_e t + \theta_{esi}(0)] \tag{4.4-8}$$

$$I_{bs} = \sqrt{2}I_s \sin[\omega_e t + \theta_{esi}(0)] \tag{4.4-9}$$

Here we see that $\tilde{I}_{bs} = -j\tilde{I}_{as}$. Actually, the set formed by (4.4-8) and (4.4-9) is but one of four possible two-phase balanced sets; that is, each expression for current could be preceded by a $\pm$ sign. The reason for selecting the set given by (4.4-8) and (4.4-9) will become apparent.

The total air gap mmf due to both stator windings, which are assumed to be sinusoidally distributed, may be expressed by adding mmf$_{as}$ and mmf$_{bs}$ as given by (4.3-4) and (4.3-5), respectively. Thus, the total air gap mmf due to the stator windings, denoted mmf$_s$, is

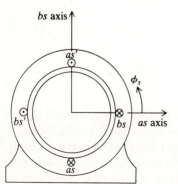

FIGURE 4.4-2
Elementary two-pole two-phase sinusoidally distributed stator windings.

$$\boxed{\begin{aligned} \text{mmf}_s &= \text{mmf}_{as} + \text{mmf}_{bs} \\ &= \frac{N_s}{2}(i_{as}\cos\phi_s + i_{bs}\sin\phi_s) \end{aligned}} \quad (4.4\text{-}10)$$

Substituting the expressions for the balanced steady-state stator currents, (4.4-8) and (4.4-9), into (4.4-10) and making use of the trigonometric relations given in Appendix A, yields

$$\text{mmf}_s = \frac{N_s}{2}\sqrt{2}I_s \cos[\omega_e t + \theta_{esi}(0) - \phi_s] \quad (4.4\text{-}11)$$

It is interesting that we have only one rotating air gap mmf or rotating magnetic field. If we set the argument equal to a constant and take the derivative with respect to time, we find that the argument is constant if $d\phi_s/dt = \omega_e$. That is, if we travel around the air gap of Fig. 4.4-2 in the counterclockwise direction at ω_e, we will always see a constant mmf$_s$ for the balanced set of stator currents given by (4.4-8) and (4.4-9). Hence, a single rotating air gap mmf is produced. The actual value that we would see as we travel around the air gap at ω_e would depend upon the selection of time zero and our position on the stator at time zero. If, for example, $\theta_{esi}(0) = 0$ and if at $t = 0$ we are at $\phi_s = 0$, the magnitude of the total air gap mmf that we would see would be $(N_s/2)\sqrt{2}I_s$, as determined from (4.4-11). [Is it clear that $I_{as}(0) = \sqrt{2}I_s$ and $I_{bs}(0) = 0$?] If now, as time increases ($t > 0$), we immediately started running in the counterclockwise direction at ω_e, we would always see this same magnitude of mmf$_s$.

We can now answer a question raised earlier regarding the selection of the balanced set given by (4.4-8) and (4.4-9). With balanced steady-state stator currents, we want mmf$_s$ to rotate in the counterclockwise direction for conventional purposes. With the assigned positive direction of current in the given arrangement of the *as* and *bs* windings shown in Fig. 4.4-1, the balanced set of stator currents given by (4.4-8) and (4.4-9) produces an mmf$_s$ that rotates counterclockwise.

It is very important not to confuse the magnetic axes of the stator windings (*as* axis and *bs* axis) which are stationary with the phasors representing the stator currents ($\tilde{I}_{as}$ and $\tilde{I}_{bs}$). However, it is interesting to note the relative position of the *as* axis and *bs* axis versus the relative position of $\tilde{I}_{as}$ and $\tilde{I}_{bs}$ for a constant, counterclockwise mmf$_s$ to occur. The *bs* axis is displaced $\frac{1}{2}\pi$ ahead of the *as* axis, as illustrated in Fig. 4.4-2; however, from (4.4-8) and (4.4-9) we see that $\tilde{I}_{bs}$ lags $\tilde{I}_{as}$ by $\frac{1}{2}\pi$. Note, that the negative of both (4.4-8) and (4.4-9) would also produce a counterclockwise mmf$_s$.

At $t = 0$ with $\theta_{esi}(0) = 0$, mmf$_s$ is a cosine function of ϕ_s with the maximum value of $(N_s/2)\sqrt{2}I_s$ directed in the positive *as* axis (Fig. 4.4-2). Recall that a north pole is established over the area where the mmf due to the current flowing in the stator windings causes flux to enter the air gap from the stator. At $t = 0$ this would occur for $\frac{1}{2}\pi < \phi_s < \frac{3}{2}\pi$. Hence, at $t = 0$, a north pole caused by the currents flowing in the stator windings exists over the range $\frac{1}{2}\pi < \phi_s < \frac{3}{2}\pi$ with the maximum strength at $\phi_s = \pi$. Similarly, a south pole caused by stator currents exists over the range $-\frac{1}{2}\pi < \phi_s < \frac{1}{2}\pi$ with the maximum strength at $\phi_s = 0$. Note that there are two poles established, hence a two-pole machine, which rotate about the air gap at ω_e with mmf$_s$. The direction of rotation of the air gap mmf (mmf$_s$) may readily be determined from the position of the poles when time has progressed to where I_{bs} is maximum and I_{as} is zero ($\omega_e t = \frac{1}{2}\pi$).

In the case of the single-phase stator winding with a sinusoidal current, the air gap mmf can be thought of as two oppositely rotating, constant-amplitude mmf's. However, the instantaneous air gap mmf is pulsating even when we are traveling with one of the rotating air gap mmf's. Unfortunately, this pulsating air gap mmf or set of poles gives rise to steady-state pulsating components of electromagnetic torque. In the case of the two-phase stator with balanced currents, only one rotating air gap mmf exists. Hence, the steady-state electromagnetic torque will not contain a pulsating or time-varying component; it will be constant with the value determined by the operating conditions.

Three-Phase Devices

Although we will focus our attention upon two-phase devices in the following chapters, there is some time devoted to the three-phase version of each device. In preparation for this, it is worthwhile to consider briefly the arrangement of the stator windings of a two-pole three-phase device shown in Fig. 4.4-3. The windings are shown connected in wye; however, a delta connection could also be used. The type of connection is irrelevant for this consideration. The windings are identical, sinusoidally distributed with N_s equivalent turns and with their magnetic axes displaced 120°; the stator is symmetrical. The positive direction of the magnetic axes is selected so as to achieve counterclockwise rotation of the rotating air gap mmf with balanced stator currents of the abc sequence. We shall see this in just a moment.

The air gap mmf's established by the stator windings may be expressed by inspection of Fig. 4.4-3. In particular,

$$\text{mmf}_{as} = \frac{N_s}{2} i_{as} \cos \phi_s \qquad (4.4\text{-}12)$$

$$\text{mmf}_{bs} = \frac{N_s}{2} i_{bs} \cos(\phi_s - \tfrac{2}{3}\pi) \qquad (4.4\text{-}13)$$

$$\text{mmf}_{cs} = \frac{N_s}{2} i_{cs} \cos(\phi_s + \tfrac{2}{3}\pi) \qquad (4.4\text{-}14)$$

As before, N_s is the number of turns of the equivalent sinusoidally distributed stator windings and ϕ_s is the angular displacement about the stator. For

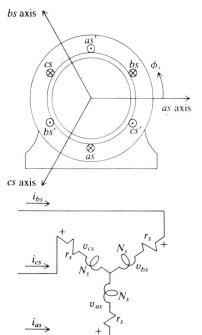

FIGURE 4.4-3
Elementary two-pole three-phase sinusoidally distributed stator windings.

balanced steady-state conditions, the stator currents for an *abc* sequence may be expressed as

$$I_{as} = \sqrt{2}I_s \cos[\omega_e t + \theta_{esi}(0)] \tag{4.4-15}$$

$$I_{bs} = \sqrt{2}I_s \cos[\omega_e t - \tfrac{2}{3}\pi + \theta_{esi}(0)] \tag{4.4-16}$$

$$I_{cs} = \sqrt{2}I_s \cos[\omega_e t + \tfrac{2}{3}\pi + \theta_{esi}(0)] \tag{4.4-17}$$

Substituting (4.4-15) through (4.4-17) into (4.4-12) through (4.4-14) and adding the resulting expression yield an expression for the rotating air gap mmf established by balanced steady-state currents flowing in the stator windings:

$$\mathrm{mmf}_s = \frac{N_s}{2}\sqrt{2}I_s \frac{3}{2}\cos[\omega_e t + \theta_{esi}(0) - \phi_s] \tag{4.4-18}$$

The trigonometric relations given in Appendix A are helpful in obtaining (4.4-18). If mmf$_s$, (4.4-18), for the three-phase device is compared with mmf$_s$ for a two-phase device given by (4.4-11), we see that they are identical except that the amplitude of the mmf for the three-phase device is $\tfrac{3}{2}$ times that of a two-phase device. Actually, it can be shown that this amplitude for multiphase devices changes from that of a two-phase device in proportion to the number of phases divided by two.

It is important to note that with the selected positive directions of the magnetic axes a counterclockwise rotating air gap mmf is obtained with a three-phase set of balanced stator currents of the *abc* sequence. As in the two-phase case, it is also important to note the relative positions of the magnetic axes versus the relative positions of the phasors representing the currents in order to establish a constant-amplitude counterclockwise rotating air gap mmf. From (4.4-12) through (4.4-14) or Fig. 4.4-3, we see that the *bs* axis is stationary at $120°$ while the *cs* axis is stationary at $-120°$. From (4.4-15) through (4.4-17), $\tilde{I}_{as}$, $\tilde{I}_{bs}$, and $\tilde{I}_{cs}$ are $120°$ out of phase; however, in order for the constant-amplitude air gap mmf to rotate in the counterclockwise direction, $\tilde{I}_{bs}$ lags $\tilde{I}_{as}$ by $120°$ and $\tilde{I}_{cs}$ leads $\tilde{I}_{as}$ by $120°$.

SP4.4-1. Assume I_{as} in Fig. 4.4-1 is $I_{as} = \sqrt{2}I_s \cos \omega_e t$. Express mmf$_{as}$ at $\phi_s = \pi$. [mmf$_{as}(\pi) = -(N_s/2)\sqrt{2}I_s \cos \omega_e t$]

SP4.4-2. When running counterclockwise at ω_e around the air gap of Fig. 4.4-1 with $I_{as} = \sqrt{2}I_s \cos[\omega_e t + \theta_{esi}(0)]$, what did the second term on the right-hand side of (4.4-3) appear as? [A one-half amplitude air gap mmf pulsating at a frequency of ω_e/π Hz]

SP4.4-3. Determine the balanced set that would produce a counterclockwise rotating mmf$_s$ if the assigned positive direction of i_{bs} in Fig. 4.4-2 were reversed. [$I_{as} = (4.4-8)$ and $I_{bs} = -(4.4-9)$; $I_{as} = -(4.4-8)$ and $I_{bs} = (4.4-9)$]

SP4.4-4. The rotor of the two-phase device shown in Fig. 4.4-2 is rotating at $0.9\omega_e$ in the counterclockwise direction. You are on the rotor. (*a*) Relative to you, what is the speed of the rotating air gap mmf given by (4.4-11)? (*b*) If now you start running clockwise on the rotor at $2\omega_e$, what would be the speed of the rotating air gap mmf relative to you? [(*a*) $0.1\omega_e$, ccw; (*b*) $2.1\omega_e$, ccw]

SP4.4-5. The steady-state stator currents flowing in the three-phase device shown in Fig. 4.4-3 are $\tilde{I}_{as} = I_s/\underline{-90°}$, $\tilde{I}_{bs} = I_s/\underline{30°}$, $\tilde{I}_{cs} = I_s/\underline{150°}$. Determine the direction of rotation of mmf$_s$. [cw]

4.5   *P*-POLE MACHINES

Thus far, we have considered only two-pole electromechanical motion devices. Actually, electromechanical devices may have any even number of poles; $2, 4, 6, 8, \ldots$, up to more than 100 in the case of large hydroturbine generators. We may at first be a little hesitant to tackle the analysis of machines with more than two poles, since it appears that this may complicate matters before we even get started. Fortunately, this is not the case. We will find that with a simple change of variables we can analyze all machines as if they were two-pole machines. We need only to modify the expression for evaluating torque and realize that, physically, the actual rotor speed of a machine with more than two poles will be a multiple less than that which we determine from our two-pole equivalent.

The characteristics of the rotating air gap mmf of a machine with more than two poles can be determined by considering the four-pole device shown in Fig. 4.5-1. Here, the outer boundary of the stator is not depicted for purposes of convenience. We shall make this omission in many future drawings of rotational electromechanical devices. In Fig. 4.5-1, each phase winding consists

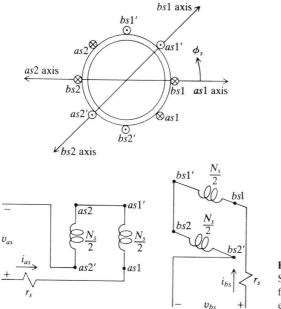

FIGURE 4.5-1
Stator winding arrangement of a four-pole two-phase symmetrical electromechanical device.

of two series-connected windings, each of which is assumed to be sinusoidally distributed. For example, $as1'$ represents a group of conductors sinusoidally distributed over $0 < \phi_s < \frac{1}{2}\pi$, etc. The phase windings consist of N_s turns, with $N_s/2$ turns in each of the series-connected windings. There may be some confusion in regard to notation. When considering the coils of a winding in Fig. 4.2-1, the notation $as_1, bs_1, \ldots$ was used. In Fig. 4.5-1, the notation $as1, bs1, \ldots$ is used to denote sinusoidally distributed windings. Thus, a subscripted number denotes a coil whereas a number which is not a subscript denotes a sinusoidally distributed winding.

Note that each of the two series windings per phase spans $\frac{1}{2}\pi$ radians and each phase winding establishes two magnetic systems. For example, at the instant when $i_{bs} = 0$ and i_{as} is a positive maximum, the $as1$-$as1'$ part of the as winding produces positive flux in the $as1$-axis direction whereas the $as2$-$as2'$ part of the as winding produces positive flux in the $as2$-axis direction. A south pole occurs for $-\frac{1}{4}\pi < \phi_s < \frac{1}{4}\pi$ and $\frac{3}{4}\pi < \phi_s < \frac{5}{4}\pi$. Now, one half of the flux which enters the stator steel for $-\frac{1}{4}\pi < \phi_s < \frac{1}{4}\pi$ reenters the air gap between $\frac{5}{4}\pi < \phi_s < \frac{7}{4}\pi$ and the other one half between $\frac{1}{4}\pi < \phi_s < \frac{3}{4}\pi$. The flux that enters the stator for $\frac{3}{4}\pi < \phi_s < \frac{5}{4}\pi$ divides similarly. Hence, two north poles occur; one for $\frac{1}{4}\pi < \phi_s < \frac{3}{4}\pi$ and one for $\frac{5}{4}\pi < \phi_s < \frac{7}{4}\pi$.

The air gap mmf established by each phase is a sinusoidal function of $2\phi_s$ for a four-pole machine or, in general, $(P/2)\phi_s$, where P is the number of poles. In particular,

$$\text{mmf}_{as} = \frac{2}{P} \frac{N_s}{2} i_{as} \cos \frac{P}{2} \phi_s \qquad (4.5\text{-}1)$$

$$\text{mmf}_{bs} = \frac{2}{P} \frac{N_s}{2} i_{bs} \sin \frac{P}{2} \phi_s \qquad (4.5\text{-}2)$$

where N_s is the total equivalent turns per stator phase. For balanced steady-state operation, the stator currents may be expressed as

$$I_{as} = \sqrt{2}I_s \cos[\omega_e t + \theta_{esi}(0)] \qquad (4.5\text{-}3)$$

$$I_{bs} = \sqrt{2}I_s \sin[\omega_e t + \theta_{esi}(0)] \qquad (4.5\text{-}4)$$

Equation (4.5-3) and (4.5-4) are the same expressions for the stator currents as given for the two-pole case. The air gap mmf established by balanced steady-state stator currents (mmf_s) may be expressed by substituting (4.5-3) and (4.5-4) into (4.5-1) and (4.5-2), respectively, and by adding the resulting equations. Thus,

$$\text{mmf}_s = \frac{N_s}{P} \sqrt{2}I_s \cos\left[\omega_e t + \theta_{esi}(0) - \frac{P}{2} \phi_s \right] \qquad (4.5\text{-}5)$$

If the argument of (4.5-5) is set equal to a constant and if the derivative of this expression is taken with respect to time, we see that the four poles established by the balanced stator currents rotate about the air gap at $(2/P)\omega_e$, that is,

$$\frac{d\phi_s}{dt} = \frac{2}{P} \omega_e \qquad (4.5\text{-}6)$$

Let us take a moment to review. With the stator arranged as in Fig. 4.5-1, balanced steady-state stator currents of frequency ω_e produce a four-pole (P-pole) magnetic system which rotates about the air gap at $\frac{2}{4}\omega_e$ or $(2/P)\omega_e$. Synchronous speed, as far as the rotor circuits are concerned, is now $(2/P)\omega_e$; however, the stator variables are unaware of this. To the electric system, ω_e is synchronous speed regardless of what the actual synchronous speed of the mechanical system is.

The four-pole single-phase reluctance machine shown in Fig. 4.5-2 can be used to establish the change of variables necessary to allow us to consider P-pole machines as two-pole devices. In Fig. 4.5-2, θ_{rm} and ω_{rm} are the angular displacement and angular velocity, respectively, of the rotor. Recall we used θ_r and ω_r for the two-pole device in our previous work. Now, let us write the self-inductance L_{asas}. If we follow the same procedure as we did to obtain (1.7-29), we will find that

$$L_{asas} = L_l + L_A - L_B \cos 2(\tfrac{2}{2}\theta_{rm}) \qquad (4.5\text{-}7)$$

Generalizing for a P-pole machine,

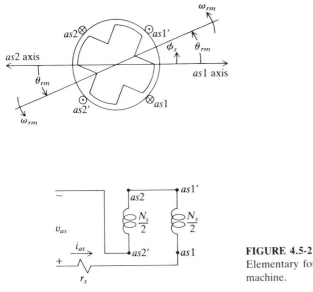

FIGURE 4.5-2
Elementary four-pole single-phase reluctance machine.

$$L_{asas} = L_l + L_A - L_B \cos 2\left(\frac{P}{2}\theta_{rm}\right) \tag{4.5-8}$$

It is apparent that if we make the substitution of

$$\boxed{\theta_r = \frac{P}{2}\theta_{rm}} \tag{4.5-9}$$

then L_{asas} for the P-pole machine, (4.5-8), is identical in form to L_{asas} for the two-pole machine, (1.7-29). Also, (4.5-9) leads to

$$\boxed{\omega_r = \frac{P}{2}\omega_{rm}} \tag{4.5-10}$$

With the substitution of θ_r and ω_r for $(P/2)\theta_{rm}$ and $(P/2)\omega_{rm}$, respectively, the voltage equations for a P-pole machine are identical to those for a two-pole machine. Actually, the electric system can tell no difference. To the electric system, the angular position of the rotor is θ_r and its velocity is ω_r. For this reason θ_r and ω_r are often called the electrical angular displacement of the rotor and the electrical angular velocity of the rotor, respectively.

One might be led to believe that the number of poles can be forgotten except when we want to know the actual physical rotor displacement θ_{rm} or angular velocity ω_{rm}. This is not quite the case. In the derivation for force (torque) in Chap. 2, the change of energy between the mechanical system and coupling fields was expressed for a rotational system as

$$dW_m = -T_e \, d\theta \qquad (4.5\text{-}11)$$

where θ must be the actual displacement of the rotating member. Hence, for a P-pole machine,

$$dW_m = -T_e \, d\theta_{rm} \qquad (4.5\text{-}12)$$

Since all electrical variables are expressed in terms of θ_r, it is convenient to replace $d\theta_{rm}$ in (4.5-12) with $d\theta_r$. Thus,

$$dW_m = -T_e \frac{2}{P} \, d\theta_r \qquad (4.5\text{-}13)$$

Therefore, to calculate the electromagnetic torque of a P-pole machine, we simply multiply all terms on the right-hand side of Table 2.5-1 by $P/2$. For all other calculations, the P-pole machine may be considered as a two-pole device. This is true regardless of the number of phases.

Example 4B. A four-pole two-phase reluctance machine is depicted in Fig. 4B-1. The stator currents are given by (4.5-3) and (4.5-4) with $\theta_{esi}(0) = 0$. The rotor is rotating in the counterclockwise direction at synchronous speed. The mechanical displacement θ_{rm} may be expressed as

$$\theta_{rm} = \omega_{rm} t + \theta_{rm}(0) \qquad (4B\text{-}1)$$

where ω_{rm} is the mechanical speed of the rotor. In a four-pole machine,

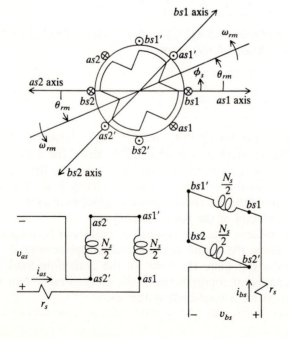

FIGURE 4B-1
Elementary four-pole two-phase reluctance machine.

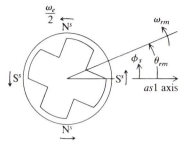

FIGURE 4B-2
Time zero location of stator north and south poles.

synchronous speed is $(2/P)\omega_e = \omega_e/2$. It is instructive to determine the time zero position of the north and south magnetic poles associated with mmf_s.

From (4.5-5) with $P = 4$ and $\theta_{esi}(0) = 0$,

$$\text{mmf}_s = \frac{N_s}{4}\sqrt{2}I_s\cos(\omega_e t - 2\phi_s) \qquad (4B\text{-}2)$$

At $t = 0$, mmf_s achieves its positive maximum value at $\phi_s = 0$ and π, which define the median locations of the stator south poles at $t = 0$. Recall that, if mmf_s is positive, the magnetic field is oriented from the rotor to the stator and magnetic flux enters the south poles. Also, from (4B-2) with $t = 0$, mmf_s is a negative maximum at $\phi_s = \frac{1}{2}\pi$ and $\frac{3}{2}\pi$, which define the median locations of the stator north poles. The time zero location of the stator north and south poles is depicted in Fig. 4B-2. Since the windings are assumed to be sinusoidally distributed, mmf_s is distributed sinusoidally. The placement of N^s and S^s in Fig. 4B-2 defines the locations of the positive and negative maximum values, respectively, of mmf_s at $t = 0$.

As time increases from zero, the stator mmf and, consequently, the stator poles (N^s and S^s) rotate at $\omega_e/2$ in the counterclockwise direction. For a constant electromagnetic torque to be produced, the rotor must also rotate at $\omega_{rm} = \omega_e/2$ (synchronous speed) in the counterclockwise direction. The electromagnetic torque acts to align the minimum reluctance paths of the rotor with the rotating stator poles.

SP4.5-1. If, during steady-state operation, $\omega_r = \omega_e$, calculate the actual rotor speed in revolutions per minute of a 120-pole 60-Hz hydroturbine synchronous generator. $[\omega_{rm} = 60\,\text{r/min}]$

SP4.5-2. Suppose $\theta_{esi}(0) = 45°$ in (4.5-3) and (4.5-4). For a six-pole two-phase stator, determine the location of the positive and negative maximum values of mmf_s at time zero. $[S^s$ at $\phi_s = 15°$, $135°$, $255°$; N^s at $\phi_s = 75°$, $195°$, $315°]$

4.6 INTRODUCTION TO SEVERAL ELECTROMECHANICAL MOTION DEVICES

We are at a point where it would be worthwhile to take a first look at the electromechanical devices which we will analyze in subsequent chapters. By introducing these devices here rather than waiting to introduce them individually at the beginning of the following appropriate chapters, you are able to study subsequent chapters in arbitrary order. For example, the chapter on induction machines is not prerequisite to the chapter on synchronous machines, even though the principles involved when starting a synchronous motor are the same principles used to explain induction motor action. However, the introductory treatment of induction machines given in this section is sufficient for explaining how synchronous machines may be started. Similarly, the introductory treatment of reluctance machines (a type of synchronous machine) given in this section enables you to study stepper motors (Chap. 8) without reading the material in Chap. 6 on synchronous machines. If, however, you choose to follow the subsequent chapters in the order that they are presented, you may wish to bypass this section.

Rotational electromechanical devices fall into three general classes—direct-current, synchronous, and induction. The dc machine was treated in Chap. 3, and we have briefly considered the single-phase reluctance machine which is a type of synchronous machine. Synchronous machines are so called because they develop an average torque only when the rotor is rotating in synchronism (synchronous speed) with the rotating air gap mmf established by currents flowing in the stator windings. Examples are reluctance machines, stepper motors, permanent-magnet machines, brushless dc motors, and the machine which has become known as simply the synchronous machine. Even though these machines are of the synchronous type, the machine commonly referred to as the synchronous machine or synchronous generator is the device used to generate electric power. All large, electric power generators are synchronous generators and are so called.

On the other hand, the induction machine, which is the principal means of converting energy from electric to mechanical, cannot develop torque at synchronous speed in its normal mode of application. The windings on the rotor are short-circuited and, in order to cause current to flow in these windings which produce torque by interacting with the air gap mmf established by the stator windings, the rotor must rotate at a speed other than synchronous speed. Although some synchronous machines are designed to start from stall and accelerate to near synchronous speed as an induction motor, the induction machine family does not include a large number of relatives as does the synchronous machine group. There is, however, some differentiation made between a large, workhorse induction motor and the small, two-phase induction motor used in low-power control systems. The latter is sometimes referred to as an *ac servomotor*.

In this section, we will show the winding arrangement for elementary versions of these electromechanical devices and describe briefly the principle of operation of each. In subsequent chapters we will analyze these devices in detail.

Reluctance Devices

Elementary, single-, and two-phase two-pole reluctance machines are shown in Fig. 4.6-1. The stator windings are assumed to be sinusoidally distributed. The

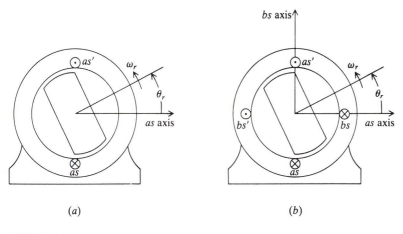

FIGURE 4.6-1
Elementary two-pole reluctance machines. (*a*) Single-phase; (*b*) two-phase.

single-phase reluctance motors are used in clocks and some turntables. The two-phase reluctance motor is often used as a control motor. Moreover, the operation of these devices with constant stator currents provides an explanation of the positioning of reluctance-type stepper motors.

The principle of operation is quite straightforward. Recall from our work in Chap. 2 that in an electromagnetic system a force (torque) is produced in an attempt to minimize the reluctance of the magnetic system. We have established that, with an alternating current flowing in the winding of the single-phase stator, two oppositely rotating mmf's are produced, (4.4-3). Therefore, once the rotor is rotating in synchronism (remember, this is a type of synchronous machine) with either of the two oppositely rotating air gap mmf's, there is a force (torque) created by the magnetic system in an attempt to align the minimum-reluctance path of the rotor with the rotating air gap mmf. When there is no load torque on the rotor, the minimum-reluctance path of the rotor is in alignment with the rotating air gap mmf. When a load torque is applied, the rotor slows ever so slightly, thereby creating a misalignment of the minimum-reluctance path and the rotating air gap mmf. When the electromagnetic torque produced in an attempt to maintain alignment is equal and opposite to the load torque on the rotor, the rotor resumes synchronous speed. It follows that, if the load torque is larger than the torque which can be produced to align, the rotor will fall out of synchronism and, since the machine cannot develop an average torque at a speed other than synchronous, it will slow to stall.

The operation of the two-phase device differs from that of the single-phase device in that only one constant-amplitude rotating air gap mmf is produced during balanced steady-state conditions. Hence, a constant torque will be developed at synchronous speed rather than a torque which pulsates about an average value as is the case with the single-phase machine. Although the reluctance motor can be started from a source which can be switched at a frequency corresponding to the rotor speed as in the case of stepper or brushless dc motors, the devices, as shown in elementary form in Fig. 4.6-1, cannot develop an average starting torque when plugged into a household power outlet.

As a youngster, the oldest of the authors recalls that his parents had a small, old, mantle-type electric clock which was not self-starting. Although it was rather annoying to the rest of the family, it was extremely interesting and entertaining to spin the rotor from stall with the thumb wheel on the back of the clock. Spinning the rotor above synchronous speed would cause it to slow to synchronous speed and then "lock in" and operate normally. With a little practice you could spin the rotor so that it would come up to just slightly less than synchronous speed, whereupon you could see and hear it get pulled into step with the rotating air gap mmf. However, the most interesting of all was the fact that the clock would operate in either direction. Moreover, once the device was operating at synchronous speed, you could pinch the thumb wheel very lightly between your thumb and index finger and feel the motor overcome the load torque you were applying. As you increased your hold on the thumb wheel, the motor would overcome this increased load and continue to run at synchronous speed until you exceeded the torque capability of the motor. Instantaneously, before you could release your hold on the thumb wheel, the motor torque would vanish and the clock would stop. Modern electric clock motors are equipped with a means to self-start without a thumb wheel and, therefore, are not nearly as intriguing or instructive to a ten-year-old.

Some stepping or stepper motors are used to convert a digital input into a mechanical motion. The stepping action from one line to the next of a line printer is accomplished with a stepper motor. Many stepper motors are of the reluctance type. In fact, some stepper motors are often called *variable-reluctance motors*. Operation of reluctance stepper motors is easily explained. For this purpose, let us assume that a constant current is flowing in the *bs* winding of Fig. 4.6-1*b* with the *as* winding open-circuited. The minimum-reluctance path of the rotor will be aligned with the *bs* axis, i.e., assume $\theta_r = 0$. Now let us reduce the *bs* winding current to zero while increasing the current in the *as* winding to a constant value. There will be forces to align the minimum-reluctance path of the rotor with the *as* axis; however, this can be satisfied with $\theta_r = \pm \frac{1}{2}\pi$. Unfortunately, it is a 50-50 chance as to which way it will rotate. Although we see how stepping action can be accomplished from this explanation, we realize that we need a device different from a single- or two-phase reluctance machine to accomplish controlled stepping. There are two common techniques used to achieve controlled, bidirectional stepping with reluctance devices: Place more than two phase windings on the stator, generally three are used, or cascade three or more single-phase reluctance machines on the same shaft with the minimum-reluctance paths of each rotor displaced from each other. The first type is called a single-stack variable-reluctance stepper motor; the second, a multistack variable-reluctance stepper motor. The two-phase reluctance machine is treated in Chap. 6 and stepper motors in Chap. 8.

Induction Machines

Elementary single- and two-phase induction machines are shown in Fig. 4.6-2. The rotors of both devices are identical in configuration; each has the equivalent of two orthogonal windings which are assumed to be sinusoidally distributed. In other words, the *ar* and *br* windings are equivalent to a symmetrical two-phase set of windings and, in the vast majority of applications, these rotor windings are short-circuited.

It is perhaps a little more convenient to explain the operation of the two-phase device first. For balanced steady-state operation, the currents flowing in the stator windings produce an air gap mmf which rotates about the air gap at an angular velocity of ω_e. With the rotor windings short-circuited, which is the only mode of operation we will consider, a voltage is induced in each of the rotor windings only if the rotor speed ω_r is different from ω_e. The currents flowing in the rotor circuits due to induction (thus, the name induction motor) will be a balanced set with a frequency equal to $\omega_e - \omega_r$, which will produce an air gap mmf that rotates at $\omega_e - \omega_r$ relative to the rotor or ω_e relative to a stationary observer. Hence, the rotating air gap mmf caused by the currents flowing in the stator windings induces currents in the short-circuited rotor windings which, in turn, establish an air gap mmf that rotates in unison with the stator rotating air gap mmf (mmf$_s$). Interaction of these magnetic systems (poles) rotating in unison provides the means of producing torque on the rotor.

Although the induction machine can operate as a motor or a generator, it is normally operated as a motor. As a motor, it can develop a torque from $0 < \omega_r < \omega_e$. At $\omega_r = \omega_e$, rotor currents are not present since the rotor is rotating at the speed of the stator rotating air gap mmf and, therefore, the rotor windings do not experience a change of flux linkages, which is, of course, necessary to induce a voltage in the rotor windings. Large induction machines

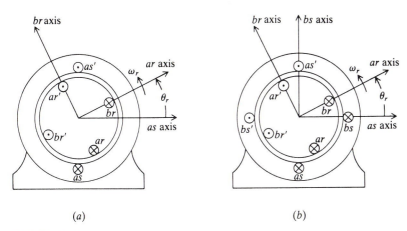

(a)   (b)

FIGURE 4.6-2
Elementary two-pole induction machines. (a) Single-phase; (b) two-phase.

are designed to operate very close to synchronous speed. On the other hand, a servomotor used in a position control system might not exceed 25 percent of synchronous speed during a positioning task.

The single-phase induction motor is, perhaps, the most widely used electromechanical device. Garbage disposals, washers, dryers, furnace fans, etc., are but a few of the many applications of fractional-horsepower single-phase induction motors. However, the device shown in Fig. 4.6-2a is not quite the whole picture of a single-phase induction motor. Recall that the single-phase stator winding produces oppositely rotating air gap mmf's of equal amplitude. If the single-phase induction motor shown in Fig. 4.6-2a is stalled, $\omega_r = 0$, and if a sinusoidal current is applied to the stator winding, the rotor will not move. This device does not develop a starting torque. Why? Well, the rotor cannot follow either of the rotating mmf's since it develops as much torque to go with one as it does to go with the other. If, however, you manually turn the rotor in either direction, it will accelerate in that direction and operate normally. Although single-phase induction motors normally operate with only one stator winding, it is necessary to use a second stator winding to start the device. Actually, the single-phase induction motors we use in our homes are two-phase induction motors with provisions to switch out one of the windings once the rotor accelerates to between 60 and 80 percent of synchronous speed. The next question is how do we get two-phase voltages from a single-phase household supply? Well, we do not actually develop a two-phase supply, but we approximate one, as far as the two-phase motor is concerned, by placing a capacitor in series with one of the stator windings. This shifts the phase of one current relative to the other, thereby producing a larger rotating air gap mmf in one direction than in the other. If you have looked closely at a single-phase induction motor, you probably noticed a cylinder 2 to 5 inches in length mounted on the housing of the motor and 9 times out of 10 it is painted black for some unknown reason. That is the capacitor, commonly called the *start capacitor*, for obvious reasons. Provisions to switch the capacitor out of the circuit is generally inside the housing of the motor. The induction machine is analyzed in Chaps. 5 and 9. In Chap. 9, attention is focused on the two-phase servomotor and the single-phase induction motor.

Synchronous Machines

Although the elementary single- and two-phase devices shown in Fig. 4.6-3 have become known as synchronous machines, they are but one of several devices which fall into the synchronous machine category. Nevertheless, we will honor convention and refer to the devices in Fig. 4.6-3 as synchronous machines.

The single-phase synchronous machine has limited application. In fact, the same can be said about the two-phase synchronous machine. It is the three-phase synchronous machine which is used to generate electric power in power systems such as in some automobiles, aircraft, utility systems, ships, and

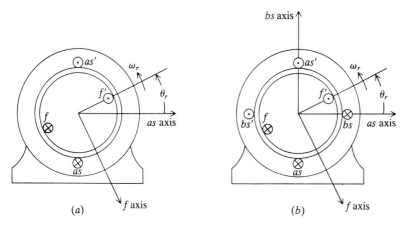

FIGURE 4.6-3
Elementary two-pole synchronous machines. (*a*) Single-phase; (*b*) two-phase.

now possibly future spacecraft. Nevertheless, the theory of operation of synchronous machines is adequately introduced by considering the two-phase version.

The elementary devices shown in Fig. 4.6-3 have only one rotor winding—the field winding (f winding). In practical synchronous machines, the rotor is equipped with short-circuited windings in addition to the f winding which help to damp oscillations about synchronous speed and, in some cases, these windings are used to start the unloaded synchronous machine from stall as an induction motor.

The principle of operation is apparent once we realize that the current flowing in the field winding is direct current. Although it may be changed in value by varying the applied field voltage, it is constant for steady-state operation of a balanced two-phase synchronous machine. When operated as a generator, the rotor is driven by some mechanical means. If the stator windings are connected to a balanced system, the stator currents produce a constant-amplitude rotating air gap mmf. A rotor air gap mmf is produced by the direct current flowing in the field winding. To produce a torque or transmit power, the air gap mmf produced by the stator and that produced by the rotor must rotate in unison about the air gap of the machine. Hence, $\omega_r = \omega_e$, that is, synchronous speed.

The synchronous machines shown in Fig. 4.6-3 are actually called round-rotor synchronous machines. A variation from this is the salient-pole synchronous machine used in low-speed multipole applications. A two-phase two-pole version is shown in Fig. 4.6-4. The principle of operation is identical to that of the round-rotor machine, except there are now two means of producing torque: One way is by the interaction of the rotating air gap mmf's, as in the case of the round-rotor device and, although normally much smaller, a reluctance torque is produced owing to the rotor configuration. Both torques

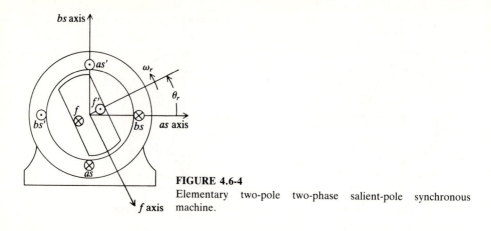

FIGURE 4.6-4
Elementary two-pole two-phase salient-pole synchronous machine.

are constant only when $\omega_r = \omega_e$. If ω_r is not equal to ω_e, the torques will pulsate with zero average value even if the stator currents are balanced. Although the synchronous machine is generally used as a generator, it can also be used as a motor. Synchronous machines are analyzed in Chap. 6.

Permanent-Magnet Devices

If we replace the rotor of the synchronous machines shown in Fig. 4.6-3 with a permanent-magnet rotor, we have the so-called permanent-magnet devices shown in Fig. 4.6-5. The operation of these devices is identical to that of the synchronous machine. Since the strength of the rotor field due to the perma-

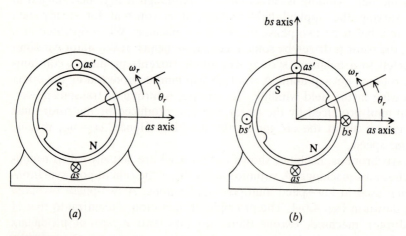

(a)   (b)

FIGURE 4.6-5
Elementary two-pole permanent-magnet devices. (a) Single-phase; (b) two-phase.

nent magnet cannot be controlled as in the case of the synchronous machine which has a field winding, it is not widely used as a means of generating power. It is, however, used widely as a drive motor. In particular, permanent-magnet motors are used as stepper motors and, extensively, as brushless dc motors, wherein the voltages applied to the stator windings are switched electronically at a frequency corresponding to the speed of the rotor. We shall analyze the brushless dc motor in detail in Chap. 7 and stepper motors in Chap. 8.

SP4.6-1. The *ar* winding of the two-phase induction machine shown in Fig. 4.6-2*b* is open-circuited and a negative direct current is supplied to the *br* winding. Find another machine in Sec. 4.6 which has the same winding configuration (active windings) and constraints and thus the same principles of operation. [Fig. 4.6-3*b*]

SP4.6-2. Repeat SP4.6-1 if the direct current in the *br* winding is held fixed by a current source. [Fig. 4.6-5*b*]

SP4.6-3. Which of the devices discussed in Sec. 4.6 can be operated as a generator supplying an *RL* static load? [Fig. 4.6-3 through Fig. 4.6-5]

SP4.6-4. We have two single-phase synchronous machines from which we want to generate two-phase voltages. What do we do? [Connect the shafts with fields displaced $\frac{1}{2}\pi$]

SP4.6-5. The current flowing in the *as* winding of the single-phase round-rotor synchronous machine shown in Fig. 4.6-3*a* is fixed at $i_{as} = \sqrt{2}I_s \cos \omega_e t$. A dc voltage is applied to the field. The rotor is rotating counterclockwise at ω_e. The steady-state field current will consist of two components. Determine the frequency of these two components. [dc; $2\omega_e$]

SP4.6-6. The two-pole two-phase permanent-magnet machine shown in Fig. 4.6-5*b* is used as a stepper motor. Initially $I_{as} = I$ and $I_{bs} = 0$. I_{as} is "stepped" to zero while I_{bs} is stepped to $-I$, where I is a positive current. Determine the initial and final values of θ_r. [$\frac{1}{2}\pi$; 0]

4.7 RECAPPING

We have seen that the electromechanical devices with rotating magnetic fields have a number of features in common. Perhaps most important of these features is the fact that the stator windings of all multiphase devices are arranged so that balanced, sinusoidal stator currents produce an air gap mmf which rotates about the air gap at $(2/P)\omega_e$ rad/s, where ω_e is the electrical angular velocity of the stator currents and P is the number of poles. Also of importance is the observation that, by simple changes in the configuration of the windings and the constraints placed upon the electrical variables, we can readily identify, in elementary form, all of the electromechanical motion devices which we will consider in the remainder of this text. If we have done our job, you now have sufficient background to proceed to any of the next four chapters. You need not read one before the other. In other words, you can now begin study of the induction, synchronous, brushless dc, and stepper motors in any order you wish.

4.8 PROBLEMS

1. A winding with five coils, each with nc_s conductors, is distributed over 90° of the stator. Sketch the configuration and indicate the direction of the positive magnetic axis of this single-phase stator winding.

2. The four coils, each with nc_s conductors, of the windings of a two-pole two-phase low-power device are concentrated so that four coil sides are placed in one slot per phase. Sketch the winding arrangement and show the as and bs axes.

3. Consider the two-phase device shown in Fig. 4.8-1. The windings are sinusoidally distributed each with N_s equivalent turns. Express (a) N_{as} and N_{bs} and (b) mmf$_{as}$ and mmf$_{bs}$.

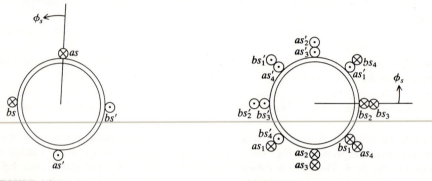

FIGURE 4.8-1
Elementary two-pole two-phase electromechanical device.

FIGURE 4.8-2
A two-pole two-phase electromechanical device.

*4. Each stator coil in Fig. 4.8-2 contains nc_s turns. Using the information given in Example 4A as a guide, sketch the developed diagram for the as and bs windings and mmf$_{as}$ and mmf$_{bs}$.

*5. Consider Fig. 4.8-3 where a winding is formed by N conductors uniformly distributed over the regions shown. Each conductor carries the current i. Sketch the air gap mmf. You will need the information in Example 4A as a guide.

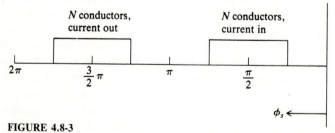

FIGURE 4.8-3
Uniformly distributed windings.

6. Consider Fig. 4.8-1.
   (a) Express mmf$_s$.
   (b) Determine the phase relationship between $\tilde{I}_{as}$ and $\tilde{I}_{bs}$ which will produce counterclockwise rotation of mmf$_s$. That is, does $\tilde{I}_{as}$ lead or lag $\tilde{I}_{bs}$ by 90°?

(c) Locate the position of N^s and S^s when $\omega_e t = 30°$ with $I_{as} = \sqrt{2}I_s \cos \omega_e t$ and $I_{bs} = -\sqrt{2}I_s \sin \omega_e t$.

7. Consider the device shown in Fig. 4.4-2. Express mmf$_s$ for (a) $I_{as} = \sqrt{2}I_s \cos \omega_e t$, $I_{bs} = \sqrt{2}I_s \cos \omega_e t$; (b) $I_{as} = I_a \cos \omega_e t$, $I_{bs} = I_b \sin \omega_e t$, where $I_a \neq I_b$. Both (a) and (b) are unbalanced sets. Set (a) is unbalanced since I_{as} and I_{bs} are not orthogonal; set (b) is unbalanced due to unequal amplitudes.

*8. The windings of the device shown in Fig. 4.8-4 are sinusoidally distributed, each with N_s equivalent turns. The bs winding may be arranged so that the bs axis is at an arbitrary angle α with the as axis. Express i_{as} and i_{bs} so that mmf$_s$ has a constant amplitude and rotates in the counterclockwise direction regardless of the value of α; in particular, mmf$_s = (N_s/2)\sqrt{2}I_s \cos(\omega_e t - \phi_s)$.

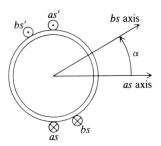

FIGURE 4.8-4
Magnetic axes displaced by an arbitrary angle α.

9. Sketch the stator windings and rotor configuration for a six-pole two-phase symmetrical-reluctance machine.

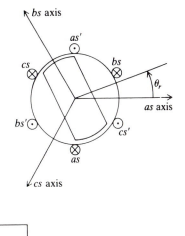

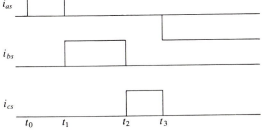

FIGURE 4.8-5
Two-pole three-phase reluctance motor with switched phase currents.

*10. Show analytically that a single-phase reluctance motor does not produce an average starting torque when $I_{as} = \sqrt{2}I_s \cos[\omega_e t + \theta_{esi}(0)]$.

*11. A two-pole three-phase reluctance machine and the currents flowing in the stator windings are shown in Fig. 4.8-5 (p. 157). Sketch θ_r versus time assuming that the rotor oscillations are sufficiently damped and, at t_0^+, θ_r has a small positive value. Repeat if θ_r has a small negative value at t_0^+. Is there any advantage to make the necessary circuit provisions so that the stator currents can be made both positive and negative? Why?

*12. A two-phase induction motor is operating at rated-load torque with balanced stator currents. The steady-state rotor speed is $\omega_r = 0.9\omega_e$. It is an eight-pole 50-Hz device. Determine (a) the actual rotor speed, (b) the angular velocity of the rotor mmf relative to the rotor, and (c) the frequency of the rotor currents.

*13. Determine the actual no-load and full-load rotor speeds (ω_{rm}) in revolutions per minute of the following devices:

(a) A four-pole three-phase 100-hp 60-Hz round-rotor synchronous machine
(b) A 120-pole three-phase 100-MW 60-Hz salient-pole synchronous machine
(c) An eight-pole two-phase 1-hp 60-Hz reluctance machine supplied from a 5-Hz voltage source
(d) A six-pole three-phase $\frac{1}{2}$-hp permanent-magnet motor supplied from a 15-Hz source

CHAPTER
5

SYMMETRICAL INDUCTION MACHINES

5.1 INTRODUCTION

Although the induction machine is used most often to convert electric power to mechanical work (motor), it can operate as either a motor or a generator. Three-phase induction motors are commonly used in large-horsepower applications. Pump drives, steel mill drives, hoist drives, and vehicle drives are examples. On a smaller scale, the two-phase servomotor (two-phase induction motor) is used extensively in position-followup control systems. Also, single-phase induction motors, which develop torque in a manner similar to the multiphase induction motor, are widely used in many household appliances as well as for bench tools. The induction motor is the workhorse of the electrical industry. As generators, induction machines have limited use; wind turbine and low-head hydroapplications are examples.

To conduct a rigorous analysis of an induction machine, it is necessary to perform a change of variables (transformation) which eliminates the time-varying mutual inductances which occur because of windings in relative motion. Although there are an infinite number of transformations which can be used in the case of the symmetrical induction machine to accomplish this goal [1], we will use only one, that is, a change of variables which, in effect, replaces the rotor windings with fictitious windings fixed in the stator. This analysis is somewhat involved. However, the resulting steady-state equivalent circuit, which is an invaluable tool, is essentially that of a transformer with two windings, one of which being short-circuited.

159

The analysis of symmetrical two- and three-phase induction machines is essentially the same. Therefore, we will focus our attention on the two-phase machine, since this enables us to become familiar with the theory of induction machines without becoming inundated with trigonometric manipulations. Once the theory has been established, the performances of a 5-hp two-phase induction motor and a $\frac{1}{10}$-hp servomotor during balanced operation are illustrated in detail by computer traces. This work sets the stage for those who have a general interest in induction machines for control and low-power applications. A section is given on the three-phase induction machine primarily for those interested in power system or large-power industrial drive applications. Single-phase and unbalanced operations of the two-phase symmetrical induction machine are covered in Chap. 9.

5.2 TWO-PHASE INDUCTION MACHINE

A two-pole two-phase induction machine is shown in Fig. 5.2-1. It is assumed that the stator windings may be portrayed by orthogonal, sinusoidally distributed windings as described in Chap. 4. It is sufficient for our present purposes

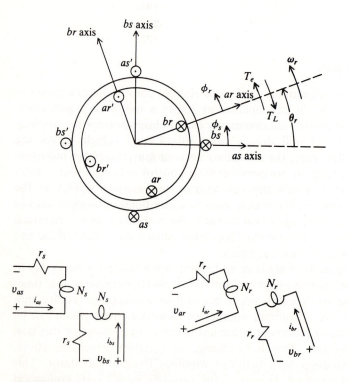

FIGURE 5.2-1
A two-pole two-phase symmetrical induction machine.

to assume that the rotor of the two-phase two-pole induction machine may be portrayed electrically by two sinusoidally distributed windings displaced 90°. We will talk about forged and squirrel cage rotors of induction motors later. Hence, for our present purpose, we will consider that the *ar* and *br* windings are sinusoidally distributed, each with the same total winding resistance. Thus, both the stator and rotor are symmetrical. For this reason, this device is often referred to as a *symmetrical induction machine*.

Note that the air gap distance between the stator and rotor is uniform. In other words, it has a uniform air gap. Also, the rotor windings of an induction machine are generally short-circuited ($v_{ar} = v_{br} = 0$). Most often, only the stator windings are connected to a source, whereupon the induction machine is said to be single-fed. In some special applications, however, both the rotor and stator windings are connected to sources, whereupon the induction machine is said to be double-fed. In this case, the rotor windings are connected to a stationary multiphase source by a brush and slip ring arrangement.

As established in Chap. 4, the angular displacement about the stator is denoted ϕ_s and it is referenced to the *as* axis. We see from Fig. 5.2-1 that the angular displacement about the rotor is denoted ϕ_r and it is referenced to the *ar* axis. The angular velocity of the rotor is ω_r and θ_r is its angular displacement. In particular, θ_r is the angular displacement between the *ar* and *as* axes. Thus, a given point on the rotor surface at the angular position ϕ_r may be related to an adjacent point on the inside stator surface with the angular position ϕ_s as

$$\boxed{\phi_s = \phi_r + \theta_r} \qquad (5.2\text{-}1)$$

The electromechanical torque T_e and the load torque T_L are also indicated in Fig. 5.2-1. We are aware from Chap. 2 that T_e is assumed to be positive in the direction of increasing θ_r whereas the load torque is positive in the opposite direction (opposing rotation).

The air gap mmf's due to the *as* and *bs* windings are given by (4.3-4) and (4.3-5). From our work in Chap. 4 we are able to write the air gap mmf's for the *ar* and *br* windings by inspection. In particular,

$$\boxed{\begin{aligned} \text{mmf}_{ar} &= \frac{N_r}{2} i_{ar} \cos \phi_r \\[2mm] \text{mmf}_{br} &= \frac{N_r}{2} i_{br} \sin \phi_r \end{aligned}} \qquad \begin{aligned} &(5.2\text{-}2) \\[2mm] &(5.2\text{-}3) \end{aligned}$$

where N_r is the equivalent number of turns of the rotor windings.

In Chap. 4, we established that balanced two-phase currents flowing in the stator windings of a two-pole device produced a constant-amplitude rotating air gap mmf during steady-state operation (4.4-11). Two rotating poles are established as a result of this rotating air gap mmf (mmf$_s$), and this magnetic system rotates about the air gap at the angular velocity ω_e of the stator currents. Let us consider the air gap mmf established by the rotor currents. During balanced steady-state operation, the rotor speed is constant and the stator currents may be expressed by (4.4-8) and (4.4-9). For now, let us assume that the rotor currents may be expressed as

$$I_{ar} = \sqrt{2}I_r \cos[(\omega_e - \omega_r)t + \theta_{eri}(0)] \tag{5.2-4}$$

$$I_{br} = \sqrt{2}I_r \sin[(\omega_e - \omega_r)t + \theta_{eri}(0)] \tag{5.2-5}$$

Once time zero is selected, $\theta_{esi}(0)$, which is the time zero position of the stator currents, and $\theta_{eri}(0)$ are determined from the instantaneous values of the stator and rotor currents at the assigned time zero. One wonders why the frequency of the rotor currents is the difference of the angular velocity ω_e of the stator currents and the angular velocity ω_r of the rotor. We will find that this must be the frequency of the rotor currents during balanced steady-state operation, regardless whether the rotor circuits are short-circuited and only the stator windings are connected to a source (single-fed), or whether both the stator and rotor windings are connected to sources (double-fed).

An expression for the air gap mmf due to the rotor currents is obtained by substituting (5.2-4) into (5.2-2) and (5.2-5) into (5.2-3) and adding the resulting expressions. Thus,

$$\text{mmf}_r = \text{mmf}_{ar} + \text{mmf}_{br}$$

$$= \frac{N_r}{2}\sqrt{2}I_r \cos[(\omega_e - \omega_r)t + \theta_{eri}(0) - \phi_r] \tag{5.2-6}$$

Setting the argument equal to a constant and taking the derivative with respect to time yields

$$\frac{d\phi_r}{dt} = \omega_e - \omega_r \tag{5.2-7}$$

Here, mmf$_r$ rotates relative to the rotor at $\omega_e - \omega_r$, and, if $\omega_e - \omega_r > 0$, the direction of rotation is counterclockwise relative to the rotor. Before proceeding, let us note the position of the poles. Although, in general, $\theta_{esi}(0)$ and $\theta_{eri}(0)$ cannot both be zero, we will assume that this is the case for a moment. Thus, at $t = 0$, mmf$_r$ is the cosine of ϕ_r positioned about the ar axis. Flux leaves the rotor and enters the air gap for $-\frac{1}{2}\pi < \phi_r < \frac{1}{2}\pi$, a north pole, and enters the rotor from the air gap for $\frac{1}{2}\pi < \phi_r < \frac{3}{2}\pi$, a south pole. Thus, we have a set of poles established by the stator currents which rotates at ω_e relative to the stator and another set of poles established by the rotor currents which rotates at $\omega_e - \omega_r$ relative to the rotor. If the "stator poles" and the "rotor poles" were to rotate at the same angular velocity about the air gap, the stage would be set for producing an electromagnetic torque due to the force exerted between the two magnetic systems rotating in unison. Are the two sets of poles rotating at the same angular velocity? Let us see. We can relate a displacement on the rotor, ϕ_r, to a displacement on the stator, ϕ_s, by (5.2-1). Taking the derivative of (5.2-1) with respect to time yields

$$\frac{d\phi_s}{dt} = \frac{d\phi_r}{dt} + \frac{d\theta_r}{dt} \tag{5.2-8}$$

Now, (5.2-7) tells us that $d\phi_r/dt = \omega_e - \omega_r$ and we know that $d\theta_r/dt = \omega_r$, thus

$$\frac{d\phi_s}{dt} = \omega_e - \omega_r + \omega_r = \omega_e \tag{5.2-9}$$

In other words, the air gap mmf established by currents flowing in the rotor, mmf$_r$, rotates about the air gap at ω_e as observed from the stator. Hence, both the stator and rotor poles rotate about the air gap at ω_e. This, of course, makes sense from another point of view. The mmf$_r$ is rotating at $\omega_e - \omega_r$ relative to the rotor, that is, if we were riding on the rotor which is moving at ω_r, we would observe mmf$_r$ rotating at $\omega_e - \omega_r$. If we now jump off the rotor and look at mmf$_r$ as a stationary observer, we would see the rotor rotating at ω_r and mmf$_r$ rotating on the rotor at $\omega_e - \omega_r$; hence, $\omega_r + (\omega_e - \omega_r) = \omega_e$. While you were on the rotor, did you look at mmf$_s$? If so, at what speed was it rotating relative to you? (*Ans.:* $\omega_e - \omega_r$.)

With mmf$_s$ and mmf$_r$ rotating at the same angular velocity an average steady-state electromagnetic torque can be produced. This is not the case if they rotate at different angular velocities. How did we cause the angular velocities to be equal? We assumed that the frequency of the steady-state rotor currents was $\omega_e - \omega_r$, and we said that this had to be the case regardless whether the machine was single- or double-fed. Now we see why. We have also said that the induction machine is nearly always operated with the rotor windings short-circuited. Hence, currents are induced in the rotor circuits (thus the name *induction motor*) by the flux established by mmf$_s$. However, to induce a current in the rotor circuits, the rotor must rotate at a speed different from that of mmf$_s$ so that the rotor circuits experience a change of flux.

We have said that an electromagnetic torque is exerted on the shaft because of the interaction of the poles established by stator currents and the poles established by the rotor currents. If the rotor currents are not present, then a torque does not exist. Thus, the induction motor has the capability of developing an average, steady-state electromagnetic torque at any rotor speed except when $\omega_r = \omega_e$. In other words, when the rotor is rotating in synchronism with the rotating magnetic field produced by the stator currents (mmf$_s$), rotor currents are not induced and, hence, an electromagnetic torque cannot be developed. We will find that this device operates as a motor when $\omega_r < \omega_e$, and as a generator when the rotor is driven above ω_e by a torque input to the shaft.

One may, at first, choose to call this device a four-pole rather than a two-pole device since two poles are established by the stator currents and two poles are established by the rotor currents. We must realize, however, that even though we have considered the stator and rotor air gap mmf's separately, they combine to form one resultant two-pole magnetic system.

A cutaway of a four-pole three-phase 7.5-hp 460-V squirrel-cage induction motor is shown in Fig. 5.2-2. It is an enclosed, fan-cooled, severe-duty motor for use in the chemical, paper, cement, and mining industries. A disassembled four-pole two-phase $\frac{1}{10}$-hp 115-V induction motor, which is used as a servomotor, is shown in Fig. 5.2-3. Also shown is the case that houses the speed-reduction gears. We will use this device as an example servomotor later in this chapter and in Chap. 9.

SP5.2-1. Assume sinusoidally distributed windings on the stator and rotor of the machine shown in Fig. 5.2-1. Express (a) mmf$_{as}$ in terms of θ_r and ϕ_r and (b) mmf$_{ar}$ in terms of θ_r and ϕ_s. [(a) mmf$_{as} = (N_s/2)i_{as} \cos(\phi_r + \theta_r)$; (b) mmf$_{ar} = (N_r/2)i_{ar} \cos(\phi_s - \theta_r)$]

SP5.2-2. The frequency of the balanced stator currents of an induction machine is 60 Hz, mmf$_s$ rotates counterclockwise. The device is operating as a motor, and the rotor of the two-pole machine is rotating counterclockwise at $0.9\omega_e$. (a) Determine the frequency of the balanced rotor currents. Determine the angular velocity of mmf$_s$ and mmf$_r$ relative to an observer sitting (b) on the rotor and (c) on the stator. [(a) 6 Hz; (b) 37.7 rad/s, ccw; (c) 377 rad/s, ccw]

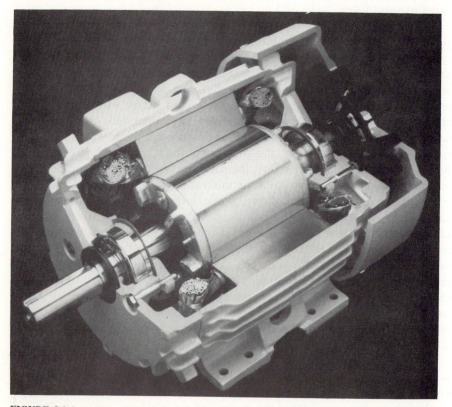

FIGURE 5.2-2
Four-pole three-phase 6.5-hp 460-V severe-duty squirrel-cage induction motor. (*Courtesy of General Electric.*)

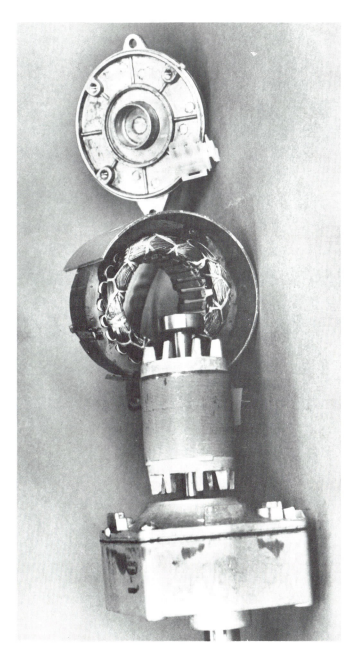

FIGURE 5.2-3
Four-pole two-phase $\frac{1}{10}$-hp 115-V servomotor with speed reduction gear.

SP5.2-3. Repeat SP5.2-2 for a six-pole induction machine operating as a generator which is being driven at $\omega_r = 1.1\omega_e$. [(a) 6 Hz; (b) 37.7/3 rad/s, cw; (c) 377/3 rad/s, ccw]

SP5.2-4. Give the phase relationship of the rotor currents for (a) SP5.2-2 and (b) SP5.2-3. [(a) $\tilde{I}_{ar} = j\tilde{I}_{br}$; (b) $\tilde{I}_{ar} = -j\tilde{I}_{br}$]

5.3 VOLTAGE EQUATIONS AND WINDING INDUCTANCES

The voltage equations for the induction machine depicted in Fig. 5.2-1 may be expressed as

$$v_{as} = r_s i_{as} + \frac{d\lambda_{as}}{dt} \tag{5.3-1}$$

$$v_{bs} = r_s i_{bs} + \frac{d\lambda_{bs}}{dt} \tag{5.3-2}$$

$$v_{ar} = r_r i_{ar} + \frac{d\lambda_{ar}}{dt} \tag{5.3-3}$$

$$v_{br} = r_r i_{br} + \frac{d\lambda_{br}}{dt} \tag{5.3-4}$$

where r_s is the resistance of the stator windings and r_r is the resistance of the rotor windings. It is convenient, for future derivations, to write (5.3-1) through (5.3-4) in matrix form as

$$\mathbf{v}_{abs} = \mathbf{r}_s \mathbf{i}_{abs} + p\boldsymbol{\lambda}_{abs} \tag{5.3-5}$$

$$\mathbf{v}_{abr} = \mathbf{r}_r \mathbf{i}_{abr} + p\boldsymbol{\lambda}_{abr} \tag{5.3-6}$$

where

$$(\mathbf{f}_{abs})^T = [f_{as} \quad f_{bs}] \tag{5.3-7}$$

$$(\mathbf{f}_{abr})^T = [f_{ar} \quad f_{br}] \tag{5.3-8}$$

In (5.3-7) and (5.3-8), f can represent voltage, current, or flux linkages and T denotes the transpose. In (5.3-5) and (5.3-6), p is the shorthand notation for the operator d/dt. Also,

$$\mathbf{r}_s = \begin{bmatrix} r_s & 0 \\ 0 & r_s \end{bmatrix} = r_s \mathbf{I} \tag{5.3-9}$$

and

$$\mathbf{r}_r = \begin{bmatrix} r_r & 0 \\ 0 & r_r \end{bmatrix} = r_r \mathbf{I} \tag{5.3-10}$$

where **I** is the identity matrix. A review of matrix algebra is given in Appendix D.

In the analysis of electric machines, we generally assume that the magnetic system is linear, whereupon the flux linkages may be expressed as linear functions of inductances and currents. In particular, we can write

$$\lambda_{as} = L_{asas}i_{as} + L_{asbs}i_{bs} + L_{asar}i_{ar} + L_{asbr}i_{br} \tag{5.3-11}$$

$$\lambda_{bs} = L_{bsas}i_{as} + L_{bsbs}i_{bs} + L_{bsar}i_{ar} + L_{bsbr}i_{br} \tag{5.3-12}$$

$$\lambda_{ar} = L_{aras}i_{as} + L_{arbs}i_{bs} + L_{arar}i_{ar} + L_{arbr}i_{br} \tag{5.3-13}$$

$$\lambda_{br} = L_{bras}i_{as} + L_{brbs}i_{bs} + L_{brar}i_{ar} + L_{brbr}i_{br} \tag{5.3-14}$$

The self- and mutual inductances given in (5.3-11) through (5.3-14) are defined by their subscripts. Reciprocity applies, thus $L_{asbs} = L_{bsas}$, $L_{asar} = L_{aras}$, etc. For future derivations, it is convenient to write (5.3-11) through (5.3-14) in matrix form as

$$\boldsymbol{\lambda}_{abs} = \mathbf{L}_s \mathbf{i}_{abs} + \mathbf{L}_{sr} \mathbf{i}_{abr} \tag{5.3-15}$$

$$\boldsymbol{\lambda}_{abr} = (\mathbf{L}_{sr})^T \mathbf{i}_{abs} + \mathbf{L}_r \mathbf{i}_{abr} \tag{5.3-16}$$

which may also be written as

$$\begin{bmatrix} \boldsymbol{\lambda}_{abs} \\ \boldsymbol{\lambda}_{abr} \end{bmatrix} = \begin{bmatrix} \mathbf{L}_s & \mathbf{L}_{sr} \\ (\mathbf{L}_{sr})^T & \mathbf{L}_r \end{bmatrix} \begin{bmatrix} \mathbf{i}_{abs} \\ \mathbf{i}_{abr} \end{bmatrix} \tag{5.3-17}$$

The inductance matrices are defined from (5.3-11) through (5.3-14) and $\boldsymbol{\lambda}$ and $\mathbf{i}$ from (5.3-7) and (5.3-8).

Our job now is to express the self- and mutual inductances of all windings. As in the case of the transformer, the self-inductance of each winding is made up of a leakage inductance caused by the flux which fails to cross the air gap and a magnetizing inductance caused by the flux which traverses the air gap and circulates through the stator and rotor steel. For symmetrical, identical stator windings, the self-inductances L_{asas} and L_{bsbs} are equal and will be denoted L_{ss}, where

$$L_{ss} = L_{ls} + L_{ms} \tag{5.3-18}$$

In (5.3-18), L_{ls} is the leakage inductance and L_{ms} the magnetizing inductance. The machine is designed to minimize the leakage inductance; it generally makes up approximately 10 percent of the self-inductance. The self-inductance of the symmetrical rotor windings may be expressed similarly,

$$L_{rr} = L_{lr} + L_{mr} \tag{5.3-19}$$

The magnetizing inductances L_{ms} and L_{mr} may be expressed in terms of turns and reluctance. In particular,

$$L_{ms} = \frac{N_s^2}{\mathcal{R}_m} \tag{5.3-20}$$

$$L_{mr} = \frac{N_r^2}{\mathcal{R}_m} \tag{5.3-21}$$

The magnetizing reluctance $\mathscr{R}_m$ is due primarily to the air gap and, since the winding is assumed to be an equivalent sinusoidally distributed winding, perhaps $\mathscr{R}_m$ should be considered an equivalent magnetizing reluctance. Nevertheless, expressions for L_{ms} and L_{mr}, and thus $\mathscr{R}_m$, as defined in (5.3-20) and (5.3-21), may be derived. In particular, it can be shown that [1]

$$L_{ms} = N_s^2 \frac{\pi \mu_0 r l}{4g} \tag{5.3-22}$$

$$L_{mr} = N_r^2 \frac{\pi \mu_0 r l}{4g} \tag{5.3-23}$$

where μ_0 is the permeability of free space, r is the mean radius of the air gap, l is the axial length of the air gap (rotor), and g is the radial length of the air gap. One must perform a rather involved and lengthy derivation to obtain (5.3-22) and (5.3-23). We will not conduct this derivation; instead, the use of an equivalent magnetizing reluctance $\mathscr{R}_m$ without evaluation will be sufficient for our purposes.

Since the stator (rotor) windings are orthogonal as depicted in Fig. 5.2-1, it would seem that coupling does not exist between the as and bs windings (L_{asbs} or L_{bsas}) or between the ar and br windings (L_{arbr} or L_{brar}). However, recall that the equivalent, sinusoidally distributed windings are depicted by one coil placed at the maximum turns density; the windings are actually distributed similar to that shown in Fig. 4.2-2. If we considered, for example, the coupling between the as and bs windings, one would be lead to believe that coupling exists since current flowing in the as_1-as_1' coil in Fig. 4.2-2 would produce a flux that couples the bs winding. However, this same current flows through the as_3-as_3' coil, which produces a flux that couples the bs winding in a direction opposite to the flux established by the as_1-as_1' coil. Therefore, if the stator (rotor) windings are distributed symmetrically about orthogonal axes, a net coupling would not exist in this uniform air gap machine. Thus, L_{asbs}, L_{bsas}, L_{arbr}, and L_{brar} are all zero. We will find that, in a three-phase machine, where the stator (rotor) windings are displaced 120° magnetically in space, a net coupling exists between the stator (rotor) windings. However, for the two-phase machine we can write

$$\mathbf{L}_s = \begin{bmatrix} L_{ss} & 0 \\ 0 & L_{ss} \end{bmatrix} = L_{ss}\mathbf{I} \tag{5.3-24}$$

$$\mathbf{L}_r = \begin{bmatrix} L_{rr} & 0 \\ 0 & L_{rr} \end{bmatrix} = L_{rr}\mathbf{I} \tag{5.3-25}$$

The stator and rotor windings are in relative motion. Therefore, coupling will occur between the stator and rotor windings and this coupling will vary with the position (θ_r) of the rotor windings relative to the stator windings. For example, when the as and ar windings are aligned, $\theta_r = 0$, the magnitude of coupling between these windings is maximum and, with the assumed direction of positive i_{as} and i_{ar}, the right-hand rule tells us that the mutual fluxes are aiding. Hence the mutual inductance at $\theta_r = 0$ is a positive maximum and can be expressed in terms of turns and $\mathscr{R}_m$ as

$$L_{asar} = \frac{N_s N_r}{\mathscr{R}_m} \qquad \text{for } \theta_r = 0 \tag{5.3-26}$$

Now, when $\theta_r = \frac{1}{2}\pi$, the as and ar windings are orthogonal and

$$L_{asar} = 0 \qquad \text{for } \theta_r = \frac{1}{2}\pi \tag{5.3-27}$$

For $\theta_r = \pi$ the windings are again aligned but now they oppose, thus

$$L_{asar} = -\frac{N_s N_r}{\mathscr{R}_m} \qquad \text{for } \theta_r = \pi \tag{5.3-28}$$

at $\theta_r = \frac{3}{2}\pi$, the windings are again orthogonal and

$$L_{asar} = 0 \qquad \text{for } \theta_r = \frac{3}{2}\pi \tag{5.3-29}$$

From (5.3-26) through (5.3-29), we see that mutual inductances might be approximated as a cosine function of θ_r. In particular, if we define L_{sr} as

$$L_{sr} = \frac{N_s N_r}{\mathscr{R}_m} \tag{5.3-30}$$

we approximate L_{asar} or L_{aras} as

$$L_{asar} = L_{sr} \cos \theta_r \tag{5.3-31}$$

If we were to carry out the derivation as in [1], we would find that (5.3-31) is, indeed, a valid expression for the mutual inductance between the as and ar windings. It follows by inspection of Fig. 5.2-1 that

$$L_{asbr} = -L_{sr} \sin \theta_r \tag{5.3-32}$$

$$L_{bsar} = L_{sr} \sin \theta_r \tag{5.3-33}$$

$$L_{bsbr} = L_{sr} \cos \theta_r \tag{5.3-34}$$

Hence,
$$\boxed{\mathbf{L}_{sr} = L_{sr} \begin{bmatrix} \cos \theta_r & -\sin \theta_r \\ \sin \theta_r & \cos \theta_r \end{bmatrix}} \tag{5.3-35}$$

One should now be able to write the mutual inductances by inspection. For practice, express the stator and rotor mutual inductances if the positive direction of i_{bs} is reversed. In this case, (5.3-31) and (5.3-32) remain unchanged; however, the sign of (5.3-33) and (5.3-34) would be changed.

Once the expressions for the mutual inductances are known, we begin to understand the complexities involved in the analysis of electric machines. The stator to rotor mutual inductances are sinusoidal functions of θ_r because of their relative motion. Hence, when in the voltage equations we take the derivative of the flux linkages with respect to time, we no longer obtain only the familiar $L(di/dt)$. Instead, two terms result—one due to the derivative of the mutual inductance, since θ_r is a function of time, and one due to the derivative of the current. For example,

$$\frac{d(L_{asar}i_{as})}{dt} = \frac{dL_{asar}}{dt}i_{as} + L_{asar}\frac{di_{as}}{dt} \tag{5.3-36}$$

Although we are going to have to deal with this problem, we shall not do it now. However, before leaving this work, let us incorporate a turns-ratio into the equations as we did for the transformer. We may not see the purpose of this at this point since it cannot yield an equivalent T circuit, as in the case of the transformer, because of the variation of the mutual inductances. However, later in this chapter, we will incorporate change of variables which will allow us to treat the induction machine from the standpoint of an equivalent T circuit with constant inductances. In preparation for this event, we will incorporate a turn-ratio at this time. In particular, we will refer the rotor variables to a winding with N_s turns by the following turns-ratios:

$$\mathbf{i}'_{abr} = \frac{N_r}{N_s}\mathbf{i}_{abr} \tag{5.3-37}$$

$$\mathbf{v}'_{abr} = \frac{N_s}{N_r}\mathbf{v}_{abr} \tag{5.3-38}$$

$$\boldsymbol{\lambda}'_{abr} = \frac{N_s}{N_r}\boldsymbol{\lambda}_{abr} \tag{5.3-39}$$

Thus (5.3-5) and (5.3-6) may be written as

$$\mathbf{v}_{abs} = \mathbf{r}_s\mathbf{i}_{abs} + p\boldsymbol{\lambda}_{abs} \tag{5.3-40}$$

$$\mathbf{v}'_{abr} = \mathbf{r}'_r\mathbf{i}'_{abr} + p\boldsymbol{\lambda}'_{abr} \tag{5.3-41}$$

where

$$\mathbf{r}'_r = \left(\frac{N_s}{N_r}\right)^2\mathbf{r}_r \tag{5.3-42}$$

Substitution of (5.3-37) and (5.3-39) into (5.3-17) yields

$$\begin{bmatrix} \boldsymbol{\lambda}_{abs} \\ \boldsymbol{\lambda}'_{abr} \end{bmatrix} = \begin{bmatrix} \mathbf{L}_s & \frac{N_s}{N_r}\mathbf{L}_{sr} \\ \frac{N_s}{N_r}(\mathbf{L}_{sr})^T & \mathbf{L}'_r \end{bmatrix} \begin{bmatrix} \mathbf{i}_{abs} \\ \mathbf{i}'_{abr} \end{bmatrix} \tag{5.3-43}$$

where

$$\mathbf{L}'_r = \left(\frac{N_s}{N_r}\right)^2\mathbf{L}_r = \begin{bmatrix} L'_{rr} & 0 \\ 0 & L'_{rr} \end{bmatrix} \tag{5.3-44}$$

Since L_{mr} and L_{ms} may be related from (5.3-20) and (5.3-21),

$$L'_{rr} = L'_{lr} + \left(\frac{N_s}{N_r}\right)^2 L_{mr} = L'_{lr} + L_{ms} \tag{5.3-45}$$

Note that

$$\frac{N_s}{N_r}\mathbf{L}_{sr} = \frac{N_s}{N_r}L_{sr}\begin{bmatrix} \cos\theta_r & -\sin\theta_r \\ \sin\theta_r & \cos\theta_r \end{bmatrix} \tag{5.3-46}$$

Comparing L_{ms}, (5.3-20), and L_{sr}, (5.3-30), we see that

$$\frac{N_s}{N_r}L_{sr} = L_{ms} \tag{5.3-47}$$

Hence, (5.3-46) may be expressed in terms of L_{ms} and, for compactness, we will define $\mathbf{L}'_{sr}$ as

$$\mathbf{L}'_{sr} = \frac{N_s}{N_r}\mathbf{L}_{sr} = L_{ms}\begin{bmatrix} \cos\theta_r & -\sin\theta_r \\ \sin\theta_r & \cos\theta_r \end{bmatrix} \tag{5.3-48}$$

Thus, (5.3-43) becomes

$$\begin{bmatrix} \boldsymbol{\lambda}_{abs} \\ \boldsymbol{\lambda}'_{abr} \end{bmatrix} = \begin{bmatrix} \mathbf{L}_s & \mathbf{L}'_{sr} \\ (\mathbf{L}'_{sr})^T & \mathbf{L}'_r \end{bmatrix}\begin{bmatrix} \mathbf{i}_{abs} \\ \mathbf{i}'_{abr} \end{bmatrix} \tag{5.3-49}$$

SP5.3-1. Assume that θ_r is positive in the clockwise direction in Fig. 5.2-1 rather than in the counterclockwise direction. Express all inductances. [$L_{asbr} = L_{sr}\sin\theta_r$; $L_{bsar} = -L_{sr}\sin\theta_r$; all others unchanged]

SP5.3-2. The as and bs windings in Fig. 5.2-1 are rotated $\frac{1}{4}\pi$ clockwise from the position shown. Express L_{asar} and L_{asbr}. [$L_{asar} = L_{sr}\cos(\theta_r + \frac{1}{4}\pi)$; $L_{asbr} = -L_{sr}\sin(\theta_r + \frac{1}{4}\pi)$]

SP5.3-3. Consider the device shown in Fig. 5.2-1. All windings are open-circuited except the as winding. $I_{as} = \sin t$, $L_{ms} = 0.1$ H, $L_{rr} = \frac{1}{4}L'_{rr}$, $\omega_r = 0$, and $\theta_r = \frac{1}{3}\pi$. Determine V_{ar}. [$V_{ar} = 0.025\cos t$]

5.4 TORQUE

From Table 2.5-1 for a P-pole machine

$$T_e(\mathbf{i}, \theta_r) = \frac{P}{2}\frac{\partial W_c(\mathbf{i}, \theta_r)}{\partial\theta_r} \tag{5.4-1}$$

In a linear magnetic system, the energy in the coupling field W_f and the coenergy W_c are equal. The field energy can be expressed as

$$W_f(\mathbf{i}, \theta_r) = \frac{1}{2}L_{ss}i_{as}^2 + \frac{1}{2}L_{ss}i_{bs}^2 + \frac{1}{2}L'_{rr}i_{ar}'^2 + \frac{1}{2}L'_{rr}i_{br}'^2$$

$$+ L_{ms}i_{as}i'_{ar}\cos\theta_r - L_{ms}i_{as}i'_{br}\sin\theta_r$$

$$+ L_{ms}i_{bs}i'_{ar}\sin\theta_r + L_{ms}i_{bs}i'_{br}\cos\theta_r \tag{5.4-2}$$

Since $W_f = W_c$ for a linear magnetic system, substituting (5.4-2) into (5.4-1) yields the electromagnetic torque for a P-pole two-phase symmetrical induction machine. In particular,

$$T_e = -\frac{P}{2} L_{ms}[(i_{as}i'_{ar} + i_{bs}i'_{br}) \sin \theta_r + (i_{as}i'_{br} - i_{bs}i'_{ar}) \cos \theta_r] \qquad (5.4\text{-}3)$$

The torque and rotor speed are related by

$$T_e = J \frac{d\omega_{rm}}{dt} + B_m \omega_{rm} + T_L \qquad (5.4\text{-}4)$$

where ω_{rm} is the actual rotor speed. For a P-pole machine, since $\omega_{rm} = (2/P)\omega_r$,

$$T_e = J \frac{2}{P} \frac{d\omega_r}{dt} + B_m \frac{2}{P} \omega_r + T_L$$

where J is the inertia of the rotor and, in some cases, the connected load. The first term on the right-hand side is the inertial torque. The units of J are kilogram $\cdot$ meter^2 (kg $\cdot$ m^2) or joules $\cdot$ second^2 (J $\cdot$ s^2). Often, the inertia is given as a quantity called WR^2 expressed in units of pound mass $\cdot$ feet^2 (lbm $\cdot$ ft^2). The load torque T_L is positive for a torque load on the shaft of the induction machine, as shown in Fig. 5.2-1. The constant B_m is a damping coefficient associated with the rotational system of the machine and mechanical load. It has the units of N $\cdot$ m $\cdot$ s/rad of mechanical rotation and it is generally small and often neglected.

SP5.4-1. Why is the sixth term on the right-hand side of (5.4-2) negative? [(5.3-32)]

5.5 MACHINE EQUATIONS IN THE STATIONARY REFERENCE FRAME

In a previous section, it was pointed out that the analysis of an induction machine was complicated because of the sinusoidal variation of the stator-to-rotor mutual inductances with respect to the rotor displacement θ_r. Fortunately, by a change of variables it is possible to eliminate the variation in the mutual inductances. For this purpose, we will define a change of variables for both the stator and rotor variables.

In the case of the stator voltages, currents, and flux linkages let

$$\begin{bmatrix} f^s_{qs} \\ f^s_{ds} \end{bmatrix} = \begin{bmatrix} 1 & 0 \\ 0 & -1 \end{bmatrix} \begin{bmatrix} f_{as} \\ f_{bs} \end{bmatrix} \qquad (5.5\text{-}1)$$

or

$$\mathbf{f}^s_{qds} = \mathbf{K}^s_s \mathbf{f}_{abs} \qquad (5.5\text{-}2)$$

where **f** can represent either voltage, current, or flux linkage. It follows that

$$\mathbf{f}_{abs} = (\mathbf{K}_s^s)^{-1}\mathbf{f}_{qds}^s \qquad (5.5\text{-}3)$$

where $(\mathbf{K}_s^s)^{-1} = \mathbf{K}_s^s$. (The reader should prove this; a review of matrix algebra is given in Appendix D.) One should question the reason for this change in variables since it only defines $f_{qs}^s = f_{as}$ and $f_{ds}^s = -f_{bs}$ and appears to be nothing more than a change of notation. This is true and, of course, it would not be necessary other than to comply with convention.

In the case of the rotor variables, let [2]

$$\begin{bmatrix} f_{qr}^{'s} \\ f_{dr}^{'s} \end{bmatrix} = \begin{bmatrix} \cos\theta_r & -\sin\theta_r \\ -\sin\theta_r & -\cos\theta_r \end{bmatrix} \begin{bmatrix} f_{ar}' \\ f_{br}' \end{bmatrix} \qquad (5.5\text{-}4)$$

or

$$\boxed{\mathbf{f}_{qdr}^{'s} = \mathbf{K}_r^s \mathbf{f}_{abr}'} \qquad (5.5\text{-}5)$$

In (5.5-4) and (5.5-5),

$$\boxed{\theta_r = \int_0^t \omega_r(\xi)\, d\xi + \theta_r(0)} \qquad (5.5\text{-}6)$$

where ξ is a dummy variable of integration and $\theta_r(0)$ is the time zero position of the rotor which is generally selected to be zero. Also,

$$\mathbf{f}_{abr}' = (\mathbf{K}_r^s)^{-1}\mathbf{f}_{qdr}^{'s} \qquad (5.5\text{-}7)$$

It is left to the reader to show that $(\mathbf{K}_r^s)^{-1} = \mathbf{K}_r^s$. Although the above changes of variables do not require any physical connotation, it may be convenient to visualize these transformations or changes of variables as trigonometric relationships between variables with direction as shown in Fig. 5.5-1. It may be helpful to note that, in the case of f_{as}, f_{bs}, f_{ar}', and f_{br}', the positive direction of the variables shown in Fig. 5.5-1 happens to be the same as the positive direction of the magnetic axes of the associated winding, as shown in Fig. 5.2-1. The change of stator variables is essentially a change in notation; however, the change in rotor variables is not as easily interpreted. We see, however, that f_{qs}^s and $f_{qr}^{'s}$ are fixed in space and are both in the same direction. Also, f_{ds}^s and $f_{dr}^{'s}$ are fixed in space 90° from f_{qs}^s and $f_{qr}^{'s}$. These changes of variables are often referred to as transformations which transform the machine variables to a stationary frame of reference. In other words, the $f_{qs}^s, f_{ds}^s, f_{qr}^{'s}$ and $f_{dr}^{'s}$ are variables associated with circuits which are stationary (fixed in the stator). The $\overset{s}{q}s$ and $\overset{s}{d}s$ circuits are the as and bs circuits, respectively, with a difference in sign in the case of the f_{bs} and f_{ds}^s variables. If one wishes to establish a similar interpretation for the $f_{qr}^{'s}$ and $f_{dr}^{'s}$ variables, one must think of these variables as being associated with fictitious stationary $\overset{s}{q}r$ and $\overset{s}{d}r$ circuits. Let us think back for a moment: What did we do when we used a turns-ratio to

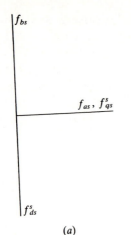

(a)

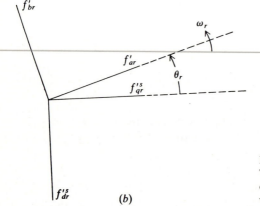

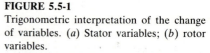

(b)

FIGURE 5.5-1
Trigonometric interpretation of the change of variables. (a) Stator variables; (b) rotor variables.

refer the rotor variables to windings with the same number of turns as the stator windings? Was this not the same thing as we are doing now? No. All we do with a turns-ratio substitution is to make the "prime" rotor variables look as if they are associated with N_s turns rather than N_r turns.

Getting back to the transformation at hand, we now start to see the significance of the subscripts s and r and the superscript s. The subscripts denote association with the stator (s) and rotor (r) variables. That is, f_{qs}^s and f_{ds}^s are related to f_{as} and f_{bs}, and $f_{qr}^{'s}$ and $f_{dr}^{'s}$ to $f_{ar}^{'}$ and $f_{br}^{'}$. Similarly, $\mathbf{K}_s^s$ is the transformation for the stator variables and $\mathbf{K}_r^s$ for the rotor variables. The superscript s indicates that the transformations are to the stationary reference frame. In this text, we will use only the stationary reference frame in the analysis of induction machines. Although we do not need to know this, it is interesting that the voltage equations of a symmetrical induction machine can be expressed in any frame of reference by a general transformation wherein an arbitrary angular displacement and velocity are used in the argument of the transformation [1]. Using the superscript s to denote stationary reference frame variables avoids any confusion as to which reference frame we are using.

The form of the voltage equations associated with these "new" circuits is of interest. For this purpose, let us write the stator and rotor voltage equations given by (5.3-40) and (5.3-41), respectively, with (5.5-3) and (5.5-7) incorporated:

$$(\mathbf{K}_s^s)^{-1}\mathbf{v}_{qds}^s = \mathbf{r}_s(\mathbf{K}_s^s)^{-1}\mathbf{i}_{qds}^s + p[(\mathbf{K}_s^s)^{-1}\boldsymbol{\lambda}_{qds}^s] \tag{5.5-8}$$

$$(\mathbf{K}_r^s)^{-1}\mathbf{v}_{qdr}^{\prime s} = \mathbf{r}_r'(\mathbf{K}_r^s)^{-1}\mathbf{i}_{qdr}^{\prime s} + p[(\mathbf{K}_r^s)^{-1}\boldsymbol{\lambda}_{qdr}^{\prime s}] \tag{5.5-9}$$

Multiplying (5.5-8) by $\mathbf{K}_s^s$ and (5.5-9) by $\mathbf{K}_r^s$ yields

$$\mathbf{v}_{qds}^s = \mathbf{K}_s^s\mathbf{r}_s(\mathbf{K}_s^s)^{-1}\mathbf{i}_{qds}^s + \mathbf{K}_s^s p[(\mathbf{K}_s^s)^{-1}\boldsymbol{\lambda}_{qds}^s] \tag{5.5-10}$$

$$\mathbf{v}_{qdr}^{\prime s} = \mathbf{K}_r^s\mathbf{r}_r'(\mathbf{K}_r^s)^{-1}\mathbf{i}_{qdr}^{\prime s} + \mathbf{K}_r^s p[(\mathbf{K}_r^s)^{-1}\boldsymbol{\lambda}_{qdr}^{\prime s}] \tag{5.5-11}$$

From which

$$\boxed{\begin{aligned} \mathbf{v}_{qds}^s &= \mathbf{r}_s\mathbf{i}_{qds}^s + p\boldsymbol{\lambda}_{qds}^s \\ \mathbf{v}_{qdr}^{\prime s} &= \mathbf{r}_r'\mathbf{i}_{qdr}^{\prime s} - \omega_r\boldsymbol{\lambda}_{dqr}^{\prime s} + p\boldsymbol{\lambda}_{qdr}^{\prime s} \end{aligned}}$$

$$\tag{5.5-12}$$
$$\tag{5.5-13}$$

where

$$(\boldsymbol{\lambda}_{dqr}^{\prime s})^T = [\lambda_{dr}^{\prime s} \quad -\lambda_{qr}^{\prime s}] \tag{5.5-14}$$

It is a little more involved to obtain (5.5-13) from (5.5-11) than (5.5-12) from (5.5-10). Consider the first term on the right-hand side of (5.5-11):

$$\mathbf{K}_r^s\mathbf{r}_r'(\mathbf{K}_r^s)^{-1}\mathbf{i}_{qdr}^{\prime s} = \mathbf{K}_r^s r_r'\mathbf{I}(\mathbf{K}_r^s)^{-1}\mathbf{i}_{qdr}^{\prime s} = r_r'\mathbf{K}_r^s(\mathbf{K}_r^s)^{-1}\mathbf{i}_{qdr}^{\prime s} = \mathbf{r}_r'\mathbf{i}_{qdr}^{\prime s} \tag{5.5-15}$$

Now, the second term on the right-hand side of (5.5-11) can be written as

$$\begin{aligned} \mathbf{K}_r^s p[(\mathbf{K}_r^s)^{-1}\boldsymbol{\lambda}_{qdr}^{\prime s}] &= \mathbf{K}_r^s[p(\mathbf{K}_r^s)^{-1}]\boldsymbol{\lambda}_{qdr}^{\prime s} + \mathbf{K}_r^s(\mathbf{K}_r^s)^{-1}p\boldsymbol{\lambda}_{qdr}^{\prime s} \\ &= \mathbf{K}_r^s\omega_r\begin{bmatrix} -\sin\theta_r & -\cos\theta_r \\ -\cos\theta_r & \sin\theta_r \end{bmatrix}\boldsymbol{\lambda}_{qdr}^{\prime s} + p\boldsymbol{\lambda}_{qdr}^{\prime s} \\ &= -\omega_r\begin{bmatrix} \cos\theta_r & -\sin\theta_r \\ -\sin\theta_r & -\cos\theta_r \end{bmatrix}\begin{bmatrix} \sin\theta_r & \cos\theta_r \\ \cos\theta_r & -\sin\theta_r \end{bmatrix}\boldsymbol{\lambda}_{qdr}^{\prime s} + p\boldsymbol{\lambda}_{qdr}^{\prime s} \\ &= -\omega_r\begin{bmatrix} 0 & 1 \\ -1 & 0 \end{bmatrix}\begin{bmatrix} \lambda_{qr}^{\prime s} \\ \lambda_{dr}^{\prime s} \end{bmatrix} + p\boldsymbol{\lambda}_{qdr}^{\prime s} \\ &= -\omega_r\boldsymbol{\lambda}_{dqr}^{\prime s} + p\boldsymbol{\lambda}_{qdr}^{\prime s} \end{aligned}$$

$$\tag{5.5-16}$$

where $\boldsymbol{\lambda}_{dqr}^{\prime s}$ is defined by (5.5-14).

At this point, one might question the logic of this change of variables since our new voltage equations have the added term $\omega_r\boldsymbol{\lambda}_{dqr}^{\prime s}$, with no apparent advantage. We shall see the facility of all this in just a moment. First, it is important to realize that the voltage equations given by (5.5-12) and (5.5-13) are valid for a linear or nonlinear magnetic system; however, for a linear magnetic system, we can express $\boldsymbol{\lambda}_{abs}$ and $\boldsymbol{\lambda}'_{abr}$ in terms of inductances as in (5.3-49):

$$\boldsymbol{\lambda}_{abs} = \mathbf{L}_s \mathbf{i}_{abs} + \mathbf{L}'_{sr} \mathbf{i}'_{abr} \tag{5.5-17}$$

$$\boldsymbol{\lambda}'_{abr} = (\mathbf{L}'_{sr})^T \mathbf{i}_{abs} + \mathbf{L}'_r \mathbf{i}'_{abr} \tag{5.5-18}$$

Substituting (5.5-3) and (5.5-7) into (5.5-17) and (5.5-18) and solving for $\boldsymbol{\lambda}^s_{qds}$ and $\boldsymbol{\lambda}'^s_{qdr}$ yields

$$\boldsymbol{\lambda}^s_{qds} = \mathbf{K}^s_s \mathbf{L}_s (\mathbf{K}^s_s)^{-1} \mathbf{i}^s_{qds} + \mathbf{K}^s_s \mathbf{L}'_{sr} (\mathbf{K}^s_r)^{-1} \mathbf{i}'^s_{qdr} \tag{5.5-19}$$

$$\boldsymbol{\lambda}'^s_{qdr} = \mathbf{K}^s_r (\mathbf{L}'_{sr})^T (\mathbf{K}^s_s)^{-1} \mathbf{i}^s_{qds} + \mathbf{K}^s_r \mathbf{L}'_r (\mathbf{K}^s_r)^{-1} \mathbf{i}'^s_{qdr} \tag{5.5-20}$$

These equations reduce to

$$\boldsymbol{\lambda}^s_{qds} = \mathbf{L}_s \mathbf{i}^s_{qds} + \mathbf{L}_{ms} \mathbf{i}'^s_{qdr} \tag{5.5-21}$$

$$\boldsymbol{\lambda}'^s_{qdr} = \mathbf{L}_{ms} \mathbf{i}^s_{qds} + \mathbf{L}'_r \mathbf{i}'^s_{qdr} \tag{5.5-22}$$

where $\mathbf{L}_s$ and $\mathbf{L}'_r$ are defined by (5.3-24) and (5.3-44), respectively, and

$$\mathbf{L}_{ms} = \begin{bmatrix} L_{ms} & 0 \\ 0 & L_{ms} \end{bmatrix} \tag{5.5-23}$$

Although the $\mathbf{L}_s \mathbf{i}^s_{qds}$ term in (5.5-21) and the $\mathbf{L}'_r \mathbf{i}'^s_{qdr}$ term in (5.5-22) are readily obtained from (5.5-19) and (5.5-20), respectively, there is some matrix multiplication necessary to obtain the $\mathbf{L}_{ms} \mathbf{i}'^s_{qdr}$ and $\mathbf{L}_{ms} \mathbf{i}^s_{qds}$ terms. The reader should take the time to perform this multiplication for at least one of these terms. However, it is more important to note that all mutual inductances are now constant and the $\overset{s}{q}s$ and $\overset{s}{q}r$ circuits are decoupled magnetically from the $\overset{s}{d}s$ and $\overset{s}{d}r$ circuits. Now we start to see the advantage of the transformation or change of variables. Although we have added the $\omega_r \boldsymbol{\lambda}'^s_{dqr}$ term, the new variables are decoupled magnetically and all inductances are constant. The latter tells us that none of the circuits are in relative motion.

Equations (5.5-12) and (5.5-13) may be written in expanded form as

$$v^s_{qs} = r_s i^s_{qs} + p\lambda^s_{qs} \tag{5.5-24}$$

$$v^s_{ds} = r_s i^s_{ds} + p\lambda^s_{ds} \tag{5.5-25}$$

$$v'^s_{qr} = r'_r i'^s_{qr} - \omega_r \lambda'^s_{dr} + p\lambda'^s_{qr} \tag{5.5-26}$$

$$v'^s_{dr} = r'_r i'^s_{dr} + \omega_r \lambda'^s_{qr} + p\lambda'^s_{dr} \tag{5.5-27}$$

For a linear magnetic system, the flux linkages are expressed by (5.5-21) and (5.5-22), which may be written as

$$\lambda_{qs}^s = L_{ls}i_{qs}^s + L_{ms}(i_{qs}^s + i_{qr}^{'s})$$

$$= L_{ss}i_{qs}^s + L_{ms}i_{qr}^{'s} \tag{5.5-28}$$

$$\lambda_{ds}^s = L_{ls}i_{ds}^s + L_{ms}(i_{ds}^s + i_{dr}^{'s})$$

$$= L_{ss}i_{ds}^s + L_{ms}i_{dr}^{'s} \tag{5.5-29}$$

$$\lambda_{qr}^{'s} = L_{lr}'i_{qr}^{'s} + L_{ms}(i_{qs}^s + i_{qr}^{'s})$$

$$= L_{rr}'i_{qr}^{'s} + L_{ms}i_{qs}^s \tag{5.5-30}$$

$$\lambda_{dr}^{'s} = L_{lr}'i_{dr}^{'s} + L_{ms}(i_{ds}^s + i_{dr}^{'s})$$

$$= L_{rr}'i_{dr}^{'s} + L_{ms}i_{ds}^s \tag{5.5-31}$$

where L_{ss} and L_{rr}' are defined by (5.3-18) and (5.3-45), respectively. Substituting (5.5-28) through (5.5-31) into (5.5-24) through (5.5-27) yields

$$\begin{bmatrix} v_{qs}^s \\ v_{ds}^s \\ v_{qr}^{'s} \\ v_{dr}^{'s} \end{bmatrix} = \begin{bmatrix} r_s + pL_{ss} & 0 & pL_{ms} & 0 \\ 0 & r_s + pL_{ss} & 0 & pL_{ms} \\ pL_{ms} & -\omega_r L_{ms} & r_r' + pL_{rr}' & -\omega_r L_{rr}' \\ \omega_r L_{ms} & pL_{ms} & \omega_r L_{rr}' & r_r' + pL_{rr}' \end{bmatrix} \begin{bmatrix} i_{qs}^s \\ i_{ds}^s \\ i_{qr}^{'s} \\ i_{dr}^{'s} \end{bmatrix} \tag{5.5-32}$$

The above voltage equations may be depicted by the equivalent circuits shown in Fig. 5.5-2. Note that when the rotor is stalled ($\omega_r = 0$) the circuits reduce to the familiar T circuits for two independent single-phase transformers.

An expression for electromagnetic torque is obtained by replacing the phase variables in (5.4-3) with the substitute variables. After some manipulation we obtain, for a P-pole machine,

$$T_e = \frac{P}{2} L_{ms}(i_{qs}^s i_{dr}^{'s} - i_{ds}^s i_{qr}^{'s}) \tag{5.5-33}$$

where T_e is positive for motor action. Equivalent expressions for torque are

$$T_e = \frac{P}{2} (\lambda_{qr}^{'s} i_{dr}^{'s} - \lambda_{dr}^{'s} i_{qr}^{'s}) \tag{5.5-34}$$

$$T_e = \frac{P}{2} (\lambda_{ds}^s i_{qs}^s - \lambda_{qs}^s i_{ds}^s) \tag{5.5-35}$$

The reader should show that (5.5-33) through (5.5-35) are equivalent. Equations (5.5-34) and (5.5-35) may be somewhat misleading since they seem to imply that the leakage inductances are involved in the energy conversion

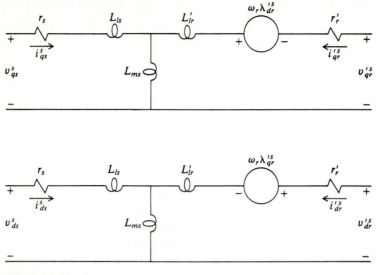

FIGURE 5.5-2
Equivalent circuit of a two-phase symmetrical induction machine in the stationary reference frame.

process. We know from Chap. 2 that this cannot be the case. Even though the flux linkages in (5.5-34) and (5.5-35) contain the leakage inductances, they are eliminated by the algebra within the parentheses.

It is important to consider what we have done and why we have done it. Due to the fact that the stator and rotor circuits are in relative motion, the mutual inductances between these circuits vary as a sinusoidal function of the angular displacement between the circuits. We defined a change of variables for the rotor circuits which we said would transform the rotor variables to fictitious stationary circuits. At that time we did not quite know why, but we went along with it anyway. Once we performed the transformation we have seen that speed voltages $(\omega_r \lambda_{qr}^{'s}$ and $\omega_r \lambda_{dr}^{'s})$ were introduced in the voltage equations, (5.5-13). Next, when we assumed a linear magnetic system and expressed the flux linkages in terms of inductances and then performed the transformation on the flux linkage equations, we discovered two important things: First, and most important, all mutual inductances became constant, as we see from (5.5-21) and (5.5-22), or (5.5-28) through (5.5-31), or (5.5-32), or the equivalent circuits given in Fig. 5.5-2. This fact gives credence to the claim that the transformation gives rise to fictitious stationary circuits. In particular, the $\overset{s}{q}r$ and $\overset{s}{d}r$ circuits must be stationary with respect to the stationary $\overset{s}{q}s$ and $\overset{s}{d}s$ circuits (which are the as and bs circuits), otherwise the mutual inductances would not be constant. In other words, the $\overset{s}{q}r$ and $\overset{s}{d}r$ circuits are not in relative motion with respect to the $\overset{s}{q}s$ and $\overset{s}{d}s$ circuits. The second important fact is that the q and d circuits are decoupled magnetically. As we can see from Fig. 5.5-2, we have two single-phase transformers which are coupled through speed voltages ($\omega_r \lambda_{qr}^{'s}$ and $\omega_r \lambda_{dr}^{'s}$) but not magnetically as coupled circuits. We will find this also to be true in the case of a three-phase symmetrical induction machine.

We might comment on something that is perhaps already obvious. In the majority of applications, the rotor windings are short-circuited and, hence, $v_{qr}^{'s}$ and $v_{dr}^{'s}$ are zero. Therefore, since the system is assumed to be magnetically linear, the frequency of the steady-state currents must be equal to the frequency of the driving voltages (v_{qs}^{s}, which is v_{as}, and v_{ds}^{s}, which is $-v_{bs}$). In other words, if the electrical angular velocity (frequency) of the applied stator voltages is ω_e, then, during steady-state operation, i_{qs}^{s}, i_{ds}^{s}, $i_{qr}^{'s}$, and $i_{dr}^{'s}$ will all vary at ω_e. That is, $i_{qr}^{'s}$ and $i_{dr}^{'s}$ do not vary at $\omega_e - \omega_r$ as do the rotor currents $i_{ar}^{'}$ and $i_{br}^{'}$.

Let us now consider two properties of the transformation, or of reference frame theory in general, which may not be obvious to us at this stage. First, one might be led to believe that the transformation we have devised may be applied to the variables of any set of windings and constant circuit parameters will result. This is not quite the case. If the transformation is to a frame of reference other than where the circuits physically exist, the windings must be symmetrical (same resistance and same number of turns) multiphase windings. In other words, if the transformation is a function of an angular displacement, the windings must be symmetrical. Example 5A, which follows, helps to illustrate this fact.

The second property of reference frame theory, as applied to symmetrical induction machines, was mentioned but not explained when we discussed the notation used for the transformation. In particular, the voltage equations of a symmetrical induction machine may be expressed in any frame of reference by a general transformation wherein an arbitrary angular displacement and velocity are used in the argument of this transformation to the arbitrary reference frame [1]. Although an extensive treatment of reference frame theory is beyond the scope of this text, let us think about this statement for a moment. First, the stator and rotor windings are symmetrical sets of windings. Therefore, according to our statement in the previous paragraph, a change of variables can be devised for the stator variables and another change of variables for the rotor variables. The change of variables will differ only in the angular displacement used in each transformation. Hence, we can transform the stator and rotor variables to any reference frame. Now, let us step back a moment and recall the reason for making the change of variables. It was to eliminate the sinusoidal variation in the mutual inductances which occurred due to windings in relative motion. The only requirement for eliminating the varying mutual inductances due to circuits in relative motion is to transform the stator and rotor variables to a common frame of reference, be it stationary, rotating, pulsating, or whatever. Since the stator and rotor windings are symmetrical and the air gap is uniform, we are not restricted as to the reference frame we can use in the analysis of an induction machine. We will find, however, that this is not the case for a synchronous machine.

Example 5A. We have mentioned that the change of variables which we introduced for the rotor variables is of benefit only if the windings are symmetrical multiphase windings. To illustrate this, let us assume that the resistances of the two rotor windings are different. Let

$$\mathbf{r}'_r = \begin{bmatrix} r'_a & 0 \\ 0 & r'_b \end{bmatrix} \tag{5A-1}$$

Now consider the first term on the right-hand side of (5.5-11):

$$\mathbf{K}^s_r \mathbf{r}'_r (\mathbf{K}^s_r)^{-1} \mathbf{i}'^s_{qdr} = \begin{bmatrix} \cos\theta_r & -\sin\theta_r \\ -\sin\theta_r & -\cos\theta_r \end{bmatrix} \begin{bmatrix} r'_a & 0 \\ 0 & r'_b \end{bmatrix} \begin{bmatrix} \cos\theta_r & -\sin\theta_r \\ -\sin\theta_r & -\cos\theta_r \end{bmatrix} \mathbf{i}'^s_{qdr} \tag{5A-2}$$

After some matrix manipulation and considerable trigonometric substitutions, we obtain

$$\mathbf{K}^s_r \mathbf{r}'_r (\mathbf{K}^s_r)^{-1} \mathbf{i}'^s_{qdr} = \begin{bmatrix} \dfrac{r'_a + r'_b}{2} + \dfrac{r'_a - r'_b}{2}\cos 2\theta_r & -\dfrac{r'_a - r'_b}{2}\sin 2\theta_r \\[2ex] -\dfrac{r'_a - r'_b}{2}\sin 2\theta_r & \dfrac{r'_a + r'_b}{2} - \dfrac{r'_a - r'_b}{2}\cos 2\theta_r \end{bmatrix} \mathbf{i}'^s_{qdr} \tag{5A-3}$$

The situation has been complicated beyond repair. We now have resistances that vary as a function of $2\theta_r$ and, as if that was not bad enough, the q and d circuits are coupled.

SP5.5-1. The induction machine shown in Fig. 5.2-1 is operating under steady-state balanced conditions, rotating counterclockwise at $\omega_r = 0.8\omega_e$. $\tilde{I}^s_{ds} = I_s \underline{/0°}$. Determine $\tilde{I}^s_{qs}$, $\tilde{I}_{as}$, and $\tilde{I}_{bs}$. [$\tilde{I}^s_{qs} = \tilde{I}_{as} = I_s \underline{/-90°}$; $\tilde{I}_{bs} = I_s \underline{/180°}$]

SP5.5-2. If $f'^s_{qr} = 0$ and $f'^s_{dr} = 1$, determine f'_{ar} and f'_{br} if $\theta_r = \frac{1}{3}\pi$. [$f'_{ar} = -\sqrt{3}/2$; $f'_{br} = -\frac{1}{2}$]

SP5.5-3. A four-pole induction motor is operating under steady-state balanced conditions with $\omega_r = 0.8\omega_e$. The frequency of i_{as} and i_{bs} is 30 Hz. Determine the frequency of (a) i_{ar} and i_{br}, (b) i^s_{qs} and i^s_{ds}, and (c) i'^s_{qr} and i'^s_{dr}. [(a) 6 Hz; (b) 30 Hz; (c) 30 Hz]

SP5.5-4. Define the conditions so that the resistance matrix given by (5A-3) (a) has all constant elements and (b) decouples the $\overset{s}{q}r$ and $\overset{s}{d}r$ circuits. [(a) $\omega_r = 0$; (b) $\omega_r = 0$ and $\theta_r(0) = 0$]

SP5.5-5. Express the rotating air gap mmf due to current flowing in the $\overset{s}{q}r$ and $\overset{s}{d}r$ windings. [mmf $= (N_r/2)(i^s_{qr}\cos\phi_s - i^s_{dr}\sin\phi_s)$]

5.6 ANALYSIS OF STEADY-STATE OPERATION

In the vast majority of applications, the induction machine is operated with the stator windings connected to a voltage source and the rotor windings short-circuited. We will not consider the double-fed induction machine since its applications are limited. During balanced steady-state operation, the variables associated with the stator circuits of the symmetrical induction machine vary at a frequency corresponding to ω_e and the variables associated with the rotor circuits vary at a frequency corresponding to $\omega_e - \omega_r$. Let us now consider the $\overset{s}{qs}$, $\overset{s}{ds}$, $\overset{s}{qr}$, and $\overset{s}{dr}$ variables. Recall that we have transformed the rotor variables to the stationary reference frame which gave rise to stationary, fictitious $\overset{s}{qr}$ and $\overset{s}{dr}$ circuits. The $\overset{s}{qs}$ and $\overset{s}{ds}$ circuits are, of course, stationary, and in this frame of reference the applied stator voltages vary at the frequency of the actual applied stator voltages since $v^s_{qs} = v_{as}$ and $v^s_{ds} = -v_{bs}$. Therefore, in the steady state, the f^s_{qs}, f^s_{ds}, f'^s_{qr}, and f'^s_{dr} variables will all vary at ω_e, the frequency of the applied stator voltages, and the rotating air gap mmf's produced by i^s_{qs} and i^s_{ds} and by i'^s_{qr} and i'^s_{dr} will rotate about the air gap at ω_e.

For the purpose of considering this from an analytical standpoint, let us substitute expressions for the steady-state stator and rotor variables into the equations of transformation. Expressions for the steady-state balanced stator and rotor currents have been set forth previously, (4.4-8), (4.4-9), (5.2-4), and (5.2-5). Since the waveform is the same for the voltages, currents, and flux linkages, we can write

$$F_{as} = \sqrt{2}F_s \cos[\omega_e t + \theta_{esf}(0)]$$
$$= \text{Re}[\sqrt{2}\tilde{F}_{as}e^{j\omega_e t}] \tag{5.6-1}$$

$$F_{bs} = \sqrt{2}F_s \sin[\omega_e t + \theta_{esf}(0)]$$
$$= \text{Re}[\sqrt{2}\tilde{F}_{bs}e^{j\omega_e t}] \tag{5.6-2}$$

$$F'_{ar} = \sqrt{2}F'_r \cos[(\omega_e - \omega_r)t + \theta_{erf}(0)]$$
$$= \text{Re}[\sqrt{2}\,\hat{F}'_{ar}e^{j(\omega_e - \omega_r)t}] \tag{5.6-3}$$

$$F'_{br} = \sqrt{2}F'_r \sin[(\omega_e - \omega_r)t + \theta_{erf}(0)]$$
$$= \text{Re}[\sqrt{2}\tilde{F}'_{br}e^{j(\omega_e - \omega_r)t}] \tag{5.6-4}$$

where uppercase F is used to denote steady-state instantaneous variables and F_s and F'_r are rms values. Also $\theta_{esf}(0)$ is the phase angle of F_{as} at time zero and $\theta_{erf}(0)$ is the phase angle of F'_{ar} at time zero (Appendix B).

From (5.5-1), $f^s_{qs} = f_{as}$ and $f^s_{ds} = -f_{bs}$ or $F^s_{qs} = F_{as}$ and $F^s_{ds} = -F_{bs}$, hence

$$\tilde{F}^s_{qs} = \tilde{F}_{as} = F_s e^{j\theta_{esf}(0)} \tag{5.6-5}$$

$$\tilde{F}^s_{ds} = -\tilde{F}_{bs} = -(-jF_s e^{j\theta_{esf}(0)}) \tag{5.6-6}$$

The above expressions for $\tilde{F}_{as}$ and $\tilde{F}_{bs}$ in terms of F_s and $\theta_{esf}(0)$ follow from (5.6-1) and (5.6-2), where we see that $\tilde{F}_{as} = j\tilde{F}_{bs}$. Hence,

$$\tilde{F}^s_{qs} = -j\tilde{F}^s_{ds} \tag{5.6-7}$$

Substituting (5.6-3) and (5.6-4) into (5.5-4) with $\theta_r(0) = 0$ transforms the rotor variables to the stationary reference frame:

$$F'^s_{qr} = \sqrt{2}F'_r \cos[\omega_e t + \theta_{erf}(0)]$$

$$= \text{Re}[\sqrt{2}\tilde{F}'_{ar}e^{j\omega_e t}] \tag{5.6-8}$$

$$F'^s_{dr} = -\sqrt{2}F'_r \sin[\omega_e t + \theta_{erf}(0)]$$

$$= \text{Re}[-\sqrt{2}\tilde{F}'_{br}e^{j\omega_e t}] \tag{5.6-9}$$

We see that F'^s_{qr} and F'^s_{dr} vary at ω_e for balanced steady-state conditions. Also, if we compare (5.6-8) and (5.6-9) with (5.6-3) and (5.6-4), we see that

$$\boxed{\tilde{F}'^s_{qr} = \tilde{F}'_{ar}} \tag{5.6-10}$$

$$\boxed{\tilde{F}'^s_{dr} = -\tilde{F}'_{br}} \tag{5.6-11}$$

From (5.6-3) and (5.6-4) we see that $\tilde{F}'_{ar} = j\tilde{F}'_{br}$. It follows that

$$\boxed{\tilde{F}'^s_{qr} = -j\tilde{F}'^s_{dr}} \tag{5.6-12}$$

We must be careful here. We are aware that F^s_{qs}, F^s_{ds}, F'^s_{qr}, and F'^s_{dr} all vary at ω_e during steady-state operation; however, (5.6-10) and (5.6-11), which equate the phasors representing F'_{ar} and F'_{br} to the phasors representing F'^s_{qr} and F'^s_{dr}, respectively, could lead us to assume that the frequency of the $\overset{s}{q}r$ and $\overset{s}{d}r$ variables is the same as the frequency of the ar and br variables. This is not the case; only the phasors, which are actually nothing more than complex numbers, are equal. In other words, the same phasor represents F'_{ar} and F'^s_{qr} (F'_{br} and $-F'^s_{dr}$). However, instantaneously F'_{ar} and F'^s_{qr} (F'_{br} and F'^s_{dr}) vary at different frequencies, in particular, $\omega_e - \omega_r$ and ω_e, respectively. Figure 5.6-1 helps to clarify these relationships.

The steady-state voltage equations may be obtained from (5.5-32) by writing all variables as phasors and replacing the operator p with $j\omega_e$ since all steady-state variables vary at a frequency corresponding to ω_e. In particular,

$$\begin{bmatrix} \tilde{V}^s_{qs} \\ \tilde{V}^s_{ds} \\ \tilde{V}'^s_{qr} \\ \tilde{V}'^s_{dr} \end{bmatrix} = \begin{bmatrix} r_s + j\omega_e L_{ss} & 0 & j\omega_e L_{ms} & 0 \\ 0 & r_s + j\omega_e L_{ss} & 0 & j\omega_e L_{ms} \\ j\omega_e L_{ms} & -\omega_r L_{ms} & r'_r + j\omega_e L'_{rr} & -\omega_r L'_{rr} \\ \omega_r L_{ms} & j\omega_e L_{ms} & \omega_r L'_{rr} & r'_r + j\omega_e L'_{rr} \end{bmatrix} \begin{bmatrix} \tilde{I}^s_{qs} \\ \tilde{I}^s_{ds} \\ \tilde{I}'^s_{qr} \\ \tilde{I}'^s_{dr} \end{bmatrix}$$

$$\tag{5.6-13}$$

Since $\tilde{F}^s_{qs} = -j\tilde{F}^s_{ds}$, and $\tilde{F}'^s_{qr} = -j\tilde{F}'^s_{dr}$, the four equations of (5.6-13) are not independent. Hence, we can use either the $\overset{s}{q}s$ and $\overset{s}{q}r$ voltage equations or the $\overset{s}{d}s$ and $\overset{s}{d}r$ voltage equations. We shall use the $\overset{s}{q}s$ and $\overset{s}{q}r$ voltage equations.

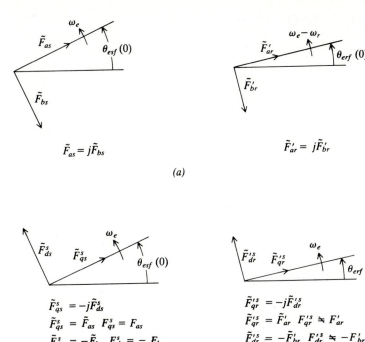

$$\tilde{F}_{as} = j\tilde{F}_{bs} \qquad\qquad \tilde{F}'_{ar} = j\tilde{F}'_{br}$$

(a)

$$\tilde{F}^s_{qs} = -j\tilde{F}^s_{ds} \qquad\qquad \tilde{F}'^s_{qr} = -j\tilde{F}'^s_{dr}$$
$$\tilde{F}^s_{qs} = \tilde{F}_{as} \quad F^s_{qs} = F_{as} \qquad \tilde{F}'^s_{qr} = \tilde{F}'_{ar} \quad F'^s_{qr} \approx F'_{ar}$$
$$\tilde{F}^s_{ds} = -\tilde{F}_{bs} \quad F^s_{ds} = -F_{bs} \qquad \tilde{F}'^s_{dr} = -\tilde{F}'_{br} \quad F'^s_{dr} \approx -F'_{br}$$

(b)

FIGURE 5.6-1
Phasor relationships between machine and substitute variables. (*a*) Machine variables; (*b*) substitute variables.

From (5.6-13),

$$\tilde{V}^s_{qs} = [r_s + j(X_{ls} + X_{ms})]\tilde{I}^s_{qs} + jX_{ms}\tilde{I}'^s_{qr} \tag{5.6-14}$$

$$\tilde{V}'^s_{qr} = jX_{ms}\tilde{I}^s_{qs} - \frac{\omega_r}{\omega_e}X_{ms}\tilde{I}^s_{ds} + [r'_r + j(X'_{lr} + X_{ms})]\tilde{I}'^s_{qr} - \frac{\omega_r}{\omega_e}(X'_{lr} + X_{ms})\tilde{I}'^s_{dr} \tag{5.6-15}$$

In the above equations we have let $L_{ss} = L_{ls} + L_{ms}$ and $L'_{rr} = L'_{lr} + L_{ms}$ and replaced all $\omega_e L$ with reactances X. Let us now incorporate the relationships between steady-state q and d variables given by (5.6-7) and (5.6-12), whereupon (5.6-15) becomes

$$\tilde{V}'^s_{qr} = jX_{ms}\frac{\omega_e - \omega_r}{\omega_e}\tilde{I}^s_{qs} + \left[r'_r + j(X'_{lr} + X_{ms})\frac{\omega_e - \omega_r}{\omega_e}\right]\tilde{I}'^s_{qr} \tag{5.6-16}$$

In (5.6-16), we see the quantity $(\omega_e - \omega_r)/\omega_e$, which is traditionally known as the slip and denoted by s, not to be confused with the Laplace operator. The slip is simply the normalized difference between the electrical angular velocity of the air gap mmf established by the stator currents and the electrical angular velocity of the rotor. However, one tends to look for the reasons behind the terms which have long been used in induction machine theory. The induction machine must run at a speed other than synchronous in order to develop a torque and, generally, it is operated as a motor. If we were to observe the rotor from the rotating magnetic field, we would see the rotor continuously "slipping back" relative to us due to the fact that $\omega_r < \omega_e$ for motor operation. The slip gives a measure to this slipping back.

Let us do two things at once: First we will divide (5.6-16) by the slip and next we will substitute $\tilde{F}_{as}$ phasors for $\tilde{F}_{qs}^s$ phasors and $\tilde{F}_{ar}'$ phasors for $\tilde{F}_{qr}'^s$ phasors. Remember we can do this in the case of $\tilde{F}_{ar}'$ and $\tilde{F}_{qr}'^s$, not because the F_{ar}' and $F_{qr}'^s$ variables have the same frequency (they do not, as we have said several times) but because they are equal as complex numbers. Taking these two steps, (5.6-14) and (5.6-16) may be written as

$$\tilde{V}_{as} = (r_s + jX_{ls})\tilde{I}_{as} + jX_{ms}(\tilde{I}_{as} + \tilde{I}_{ar}')$$ (5.6-17)

$$\frac{\tilde{V}_{ar}'}{s} = \left(\frac{r_r'}{s} + jX_{lr}'\right)\tilde{I}_{ar}' + jX_{ms}(\tilde{I}_{as} + \tilde{I}_{ar}')$$ (5.6-18)

where

$$s = \frac{\omega_e - \omega_r}{\omega_e}$$ (5.6-19)

$$X_{ls} + X_{ms} = \omega_e L_{ss} = \omega_e L_{ls} + \omega_e L_{ms}$$ (5.6-20)

$$X_{lr}' + X_{ms} = \omega_e L_{rr}' = \omega_e L_{lr}' + \omega_e L_{ms}$$ (5.6-21)

Equations (5.6-17) and (5.6-18) suggest the equivalent circuit shown in Fig. 5.6-2. This is the standard equivalent circuit used to calculate the steady-state performance of symmetrical two-phase induction machines. We will find that,

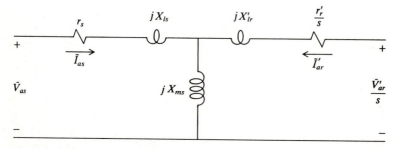

FIGURE 5.6-2
Equivalent circuit of a two-phase symmetrical induction machine for balanced steady-state operation.

with a slight modification (X_{ms} becomes $\frac{3}{2}X_{ms}$), this circuit may also be used to calculate the steady-state performance of a three-phase symmetrical induction machine.

An expression for the steady-state electromagnetic torque may be obtained by substituting the expressions for the instantaneous currents into (5.5-33). In particular,

$$T_e = \frac{P}{2} L_{ms}\{-\sqrt{2}I_s \cos[\omega_e t + \theta_{esi}(0)]\sqrt{2}I_r' \sin[\omega_e t + \theta_{eri}(0)]$$

$$+ \sqrt{2}I_s \sin[\omega_e t + \theta_{esi}(0)]\sqrt{2}I_r' \cos[\omega_e t + \theta_{eri}(0)]\} \qquad (5.6\text{-}22)$$

This expression may be reduced to

$$\boxed{T_e = 2\left(\frac{P}{2}\right) L_{ms} \operatorname{Re}[j\tilde{I}_{as}^* \tilde{I}_{ar}']} \qquad (5.6\text{-}23)$$

where $\tilde{I}_{as}^*$ is the conjugate of $\tilde{I}_{as}$. Obtaining (5.6-23) from (5.6-22) is given as a problem at the end of this chapter.

The balanced steady-state torque-speed or torque-slip characteristic of a single-fed induction machine warrants discussion. The vast majority of induction machines in use today are single-fed, wherein electric power is transferred to or from the induction machine via the stator circuits with the rotor windings short-circuited. Moreover, the majority of the single-fed induction machines are of the squirrel-cage rotor type. In this type of rotor construction, copper or aluminum bars are uniformly distributed and embedded in a ferromagnetic material with all bars terminated in a common ring at each end of the rotor. It may at first appear that the mutual inductance between a uniformly distributed rotor winding and a sinusoidally distributed stator winding would include more than just the fundamental component. In most cases, however, a uniformly distributed winding is adequately described by its fundamental sinusoidal component.

For single-fed machines, $\tilde{V}_{ar}'$ is zero, whereupon (5.6-18) may be written as

$$\tilde{I}_{ar}' = -\frac{jX_{ms}}{r_r'/s + jX_{rr}'}\tilde{I}_{as} \qquad (5.6\text{-}24)$$

Substituting (5.6-24) into (5.6-23) yields the following expression for electromagnetic torque of a single-fed two-phase symmetrical induction machine during balanced operation:

$$T_e = \frac{2(P/2)(X_{ms}^2/\omega_e)(r_r'/s)|\tilde{I}_{as}|^2}{(r_r'/s)^2 + X_{rr}'^2} \qquad (5.6\text{-}25)$$

It is important to note from (5.6-25) that torque is positive (motor action) when slip is positive which occurs when $\omega_r < \omega_e$, and negative (generator action) when the slip is negative which occurs when $\omega_r > \omega_e$, and zero when the slip is zero ($\omega_r = \omega_e$). (Prove the last of these three statements.) In other words, the single-fed induction machine develops torque at all speeds except at synchronous speed.

With $\tilde{V}'_{ar} = 0$, the input impedance of the equivalent circuit shown in Fig. 5.6-2 is

$$Z = \frac{r_s r'_r/s + (X^2_{ms} - X_{ss}X'_{rr} + j[(r'_r/s)X_{ss} + r_s X'_{rr}]}{r'_r/s + jX'_{rr}} \tag{5.6-26}$$

Now $|\tilde{I}_{as}|^2$ is I_s^2 and

$$I_s = \frac{|\tilde{V}_{as}|}{|Z|} \tag{5.6-27}$$

Hence, the expression for the steady-state electromagnetic torque for a single-fed two-phase symmetrical induction machine becomes

$$T_e = \frac{2(P/2)(X^2_{ms}/\omega_e)r'_r s|\tilde{V}_{as}|^2}{[r_s r'_r + s(X^2_{ms} - X_{ss}X'_{rr})]^2 + (r'_r X_{ss} + sr_s X'_{rr})^2} \tag{5.6-28}$$

Thus, for a given set of parameters and source frequency ω_e, the steady-state torque varies as the square of the magnitude of the applied voltages. The steady-state torque-speed characteristics typical of many single-fed two-phase induction machines are shown in Fig. 5.6-3. The parameters of the machine are often selected so that, for rated frequency operation, the maximum torque occurs near synchronous (80 to 90 percent of synchronous speed). Generally, the maximum torque is two or three times the rated torque of the machine.

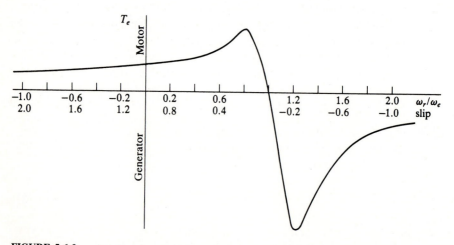

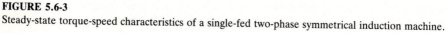

FIGURE 5.6-3
Steady-state torque-speed characteristics of a single-fed two-phase symmetrical induction machine.

Although we are considering a two-phase machine, the general shape of the torque-speed characteristic is similar for multiphase induction machines.

Let us take a moment to consider stable, steady-state operation of an induction machine. Consider (5.4-5); for steady-state operation $d\omega_r/dt$ is zero since the speed is constant. Thus $T_e = T_L$. From Fig. 5.6-3, it would appear that, for a given load torque T_L, there could be two operating points on the T_e versus ω_r plot. When we calculate the steady-state torque we are assuming, in the calculation, that the speed is constant without regard to whether or not it is a stable operating point. Let us consider the T_e versus ω_r plot shown in Fig. 5.6-4. Since induction machines are used primarily as motors, the characteristic shown in Fig. 5.6-4 is the range of speeds over which motors operate. For the load torque T_L, shown in Fig. 5.6-4, there appear to be two operating points, 1 and 1'. If we assume stable steady-state operation occurs at either point 1 or 1', then, if the system is displaced from this operating point, there will be a torque established to return the system to this operating point. This, of course, is the same procedure used to determine the stable operating points of the relay in Chap. 2.

Consider operation at point 1. If the speed ω_r were to increase ever so slightly, T_e would become less than T_L. In other words, the load torque (T_L) requirements are larger than the electromagnetic torque (T_e) which can be developed by the machine. Consequently, the machine will slow down due to the fact that $T_L > T_e$. Hence, the machine will return to operation at point 1. If, instead, the speed decreased from operating point 1, T_e is greater than T_L. Hence, the machine will accelerate back to point 1. It follows that point 1 is a stable steady-state operating point.

Now consider operation at point 1'. An increase in ω_r causes T_e to become greater than T_L, hence, the machine will accelerate and move away from point 1'. In fact, the machine will accelerate to point 1 where it will operate stably. If ω_r decreases slightly from point 1', T_e becomes less than T_L. The machine decelerates to stall and, if T_L remains applied to the shaft, the rotor will reverse direction and accelerate forever in the opposite direction. Point 1' is an unstable operating point. It follows that stable operation, for either motor or generation operation, occurs on the negative slope portion of the T_e versus ω_r characteristic.

Looking at Fig. 5.6-4, one may wonder how operation could be established at point 1 since T_L is larger than T_e at stall and the machine could not

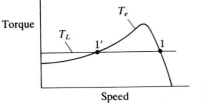

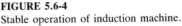

FIGURE 5.6-4
Stable operation of induction machine.

develop a large enough starting torque to accelerate the machine up to operation at point 1. In most cases, the load torque is a function of ω_r, say $T_L = K\omega_r^2$, for example. In these cases, the machine can develop sufficient starting torque and, if T_L and T_e match on the negative slope portion, stable operation will occur. If, on the other hand, T_L is constant and greater than T_e at $\omega_r = 0$, we have at least three choices: (1) increase the stator voltage; (2) increase the rotor resistance; or (3) use a different machine. Increasing the rotor resistance to increase the starting torque is something that we have not yet discussed. We will now.

An expression for the slip at maximum torque may be obtained by taking the derivative of (5.6-28) with respect to slip and setting the result equal to zero. In particular,

$$s_m = r_r'G \tag{5.6-29}$$

where

$$G = \pm\sqrt{\frac{r_s^2 + X_{ss}^2}{(X_{ms}^2 - X_{ss}X_{rr}')^2 + r_s^2 X_{rr}'^2}} \tag{5.6-30}$$

Two values of slip at maximum torque, s_m, are obtained, one for motor action and one for generator action. It is important to note that G is not a function of r_r'; thus, the slip at maximum torque, (5.6-29), is directly proportional to r_r'. Consequently, since all other machine parameters are constant, the speed at which maximum steady-state torque occurs may be varied by inserting external rotor resistance. This feature is often used when starting large motors which have coil-wound rotor windings with slip rings. In this application, balanced external rotor resistances are placed across the terminals of the rotor windings so that maximum torque occurs near stall. As the machine speeds up, the external resistors are short-circuited. On the other hand, two-phase induction machines used as servomotors are designed with high-resistance rotor windings. Since they are used in positioning devices, the rotor is generally stationary and rotates only when an error signal causes the applied voltage(s) to increase from zero. Consequently, maximum torque must be available at or near stall to provide fast response of the position-followup system.

It may at first appear that the magnitude of the maximum torque would be influenced by r_r'. However, if (5.6-29) is substituted into (5.6-28), the maximum torque may be expressed as

$$T_{e,max} = \frac{2(P/2)(X_{ms}^2/\omega_e)G|\tilde{V}_{as}|^2}{[r_s + G(X_{ms}^2 - X_{ss}X_{rr}')]^2 + (X_{ss} + Gr_s X_{rr}')^2} \tag{5.6-31}$$

Equation (5.6-31) is independent of r_r'. Thus, the maximum torque remains constant if only r_r' is varied; however, the slip at which maximum torque is produced varies in accordance with (5.6-29). Figure 5.6-5 illustrates the effect of changing r_r'. Therein $r_{r3}' > r_{r2}'$ and $r_{r2}' > r_{r1}'$.

In variable-frequency drive systems, the operating speed of the electromechanical device (reluctance, synchronous, or induction machine) is con-

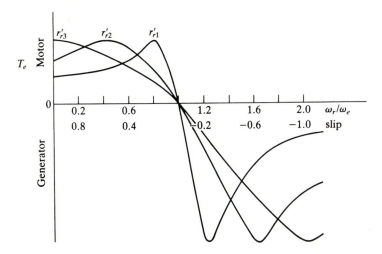

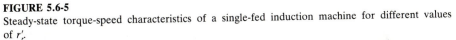

FIGURE 5.6-5
Steady-state torque-speed characteristics of a single-fed induction machine for different values of r'_r.

trolled by changing the frequency of the applied voltages by either an inverter (solid-state dc-to-ac converter) or a cycloconverter (ac frequency changer) arrangement. The phasor voltage equations are applicable regardless of the frequency of operation. It is only necessary to keep in mind that the reactances given in the steady-state equivalent circuit, Fig. 5.6-2, are defined as the product of ω_e and the inductances. As the frequency is decreased, the time rate of change of the steady-state variables is decreased proportionally. Thus, the reactances decrease linearly with frequency. If the amplitude of the applied voltages is maintained at the rated value, the currents will become excessive. To prevent large currents, the magnitude of the stator voltages is decreased as the frequency is decreased. In many applications, the voltage magnitude is reduced linearly with frequency until a low frequency is reached, whereupon the decrease in voltage is programmed in a manner to compensate for the effects of the stator resistance.

The influence of frequency upon the steady-state torque-speed characteristics is illustrated in Fig. 5.6-6. These characteristics are for a linear relationship between the magnitude of the applied voltages and frequency. This machine is designed to operate at $\omega_e = \omega_b$, where ω_b corresponds to the rated frequency. Rated voltage is applied at rated frequency, that is, when $\omega_e = \omega_b$, $|\tilde{V}_{as}| = V_B$, where V_B is the base or rated voltage. Since the reactances $(\omega_e L)$ decrease with frequency, the voltage is reduced as frequency is reduced to avoid large stator currents. The maximum torque is reduced markedly at $\omega_e/\omega_b = 0.1$. At this frequency, the voltage would probably be increased somewhat so as to obtain a higher torque. Perhaps a voltage of, say, $0.15V_B$ or $0.2V_B$ would be used rather than $0.1V_B$. Saturation of the stator or rotor steel

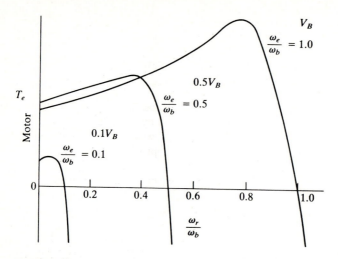

FIGURE 5.6-6
Steady-state torque-speed characteristics of a single-fed symmetrical induction machine for different operating frequencies.

may, however, cause the stator currents to be excessive at this higher voltage. These practical considerations of variable-frequency drives are of major importance but well beyond the scope of this text.

Example 5B. The parameters for the equivalent circuit shown in Fig. 5.6-2 may be calculated by using electromagnetic field theory or determined from tests. The tests generally performed are a dc test, a no-load test and a blocked-rotor test. The following test data are given for a 3-hp four-pole 110-V (rms) two-phase 60-Hz induction machine, where all ac voltages and currents are in rms values:

DC test	No-load test	Blocked-rotor test
$V_{dc} = 6.9$ V	$V_{nl} = 110$ V	$V_{br} = 23.5$ V
$I_{dc} = 13.0$ A	$I_{nl} = 3.86$ A	$I_{br} = 16.1$ A
	$P_{nl} = 134$ W	$P_{br} = 469$ W
	$f = 60$ Hz	$f = 15$ Hz

During the dc test, a dc voltage is applied to one phase while the machine is at standstill. Thus,

$$r_s = \frac{V_{dc}}{I_{dc}} = \frac{6.9}{13} = 0.531 \ \Omega \tag{5B-1}$$

The no-load test, which is analogous to the transformer open-circuit test, is performed with balanced two-phase 60-Hz voltages applied to the stator windings without mechanical load on the rotor (no load). The total power input during this test is the sum of the stator ohmic losses per phase, the core losses due to hysteresis and eddy currents, and rotational losses due to friction and windage.

The total stator ohmic losses (two phases) are

$$P_r = 2I_{nl}^2 r_s = 2(3.86)^2(0.531) = 15.8 \text{ W} \tag{5B-2}$$

Therefore, the power loss due to friction, windage, and core losses is

$$P_{fWC} = P_{nl} - P_r = 134 - 15.8 = 118.2 \text{ W} \tag{5B-3}$$

In the equivalent circuit shown in Fig. 5.6-2, the core loss is neglected. It is generally small and, in most cases, little error is introduced by neglecting it. It can be taken into account by placing a resistor in shunt with the magnetizing reactance X_{ms}. The friction and windage losses may be approximated with B_m in (5.4-5).

It is noted from the no-load test data that the power factor is very small since the total apparent power input to the motor is

$$|S_{nl}| = 2V_{nl}I_{nl} = 2(110)(3.86) = 849.2 \text{ VA} \tag{5B-4}$$

Therefore, the no-load impedance is highly inductive, and its magnitude is assumed to be the sum of the stator leakage reactance and the magnetizing reactance since the rotor speed is essentially synchronous, whereupon r_r'/s is much larger than X_{ms} in Fig. 5.6-2. Thus,

$$X_{ls} + X_{ms} \cong \frac{V_{nl}}{I_{nl}} = \frac{110}{3.86} = 28.5 \, \Omega \tag{5B-5}$$

During the blocked-rotor test, which is analogous to the transformer short-circuit test, the rotor is locked by some mechanical means and balanced two-phase stator voltages are applied. The frequency of the applied voltages is often less than rated in order to obtain a representative value of r_r', since during normal operation the frequency of the rotor currents is low and the rotor resistance of some induction machines vary considerably with frequency. During stall ($s = 1$), the rotor impedance $r_r'/s + jX_{lr}'$ is much smaller in magnitude than X_{ms}, whereupon the current flowing in the magnetizing reactance may be neglected in these calculations. Hence, the total power input to the motor during the blocked-rotor test is

$$P_{br} = 2I_{br}^2(r_s + r_r') \tag{5B-6}$$

From which

$$r_r' = \frac{P_{br}}{2I_{br}^2} - r_s = \frac{469}{(2)(16.1)^2} - 0.531 = 0.374 \, \Omega \tag{5B-7}$$

The magnitude of the blocked-rotor input impedance is

$$|Z_{br}| = \frac{V_{br}}{I_{br}} = \frac{23.5}{16.1} = 1.46 \, \Omega \tag{5B-8}$$

Thus,

$$\left| (r_s + r_r') + j\tfrac{15}{60}(X_{ls} + X_{lr}') \right| = 1.46 \, \Omega \tag{5B-9}$$

From which

$$[\tfrac{15}{60}(X_{ls} + X_{lr}')]^2 = (1.46)^2 - (r_s + r_r')^2$$
$$= (1.46)^2 - (0.531 + 0.374)^2 = 1.31 \, \Omega^2 \tag{5B-10}$$

Thus,

$$X_{ls} + X_{lr}' = 4.58 \, \Omega \tag{5B-11}$$

Generally, X_{ls} and X'_{lr} are assumed equal; however, in some types of induction machines a different ratio is suggested. We will assume $X_{ls} = X'_{lr}$, whereupon we have determined the machine parameters. In particular, for $\omega_e = 377$ rad/s, the parameters are $r_s = 0.531\,\Omega$, $X_{ls} = 2.29\,\Omega$, $X_{ms} = 26.2\,\Omega$, $r'_r = 0.374\,\Omega$, $X'_{lr} = 2.29\,\Omega$.

Example 5C. A four-pole 110-V (rms) 28-A 7.5-hp two-phase induction motor has the following parameters: $r_s = 0.3\,\Omega$, $L_{ls} = 0.0015$ H, $L_{ms} = 0.035$ H, $r'_r = 0.15\,\Omega$, $L'_{lr} = 0.0007$ H. The machine is supplied from a 110-V 60-Hz source. Calculate the starting torque and starting current.

It would be necessary to use a computer to solve for the starting current and torque if the electrical and mechanical transients are to be considered. However, an approximation of the actual starting characteristics may be obtained from a constant-speed steady-state analysis. For this purpose, it is assumed that the speed is fixed at zero and the electric system has established steady-state operation.

$$X_{ss} = \omega_e(L_{ls} + L_{ms}) = 377(0.0015 + 0.035) = 13.76\,\Omega \tag{5C-1}$$

$$X'_{rr} = \omega_e(L'_{lr} + L_{ms}) = 377(0.0007 + 0.035) = 13.46\,\Omega \tag{5C-2}$$

$$X_{ms} = \omega_e L_{ms} = 377(0.035) = 13.2\,\Omega \tag{5C-3}$$

The steady-state torque with $\omega_r = 0$ ($s = 1$) may be calculated from (5.6-28).

$$T_e = \frac{2(P/2)(X_{ms}^2/\omega_e)r'_s s|\tilde{V}_{as}|^2}{[r_s r'_r + s(X_{ms}^2 - X_{ss}X'_{rr})]^2 + (r'_r X_{ss} + s r_s X'_{rr})^2}$$

$$= \frac{2(\tfrac{4}{2})(13.2^2/377)(0.15)(1)(110)^2}{\{(0.3)(0.15) + (1)[13.2^2 - (13.76)(13.46)]\}^2 + [(0.15)(13.76) + (1)(0.3)(13.46)]^2}$$

$$= 21.4\ \text{N}\cdot\text{m} \tag{5C-4}$$

Since $s = 1$, the rotor impedance in parallel with X_{ms} is much smaller than X_{ms}. Thus, for this mode of operation the input impedance is approximately

$$Z = (r_s + r'_r) + j(X_{ls} + X'_{lr})$$

$$= (0.30 + 0.15) + j377(0.0015 + 0.0007)$$

$$= 0.45 + j0.83\,\Omega \tag{5C-5}$$

Assuming $\tilde{V}_{as}$ as the reference phasor, then

$$\tilde{I}_{as} = \frac{\tilde{V}_{as}}{Z} = \frac{110\underline{/0^\circ}}{0.944\underline{/61.5^\circ}} = 117\underline{/-61.5^\circ}\ \text{A} \tag{5C-6}$$

The stall or starting current is over four times larger than the rated current. In some large machines, the starting current with rated voltage applied may be ten times the rated current. This high value of current may cause overheating and damage to the windings. Consequently, reduced voltage is applied to many large machines during the starting period, and rated voltage is not applied until the machine has accelerated to near rated speed. This is generally accomplished by closed-transition transformer switching wherein the voltage is increased by switching from a lower to a higher transformer tap without opening the stator circuits.

SP5.6-1. The rotor speed ω_{rm} of a six-pole two-phase induction motor is $0.3\omega_e$. Express (a) I'_{ar} (b) I'^s_{qr}, and (c) $\tilde{I}'_{ar}$ for balanced steady-state operation with $\tilde{I}'^s_{qr} = I'_r \underline{/30°}$. [(a) $I'_{ar} = \sqrt{2}I'_r \cos(0.1\omega_e t + 30°)$; (b) $I'^s_{qr} = \sqrt{2}I'_r \cos(\omega_e t + 30°)$; (c) $\tilde{I}'_{ar} = \tilde{I}'^s_{qr}$]

SP5.6-2. Neglecting the current flowing in X_{ms} is generally an acceptable approximation when calculating the machine parameters from the blocked-rotor test (Example 5B). However, this approximation is not valid when calculating the blocked-rotor torque. Use (5.6-23) to show that T_e is zero regardless of the rotor speed if the current flowing in X_{ms} is assumed to be negligibly small. [Let $\tilde{I}_{as} = a + jb$]

SP5.6-3. Neglect the core losses and assume that the friction and windage losses P_{fWC}, calculated in Example 5B, are to be represented by $B_m \omega_{rm}$, with B_m selected so that a load equivalent to 118.2 W occurs at $\omega_r = 0.9 \omega_e$. Determine B_m. [$B_m = 4.11 \times 10^{-3}$ N·m·s/rad]

SP5.6-4. The parameters of a 60-Hz four-pole two-phase servomotor are $r_s = r'_r = 20\,\Omega$, $L_{ls} = L'_{lr} = 25$ mH, and $L_{ms} = 0.3$ H. Calculate $\tilde{I}_{as}$ for (a) no-load conditions and (b) blocked-rotor conditions ($\omega_r = 0$), with $\tilde{V}_{as} = 115\underline{/0°}$. Approximate the blocked-rotor current. [(a) $\tilde{I}_{as} = 0.927\underline{/-80.75°}$ A; (b) $\tilde{I}_{as} = 2.6\underline{/-25.2°}$ A]

5.7 DYNAMIC AND STEADY-STATE PERFORMANCE—MACHINE VARIABLES

It is instructive to observe the machine variables during transient and steady-state operation. For this purpose, the nonlinear differential equations which describe the induction machine were programmed on a computer. Two induction machines are considered; a 5-hp general-purpose machine with characteristics typical of many two- and three-phase squirrel-cage industrial-type induction motors and a $\frac{1}{10}$-hp servomotor which is a two-phase induction motor with high-resistance rotor windings to produce a relatively large starting torque. The majority of the performance characteristics given in this and the following section are for the single-fed two-pole two-phase 5-hp 110-V (rms) 60-Hz induction machine with the following parameters: $r_s = 0.295\,\Omega$, $L_{ls} = 0.944$ mH, $L_{ms} = 35.15$ mH, $r'_r = 0.201\,\Omega$, $L'_{lr} = 0.944$ mH. The inertia of the rotor and connected mechanical load is $J = 0.026$ kg·m^2.

The servomotor is a single-fed four-pole two-phase $\frac{1}{10}$-hp 115-V(rms) 60-Hz induction motor with the following parameters: $r_s = 24.5\,\Omega$, $L_{ls} = 27.06$ mH, $L_{ms} = 273.7$ mH, $r'_r = 23\,\Omega$, $L'_{lr} = 27.06$ mH. The inertia of the rotor and connected mechanical load is $J = 1 \times 10^{-3}$ kg·m^2. This device is shown in Fig. 5.2-3.

Free Acceleration from Stall

The free-acceleration characteristics are depicted in Figs. 5.7-1 and 5.7-2 for the 5-hp machine and in Figs. 5.7-3 and 5.7-4 for the $\frac{1}{10}$-hp servomotor. The variables v_{as}, i_{as}, v_{bs}, i_{bs}, i'_{ar}, i'_{br}, T_e, and rotor speed (ω_{rm}) are plotted in Figs. 5.7-1 and 5.7-3, whereas T_e versus rotor speed is shown in Figs. 5.7-2 and 5.7-4. At $t = 0$, rated voltages are applied of the form $v_{as} = \sqrt{2}V_s \cos 377t$ and

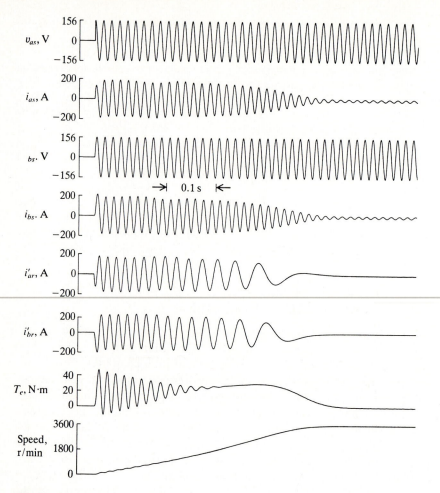

FIGURE 5.7-1

Free-acceleration characteristics of a two-pole two-phase 5-hp induction motor—machine variables.

$v_{bs} = \sqrt{2}V_s \sin 377t$, where $V_s = 110$ V for the 5-hp machine and $V_s = 115$ V for the $\frac{1}{10}$-hp servomotor. The machines accelerate from stall with zero load torque and, since friction and windage losses are not taken into account, the simulated machine accelerates to synchronous speed. In practice, friction and windage losses would exist and the machines would not reach synchronous speed. Instead, they will operate at a speed slightly less than synchronous, developing an electromagnetic torque T_e sufficient to satisfy the small torque load due to friction and windage. For the two-pole machine (5 hp), synchronous speed is 3600 r/min, where $\omega_{rm} = 377$ rad/s. For the four-pole servomotor, synchronous speed is 1800 r/min, where $\omega_{rm} = 188.5$ rad/s. In both cases, ω_r, the electrical angular velocity of the rotors, is 377 rad/s at synchronous speed.

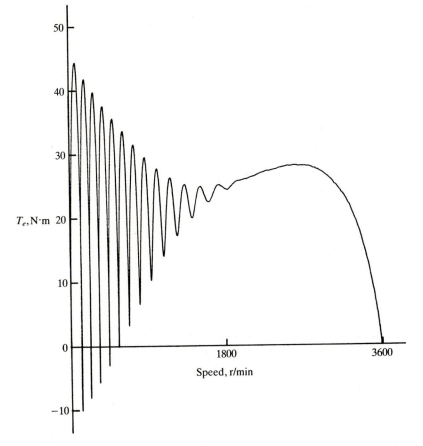

FIGURE 5.7-2
Torque versus speed during free acceleration shown in Fig. 5.7-1.

We will find that rated current (rms) for the 5-hp machine is on the order of 20 A and 1.5 A for the $\frac{1}{10}$-hp servomotor. The starting current for the 5-hp machine (Fig. 5.7-1) is approximately 120 A (rms) or on the order of six times the rated current. This ratio is typical of industrial-type induction motors which, depending upon the load, may require reduced-voltage starting, as mentioned in Example 5C. The starting current for the servomotor is about 2 A (rms) or less than twice the rated current. These devices are designed to withstand full-voltage starting since they are often used in position- or motion-control systems requiring rapid, short-duration acceleration from stall and, depending upon the application, may seldom reach rated speed.

The transient offset of the stator and rotor currents, characteristic of rL circuits, is evident in Figs. 5.7-1 and 5.7-3. This transient or dc offset in the stator and rotor currents give rise to the transient pulsation in the electromagnetic torque. Note that the pulsation in torque, which is the frequency

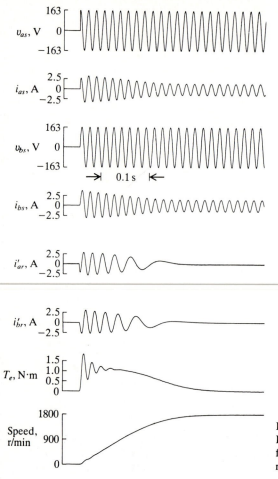

FIGURE 5.7-3

Free-acceleration characteristics of a four-pole two-phase $\frac{1}{10}$-hp servomotor—machine variables.

of the stator source voltages (60 Hz), disappears when the transient offsets in the stator and rotor currents disappear. Also apparent in the stator and rotor currents, especially in the case of the 5-hp machine, is the variation in the envelope of the currents during the transient period. This is caused by the interaction of the transient offset in the stator currents with the transient offset in the rotor currents and the fact that the stator and rotor circuits are in relative motion [1].

For small-horsepower induction machines, the steady-state torque-speed characteristic is essentially the average of the transient torque-speed curve during the time the transient pulsating torque exists. The inertia of the rotor and connected load is generally large enough to prevent this pulsating torque

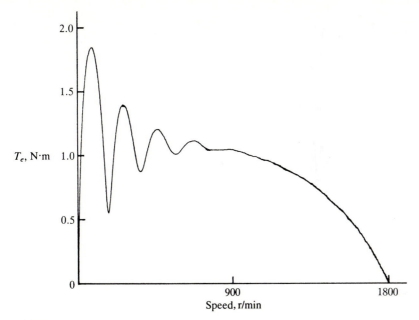

FIGURE 5.7-4
Torque versus speed during free acceleration shown in Fig. 5.7-3.

from causing significant variations in rotor speed. This is not the case, however, if the induction machine is operated with stator voltages of low frequencies as may occur in variable-frequency drive systems. Although we will not explore a detailed comparison between transient and steady-state torque-speed characteristics for rated frequency, it is worth mentioning in passing that the transient torque-speed characteristic of large-horsepower machines (larger than, say, 500 hp) may differ considerably from the steady-state characteristics particularly at rotor speeds above 60 to 80 percent of synchronous speed [1].

Acceleration from Stall with Load Torque

The acceleration characteristics of the 5-hp motor with a load torque are shown in Figs. 5.7-5 and 5.7-6. Here the load torque is $T_L = K\omega_{rm}^2$, where K is selected as $10(377)^{-2}\,\mathrm{N \cdot m \cdot s^2/rad^2}$. A load torque characteristic of this type is typical of a fan load. Rated torque load is approximately $10\,\mathrm{N \cdot m}$. The constant K is selected so that a load torque of $10\,\mathrm{N \cdot m}$ occurs at synchronous speed. In Fig. 5.7-6, the difference between the average value of the electromagnetic torque T_e and the load torque T_L is the torque that accelerates the rotor, commonly referred to as the accelerating torque. Acceleration occurs until $T_e = T_L$.

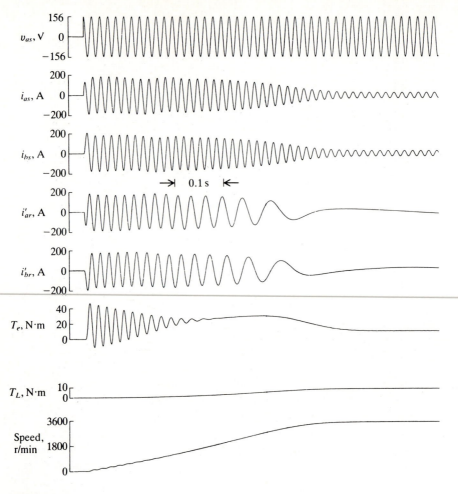

FIGURE 5.7-5
Acceleration from stall of a 5-hp induction machine with $T_L = K\omega_{rm}^2$, where $K = 10(377)^{-2}$ N·m·s²/rad².

The acceleration characteristics of the servomotor with a constant load torque are shown in Fig. 5.7-7. A servomotor is often started and operated for brief periods under a torque load which may be larger than rated. Rated torque of the servomotor is approximately 0.4 N·m. (How did we arrive at this value?) In Fig. 5.7-7, the load torque is held constant at 0.65 N·m or approximately 160 percent rated torque. The capability of a servomotor to accelerate rapidly with larger than rated torque is clearly illustrated in Fig. 5.7-7.

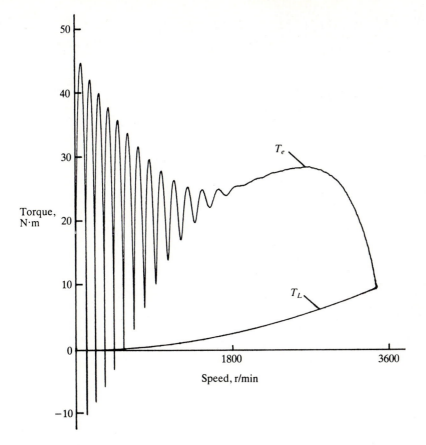

FIGURE 5.7-6
Torque versus speed during acceleration shown in Fig. 5.7-5.

Step Changes in Load Torque

The dynamic performance of the 5-hp motor during load torque changes are shown in Fig. 5.7-8. Initially the machine is operating with a constant load torque of $5 \, \text{N} \cdot \text{m}$. The load torque is stepped to $10 \, \text{N} \cdot \text{m}$ and the machine is allowed to reach steady-state operation whereupon the load torque is stepped back to $5 \, \text{N} \cdot \text{m}$. You should be able to calculate the rotor speed for $T_L = 5 \, \text{N} \cdot \text{m}$ and $T_L = 10 \, \text{N} \cdot \text{m}$ from the frequency of the rotor currents shown in Fig. 5.7-8. (For $T_L = 5 \, \text{N} \cdot \text{m}$, $\omega_{rm} \cong 370 \, \text{rad/s}$; $T_L = 10 \, \text{N} \cdot \text{m}$, $\omega_{rm} \cong 362 \, \text{rad/s}$.)

Step Changes in Stator Frequency

In Fig. 5.7-9, the 5-hp induction machine is operating in the steady state with $T_L = 10 \, \text{N} \cdot \text{m}$. The frequency of the stator voltages is stepped to 50 Hz, and, at

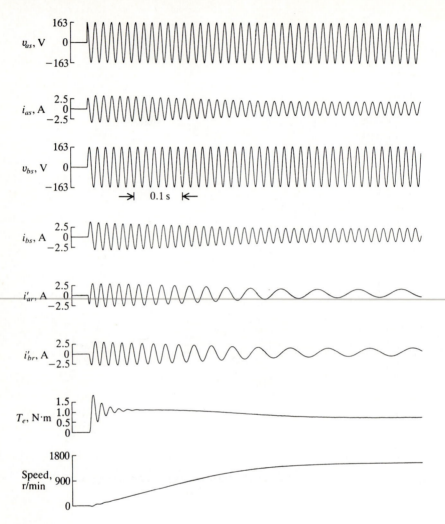

FIGURE 5.7-7
Acceleration of $\frac{1}{10}$-hp servomotor with $T_L = 0.65\,\text{N}\cdot\text{m}$.

the same time, the amplitude of the stator voltages is decreased to $(\frac{5}{6})(110)$ V. Once the system has reached steady-state operation, the frequency and amplitude of the stator voltages are stepped back to rated (60 Hz and 110 V). The load torque was held fixed at $10\,\text{N}\cdot\text{m}$ during this change in stator frequency. This type of operation is possible when the induction machine is supplied from a variable-frequency inverter where the frequency and amplitude of the fundamental component of the stator voltages may be rapidly changed by controlling the switching of the inverter which converts a dc voltage to a variable-frequency ac voltage.

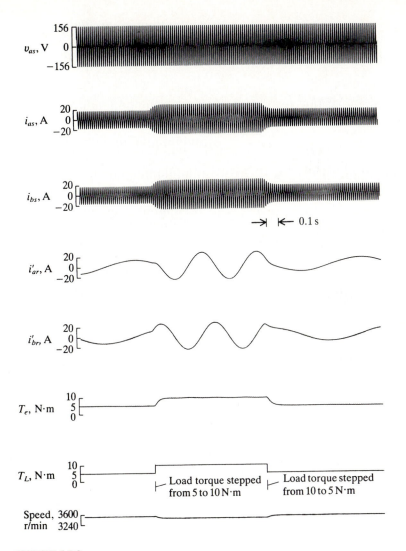

FIGURE 5.7-8
Step changes in load torque of a 5-hp induction motor.

The dynamic torque-speed characteristics during these step changes in frequency are shown in Fig. 5.7-10. Therein, the steady-state torque-speed characteristics for 50- and 60-Hz operations are also shown. Initially, operation is at point 1. When the frequency and voltage amplitude are changed, the instantaneous torque decreases, whereupon the rotor slows down and steady-state operation finally occurs at point 2. When the stator voltages are stepped back to rated conditions, the instantaneous torque increases, the rotor accelerates, and the original operating condition is reestablished.

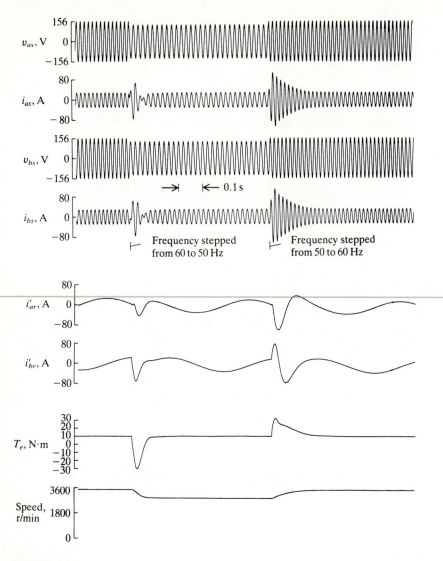

FIGURE 5.7-9
Step changes in the frequency of the stator voltages of a 5-hp induction motor.

SP5.7-1. From the plot of the rotor currents, approximate the final rotor speed of the servomotor depicted in Fig. 5.7-7. [$\omega_{rm} \cong 152$ rad/s]

SP5.7-2. Approximate the accelerating torque from Fig. 5.7-6 when the rotor speed is 280 rad/s. [$\cong 23$ N·m]

SP5.7-3. In going from point 1 to point 2 in Fig. 5.7-10, the induction machine acts as a generator supplying energy to the electric system. What is the source of this energy. [Rotor]

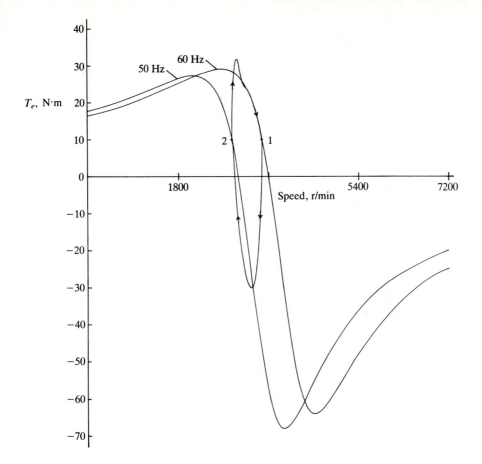

FIGURE 5.7-10
Torque versus speed for step changes in stator frequency shown in Fig. 5.7-9.

5.8 DYNAMIC AND STEADY-STATE PERFORMANCE—STATIONARY REFERENCE FRAME VARIABLES

The change of variables transforms the rotor variables to the stationary reference frame. In particular, v_{as} and i_{as} became v_{qs}^s and i_{qs}^s; v_{bs} and i_{bs} became $-v_{ds}^s$ and $-i_{ds}^s$; and i'_{ar} and i'_{br} are transformed by (5.5-4) to $i_{qr}^{'s}$ and $i_{dr}^{'s}$. Because of this change of variables, all machine variables are in a common frame of reference and, therefore, will vary at the same frequency during steady-state operation. To provide an instructive comparison between machine variables and stationary reference frame variables, Figs. 5.8-1, 5.8-2, and 5.8-3 show the stationary reference frame variables for the same mode of operation as depicted in Figs. 5.7-1, 5.7-8, and 5.7-9, respectively.

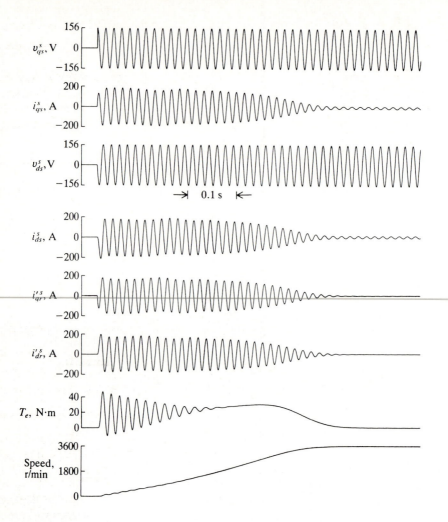

FIGURE 5.8-1
Same as Fig. 5.7-1—stationary reference frame variables.

SP5.8-1. Determine the frequency of $I_{qr}^{\prime s}$ and $I_{dr}^{\prime s}$ in Figs. 5.8-1 through 5.8-3. [60 Hz; 60 Hz; 60-50-60 Hz]

SP5.8-2. Approximate the power dissipated in the rotor resistors in Figs. 5.7-8 and 5.8-2 when $T_L = 10 \, \text{N} \cdot \text{m}$. [$P_r \cong 146 \, \text{W}$]

SP5.8-3. Approximate the peak value of $I_{ar}^{\prime}$ in Fig. 5.7-9 for 50-Hz operation and $I_{qr}^{\prime s}$ in Fig. 5.8-3 for the same mode of operation. [23 A]

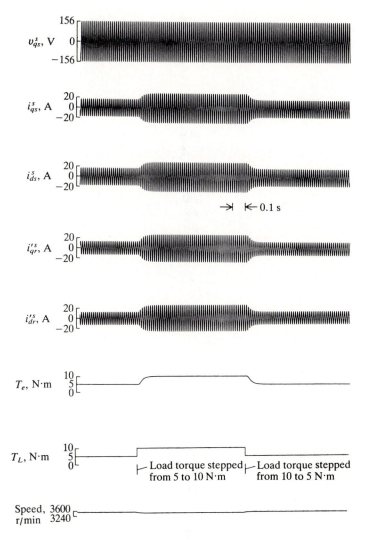

FIGURE 5.8-2

Same as Fig. 5.7-8—stationary reference frame variables.

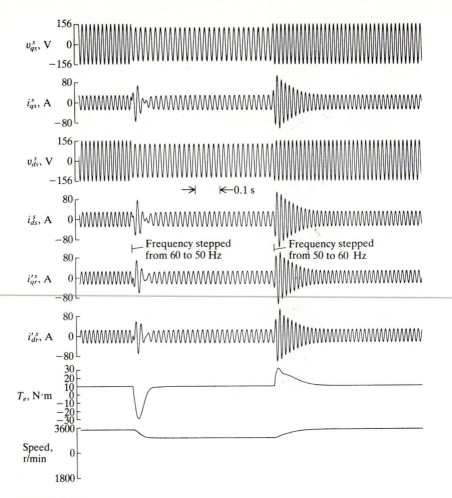

FIGURE 5.8-3
Same as Fig. 5.7-9—stationary reference frame variables.

5.9 THREE-PHASE INDUCTION MACHINE

A two-pole three-phase symmetrical induction machine is shown in Fig. 5.9-1. The three-phase machine has three identical, sinusoidally distributed stator (rotor) windings the magnetic axes of which are displaced 120° from each other. For the most part, the analysis of a three-phase machine is a direct extension of that of a two-phase machine. Therefore, it is unnecessary to repeat many of the details set forth thus far in this chapter. In fact, once we are acquainted with the two-phase symmetrical induction machine, the analysis of the three-phase symmetrical induction machine is straightforward and is more involved only because of the trigonometry necessary to deal with three rather than two variables. Perhaps the only aspect which may cause some study is the

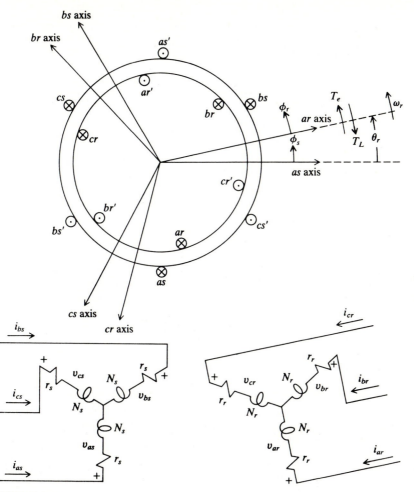

FIGURE 5.9-1
A two-pole three-phase symmetrical induction machine.

addition of the third fictitious variable in the transformation. However, these so-called zero variables or zero quantities play no role in the analysis of balanced operation of three-phase symmetrical induction machines. We will not consider unbalanced operation; a detailed analysis of unbalanced operation is found in [1].

Voltage Equations and Winding Inductances

For the three-phase symmetrical induction machine

$$v_{as} = r_s i_{as} + \frac{d\lambda_{as}}{dt} \tag{5.9-1}$$

$$v_{bs} = r_s i_{bs} + \frac{d\lambda_{bs}}{dt} \tag{5.9-2}$$

$$v_{cs} = r_s i_{cs} + \frac{d\lambda_{cs}}{dt} \tag{5.9-3}$$

$$v_{ar} = r_r i_{ar} + \frac{d\lambda_{ar}}{dt} \tag{5.9-4}$$

$$v_{br} = r_r i_{br} + \frac{d\lambda_{br}}{dt} \tag{5.9-5}$$

$$v_{cr} = r_r i_{cr} + \frac{d\lambda_{cr}}{dt} \tag{5.9-6}$$

where $r_s(r_r)$ is the resistance of the stator (rotor) windings. In matrix form,

$$\mathbf{v}_{abcs} = \mathbf{r}_s \mathbf{i}_{abcs} + p\boldsymbol{\lambda}_{abcs} \tag{5.9-7}$$

$$\mathbf{v}_{abcr} = \mathbf{r}_r \mathbf{i}_{abcr} + p\boldsymbol{\lambda}_{abcr} \tag{5.9-8}$$

where

$$(\mathbf{f}_{abcs})^T = [f_{as} \quad f_{bs} \quad f_{cs}] \tag{5.9-9}$$

$$(\mathbf{f}_{abcr})^T = [f_{ar} \quad f_{br} \quad f_{cr}] \tag{5.9-10}$$

and

$$\mathbf{r}_s = r_s \mathbf{I} \tag{5.9-11}$$

$$\mathbf{r}_r = r_r \mathbf{I} \tag{5.9-12}$$

As in the two-phase case, f can represent voltage, current, or flux linkage, the superscript T denotes the transpose, p is d/dt, and $\mathbf{I}$ a 3×3 identity matrix.

The flux linkage equations may be written as

$$\begin{bmatrix} \boldsymbol{\lambda}_{abcs} \\ \boldsymbol{\lambda}_{abcr} \end{bmatrix} = \begin{bmatrix} \mathbf{L}_s & \mathbf{L}_{sr} \\ (\mathbf{L}_{sr})^T & \mathbf{L}_r \end{bmatrix} \begin{bmatrix} \mathbf{i}_{abcs} \\ \mathbf{i}_{abcr} \end{bmatrix} \tag{5.9-13}$$

Although the same notation is used for the inductance matrices as for the two-phase case, they are not the same. The self-inductances are all constant and can be expressed as in the case of the two-phase machine; however, a mutual coupling exists between stator (rotor) phases. To determine this mutual inductance, let us consider the coupling between the *as* and *bs* winding, and let us imagine that we can take a hold of the *bs* winding and rotate it 120° in a clockwise direction. If this were possible, then the *as* and *bs* windings would be on top of each other and, thus, tightly coupled with a mutual inductance of

$$L = \frac{N_s N_s}{\mathcal{R}_m} = L_{ms} \tag{5.9-14}$$

If now we rotate the *bs* winding back to its original position and if we assume that the mutual inductance varies in proportion to $\cos \phi_s$ as the *bs* winding is rotated, then

$$L_{asbs} = L_{ms} \cos \phi_s \big|_{\phi_s = 2\pi/3} = -\tfrac{1}{2} L_{ms} \tag{5.9-15}$$

It follows that

$$\mathbf{L}_s = \begin{bmatrix} L_{ss} & -\tfrac{1}{2} L_{ms} & -\tfrac{1}{2} L_{ms} \\ -\tfrac{1}{2} L_{ms} & L_{ss} & -\tfrac{1}{2} L_{ms} \\ -\tfrac{1}{2} L_{ms} & -\tfrac{1}{2} L_{ms} & L_{ss} \end{bmatrix} \tag{5.9-16}$$

$$\mathbf{L}_r = \begin{bmatrix} L_{rr} & -\tfrac{1}{2} L_{mr} & -\tfrac{1}{2} L_{mr} \\ -\tfrac{1}{2} L_{mr} & L_{rr} & -\tfrac{1}{2} L_{mr} \\ -\tfrac{1}{2} L_{mr} & -\tfrac{1}{2} L_{mr} & L_{rr} \end{bmatrix} \tag{5.9-17}$$

where $L_{ss} = L_{ls} + L_{ms}$ and $L_{rr} = L_{lr} + L_{mr}$. Also,

$$\mathbf{L}_{sr} = L_{sr} \begin{bmatrix} \cos \theta_r & \cos(\theta_r + \tfrac{2}{3}\pi) & \cos(\theta_r - \tfrac{2}{3}\pi) \\ \cos(\theta_r - \tfrac{2}{3}\pi) & \cos \theta_r & \cos(\theta_r + \tfrac{2}{3}\pi) \\ \cos(\theta_r + \tfrac{2}{3}\pi) & \cos(\theta_r - \tfrac{2}{3}\pi) & \cos \theta_r \end{bmatrix} \tag{5.9-18}$$

where

$$L_{sr} = \frac{N_s N_r}{\mathcal{R}_m} \tag{5.9-19}$$

All rotor variables may be referred to the stator windings by the following turns-ratios:

$$\mathbf{i}'_{abcr} = \frac{N_r}{N_s} \mathbf{i}_{abcr} \tag{5.9-20}$$

$$\mathbf{v}'_{abcr} = \frac{N_s}{N_r} \mathbf{v}_{abcr} \tag{5.9-21}$$

$$\boldsymbol{\lambda}'_{abcr} = \frac{N_s}{N_r} \boldsymbol{\lambda}_{abcr} \tag{5.9-22}$$

Thus, (5.9-7) and (5.9-8) become

$$\mathbf{v}_{abcs} = \mathbf{r}_s \mathbf{i}_{abcs} + p\boldsymbol{\lambda}_{abcs} \tag{5.9-23}$$

$$\mathbf{v}'_{abcr} = \mathbf{r}'_r \mathbf{i}'_{abcr} + p\boldsymbol{\lambda}'_{abcr} \tag{5.9-24}$$

where

$$\mathbf{r}'_r = \left(\frac{N_s}{N_r} \right)^2 \mathbf{r}_r \tag{5.9-25}$$

The flux linkage equations may now be written as

$$\begin{bmatrix} \boldsymbol{\lambda}_{abcs} \\ \boldsymbol{\lambda}'_{abcr} \end{bmatrix} = \begin{bmatrix} \mathbf{L}_s & \mathbf{L}'_{sr} \\ (\mathbf{L}'_{sr})^T & \mathbf{L}'_r \end{bmatrix} \begin{bmatrix} \mathbf{i}_{abcs} \\ \mathbf{i}'_{abcr} \end{bmatrix} \tag{5.9-26}$$

where, by definition,

$$\mathbf{L}'_{sr} = \frac{N_s}{N_r}\mathbf{L}_{sr} = \frac{L_{ms}}{L_{sr}}\mathbf{L}_{sr} \tag{5.9-27}$$

and
$$\mathbf{L}'_r = \begin{bmatrix} L'_{lr} + L_{ms} & -\frac{1}{2}L_{ms} & -\frac{1}{2}L_{ms} \\ -\frac{1}{2}L_{ms} & L'_{lr} + L_{ms} & -\frac{1}{2}L_{ms} \\ -\frac{1}{2}L_{ms} & -\frac{1}{2}L_{ms} & L'_{lr} + L_{ms} \end{bmatrix} \tag{5.9-28}$$

In (5.9-28),

$$L'_{lr} = \left(\frac{N_s}{N_r}\right)^2 L_{lr} \tag{5.9-29}$$

Torque

The electromagnetic torque, positive for motor action, may be expressed by using the second entry in Table 2.5-1. In particular,

$$T_e = -\frac{P}{2}L_{ms}\Big\{[i_{as}(i'_{ar} - \tfrac{1}{2}i'_{br} - \tfrac{1}{2}i'_{cr}) + i_{bs}(i'_{br} - \tfrac{1}{2}i'_{ar} - \tfrac{1}{2}i'_{cr})$$

$$+ i_{cs}(i'_{cr} - \tfrac{1}{2}i'_{br} - \tfrac{1}{2}i'_{ar})]\sin\theta_r$$

$$+ \frac{\sqrt{3}}{2}[i_{as}(i'_{br} - i'_{cr}) + i_{bs}(i'_{cr} - i'_{ar}) + i_{cs}(i'_{ar} - i'_{br})]\cos\theta_r\Big\} \tag{5.9-30}$$

The torque and rotor speeds are related by (5.4-5), which is repeated for convenience

$$\boxed{T_e = J\left(\frac{2}{P}\right)\frac{d\omega_r}{dt} + B_m\left(\frac{2}{P}\right)\omega_r + T_L} \tag{5.9-31}$$

where J is the inertia of the rotor and, in some cases, the connected load. The units of J and the damping coefficient B_m are discussed following (5.4-5). The load torque T_L is positive for a torque load on the shaft of the induction machine, as indicated in Fig. 5.9-1.

Machine Equations in the Stationary Reference Frame

Since there are three stator variables (f_{as}, f_{bs}, and f_{cs}) and three rotor variables (f'_{ar}, f'_{br}, and f'_{cr}) in the case of the three-phase induction machine, it is necessary that we replace both the stator and rotor variables with three substitute variables. The change of variables for the stator variables is

$$\boxed{\mathbf{f}^s_{qd0s} = \mathbf{K}^s_s\mathbf{f}_{abcs}} \tag{5.9-32}$$

where
$$(\mathbf{f}^s_{qd0s})^T = [f^s_{qs} \quad f^s_{ds} \quad f_{0s}] \tag{5.9-33}$$

$$(\mathbf{f}_{abcs})^T = [f_{as} \quad f_{bs} \quad f_{cs}] \tag{5.9-34}$$

$$\mathbf{K}_s^s = \frac{2}{3}\begin{bmatrix} 1 & -\frac{1}{2} & -\frac{1}{2} \\ 0 & -\frac{\sqrt{3}}{2} & \frac{\sqrt{3}}{2} \\ \frac{1}{2} & \frac{1}{2} & \frac{1}{2} \end{bmatrix} \quad (5.9\text{-}35)$$

The inverse of $\mathbf{K}_s^s$ is

$$(\mathbf{K}_s^s)^{-1} = \begin{bmatrix} 1 & 0 & 1 \\ -\frac{1}{2} & -\frac{\sqrt{3}}{2} & 1 \\ -\frac{1}{2} & \frac{\sqrt{3}}{2} & 1 \end{bmatrix} \quad (5.9\text{-}36)$$

The change of variables which transforms the rotor variables to the stationary reference frame is

$$\boxed{\mathbf{f}_{qd0r}'^s = \mathbf{K}_r^s \mathbf{f}_{abcr}'} \quad (5.9\text{-}37)$$

where

$$(\mathbf{f}_{qd0r}'^s)^T = [f_{qr}'^s \quad f_{dr}'^s \quad f_{0r}'] \quad (5.9\text{-}38)$$

$$(\mathbf{f}_{abcr}')^T = [f_{ar}' \quad f_{br}' \quad f_{cr}'] \quad (5.9\text{-}39)$$

$$\mathbf{K}_r^s = \frac{2}{3}\begin{bmatrix} \cos\theta_r & \cos(\theta_r + \frac{2}{3}\pi) & \cos(\theta_r - \frac{2}{3}\pi) \\ -\sin\theta_r & -\sin(\theta_r + \frac{2}{3}\pi) & -\sin(\theta_r - \frac{2}{3}\pi) \\ \frac{1}{2} & \frac{1}{2} & \frac{1}{2} \end{bmatrix} \quad (5.9\text{-}40)$$

The inverse is

$$(\mathbf{K}_r^s)^{-1} = \begin{bmatrix} \cos\theta_r & -\sin\theta_r & 1 \\ \cos(\theta_r + \frac{2}{3}\pi) & -\sin(\theta_r + \frac{2}{3}\pi) & 1 \\ \cos(\theta_r - \frac{2}{3}\pi) & -\sin(\theta_r - \frac{2}{3}\pi) & 1 \end{bmatrix} \quad (5.9\text{-}41)$$

In the above equations, θ_r is defined by (5.5-6). It is important that the same notation ($\mathbf{K}_s^s$ and $\mathbf{K}_r^s$) is used for the transformation for both two- and three-phase change of variables.

A trigonometric interpretation of the above change of variables is shown in Fig. 5.9-2. The variables that have been added are the $0s$ and $0r$ variables, commonly referred to as the zero quantities. One property of the zero quantities can be noted from the transformation; both f_{0s} and f_{0r}' are identically equal to zero if the three-phase sets are balanced. A raised index is generally not associated with the zero quantities since these variables are always in the same frame of reference as the actual phase variables.

If the above change of variables is substituted into the three-phase stator and rotor voltage equations, we will obtain q and d voltage equations identical in form to those given by (5.5-24) through (5.5-27). We must add, however, the voltage equations for the zero quantities. In particular, we find that

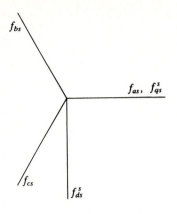

(a)

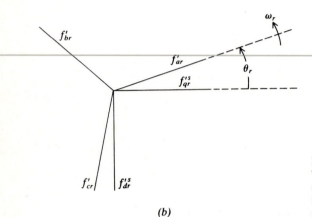

(b)

FIGURE 5.9-2
Trigonometric interpretation of the change of variables for a three-phase induction machine. (a) Stator variables; (b) rotor variables.

$$v_{0s} = r_s i_{0s} + p\lambda_{0s} \tag{5.9-42}$$

$$v'_{0r} = r'_r i'_{0r} + p\lambda'_{0r} \tag{5.9-43}$$

For a magnetically linear system, the q and d flux linkage equations are identical to those given by (5.5-28) through (5.5-31), with the exception that L_{ms} should be replaced by $\frac{3}{2}L_{ms}$, which is often referred to as M. Thus,

$$\lambda^s_{qs} = L_{ls}i^s_{qs} + M(i^s_{qs} + i'^s_{qr}) \tag{5.9-44}$$

and, similarly, for λ^s_{ds}, λ'^s_{qr}, and λ'^s_{dr}, where

$$M = \frac{3}{2}L_{ms} \tag{5.9-45}$$

In (5.9-42) and (5.9-43) we can show that

$$\lambda_{0s} = L_{ls}i_{0s} \tag{5.9-46}$$

$$\lambda'_{0r} = L'_{lr}i'_{0r} \tag{5.9-47}$$

Hence, for a three-phase machine, the voltage equations in the stationary reference frame may be written in terms of inductances as

$$
\begin{bmatrix} v^s_{qs} \\ v^s_{ds} \\ v_{0s} \\ v'^s_{qr} \\ v'^s_{dr} \\ v'_{0r} \end{bmatrix} =
\begin{bmatrix}
r_s + pL_{ss} & 0 & 0 & pM & 0 & 0 \\
0 & r_s + pL_{ss} & 0 & 0 & pM & 0 \\
0 & 0 & r_s + pL_{ls} & 0 & 0 & 0 \\
pM & -\omega_r M & 0 & r'_r + pL'_{rr} & -\omega_r M & 0 \\
\omega_r M & pM & 0 & \omega_r M & r'_r + pL'_{rr} & 0 \\
0 & 0 & 0 & 0 & 0 & r'_r + pL'_{lr}
\end{bmatrix}
\begin{bmatrix} i^s_{qs} \\ i^s_{ds} \\ i_{0s} \\ i'^s_{qr} \\ i'^s_{dr} \\ i'_{0r} \end{bmatrix}
$$

$$\tag{5.9-48}$$

where in the case of the three-phase machine

$$
\begin{aligned}
L_{ss} &= L_{ls} + M \\
L'_{rr} &= L'_{lr} + M
\end{aligned}
\qquad
\begin{matrix} (5.9\text{-}49) \\ (5.9\text{-}50) \end{matrix}
$$

The q and d equivalent circuits given in Fig. 5.5-2 for a two-phase machine are valid for the three-phase machine if L_{ms} is replaced by $M(=\frac{3}{2}L_{ms})$. Equivalent circuits for the $0s$ and $0r$ quantities are series rL circuits.

Expressions of the electromagnetic torque for a three-phase induction machine are identical in form to (5.5-33) through (5.5-35) with the following exceptions. All equations must be multiplied by $\frac{3}{2}$ and, in the case of (5.5-33), L_{ms} must be replaced by M. Also, it follows that the steady-state equivalent circuit given for a two-phase machine (Fig. 5.6-1) is valid for a three-phase machine if X_{ms} is replaced by X_M ($\omega_e M$).

SP5.9-1. The operating characteristics shown in Figs. 5.8-1 through 5.8-3 are for a balanced two-phase induction machine with the variables in the stationary reference frame. Would the waveform of the variables be the same for a balanced three-phase induction machine? Unbalanced? [Yes; no]

SP5.9-2. Consider the three-phase device shown in Fig. 5.9-1. Assume the rotor windings are open-circuited. The terminal of the cs winding is open. Determine (a) the resistance between the as and bs terminals; (b) the inductance. [(a) $2r_s$; (b) $2L_{ls} + 3L_{ms}$]

SP5.9-3. A two-phase induction motor and a three-phase induction motor have identical r_s, r'_r, L_{ss}, and L'_{rr}. The amplitude and frequency of the applied phase voltages are equal. If the free-acceleration characteristics are identical, what must be the relationship between the inertia of these devices? [$J_3 = \frac{3}{2}J_2$]

5.10 RECAPPING

There are at least two aspects of the material presented in this chapter which are worth reemphasizing. First, to analyze a symmetrical induction machine, it is necessary to incorporate a change of variables so that the resulting variables (windings) are not in relative motion. Secondly, even though this change of variables is a new concept to us and somewhat involved, the resulting steady-state equivalent circuit is very easy to work with, making it an invaluable tool for predicting and understanding the operation of symmetrical induction machines. It has also become very apparent that, in order to investigate the dynamic characteristics of the induction machine, it is necessary to use a computer. Although the implementation of a computer simulation of an induction machine is beyond the scope of this text, the computer traces obtained from such an implementation are instructive not only to portray the dynamic characteristics of this device but also to illustrate the change of variables used in the analysis.

In Chap. 9, unbalanced operation of two-phase symmetrical induction motors is analyzed. This sets the stage for the analysis of the servomotor used in position-control systems and for single-phase operation of a symmetrical two-phase induction machine. The material which has been presented in this chapter is prerequisite for Chap. 9.

5.11 REFERENCES

1. P. C. Krause, *Analysis of Electric Machinery*, McGraw-Hill Book Company, New York, 1986.
2. H. C. Stanley, "An Analysis of the Induction Motor," *AIEE Trans.*, vol. 57 (Supplement), 1938, pp. 751–755.

5.12 PROBLEMS

1. Consider the two-pole two-phase induction machine shown in Fig. 5.2-1. The device is operating as a motor at $\omega_r = 95\pi$ rad/s with $I'_{ar} = \cos 5\pi t$ and $I'_{br} = -\sin 5\pi t$. Determine the angular velocity and direction of mmf$_r$ relative to (a) an observer on the rotor and (b) an observer on the stator. Also determine (c) the angular velocity of the stator currents and (d) the direction of rotation of the rotor.

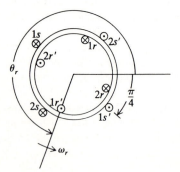

FIGURE 5.12-1
Winding arrangement for a two-pole two-phase electromechanical device.

2. The windings shown in Fig. 5.12-1 are sinusoidally distributed and the device is symmetrical. The amplitude of the stator to rotor mutual inductance is L_{sr}. Express all mutual inductances as functions of L_{sr} and θ_r.

3. Consider the two-pole two-phase symmetrical induction machine shown in Fig. 5.2-1. For steady-state operation let $I_{as} = \sqrt{2}I_s \cos \omega_e t$, $I_{bs} = \sqrt{2}I_s \sin \omega_e t$, $I'_{ar} = \sqrt{2}I'_r \cos[(\omega_e - \omega_r)t + \alpha]$, and $I'_{br} = \sqrt{2}I'_r \sin[(\omega_e - \omega_r)t + \alpha]$, where $\alpha = \theta_{eri}(0)$. Assume the rotor speed is constant and $V'_{ar} = V'_{br} = 0$. Use v_{as} of (5.3-40) and λ_{as} of (5.3-49) to express V_{as} in the form $A \cos \omega_e t + B \sin \omega_e t$.

4. The rotor windings of the two-pole two-phase symmetrical induction machine shown in Fig. 5.2-1 are open-circuited. $I_{as} = \sqrt{2}I_s \cos \omega_e t$ and $I_{bs} = -\sqrt{2}I_s \sin \omega_e t$. The rotor is driven at $\omega_r = \omega_e$ in the counterclockwise direction. Express V_{ar}.

*5. For the device shown in Fig. 5.2-1, let $r_s = r'_r = 0.5\,\Omega$, $L_{ss} = L'_{rr} = 0.1\,\text{H}$, and $L_{ms} = 0.09\,\text{H}$. With $V_{as} = \sqrt{2}\cos t$, $\omega_r = 0$, $\theta_r = \frac{1}{4}\pi$, and the bs and br windings are open-circuited. The ar winding is short-circuited. Calculate $\tilde{I}_{as}$.

6. Derive the expression for the electromagnetic torque T_e given by (5.4-3), starting with (5.4-1) and (5.4-2).

7. Obtain (5.5-21) and (5.5-22) from (5.5-19) and (5.5-20), respectively.

8. Write the voltage equations given by (5.5-32) in terms of flux linkages instead of currents. You need only formulate the equations in matrix form; you need not perform the multiplication of the matrices.

9. It is often convenient to express the voltage and flux linkage equations in the stationary reference frame, and the torque in terms of flux linkages per second rather than flux linkages. To do this, a base electrical angular velocity (ω_b) is defined, say 377 rad/s for a 60-Hz machine, and the flux linkage equations are multiplied by ω_b. For example, (5.5-28) would become

$$\psi^s_{qs} = X_{ls}i^s_{qs} + X_{ms}(i^s_{qs} + i'^s_{qr})$$

where $X_{ls} = \omega_b L_{ls}$ and $X_{ms} = \omega_b L_{ms}$. Rewrite (5.5-24) through (5.5-27) and (5.5-34) and (5.5-35) in terms of flux linkages per second.

*10. Obtain (5.6-23) from (5.6-22).

11. Obtain (5.6-28) from (5.6-25).

*12. Verify (5.6-29) and (5.6-31).

13. Repeat SP5.6-4b without approximation.

14. A four-pole two-phase induction machine has the following parameters: $r_s = 0.3\,\Omega$, $L_{ls} = 1\,\text{mH}$, $L_{ms} = 20\,\text{mH}$, $r'_r = 0.2\,\Omega$, $L'_{lr} = 1\,\text{mH}$. The device is supplied from a 60-Hz source; the rotor speed is $\omega_r = 360\,\text{rad/s}$. In this mode of operation $\tilde{I}_{as} = 28.8\underline{/-36.1°}$ and $\tilde{I}'_{ar} = 23.9\underline{/173.2°}$. Calculate (a) T_e, (b) the total ohmic loss in the rotor windings, (c) the mechanical power delivered to the load, and (d) express I_{as}, I'_{ar}, I^s_{qs}, and I'^s_{qr}.

*15. Calculate the actual rotor speed, in rad/s, at maximum steady-state torque for the machine given in Example 5B for operation as a motor when connected to an electric source of (a) 120 Hz with twice rated voltage, (b) 60 Hz with rated voltage, (c) 30 Hz with one-half rated voltage, and (d) 6 Hz with 10 percent rated voltage.

*16. The induction machine described in Example 5C is operating steadily at no load. The polarity of the applied voltage of one stator winding is suddenly reversed. Assume the electric system establishes steady-state operation before the speed of the rotor has changed appreciably. Calculate the torque. Describe the behavior of the machine.

17. Select two identical capacitors so that when they are connected in parallel with each phase (one capacitor per phase) of the induction machine described in Example 5C, the no-load capacitor induction machine combination operates at unity power factor. Assume the capacitors are ideal (zero resistance).

*18. Assume that the current at rated load of a 110-V (rms) 60-Hz machine is $20 - j10$ A (rms). Select two identical capacitors so that the parallel combination of the capacitors and the induction machine (one capacitor per phase) has a power factor of 0.95 (current lagging the voltage) when the machine is operating at rated load. Assume the capacitors are ideal (zero resistance).

*19. Consider the 5- and $\frac{1}{10}$-hp machines described in Sec. 5.7. Make reasonable approximations and determine the blocked-rotor and no-load steady-state stator phase currents $\tilde{I}_{as}$, and compare with the starting current and current at synchronous speed depicted in Figs. 5.7-1 and 5.7-3 for the 5- and $\frac{1}{10}$-hp machines, respectively.

20. Assume the torque load on the 5-hp machine described in Sec. 5.7 is $T_L = K\omega_r^2$. At a rotor speed of 1800 r/min, T_L is 22 N·m. Calculate K. Use the plot of T_e for 60 Hz in Fig. 5.7-10 to approximate the speed at which $T_L = T_e$. Will the machine operate at this speed? Explain. Use Fig. 5.7-1 to approximate $\tilde{I}_{as}$ for this speed.

21. Obtain (5.9-30).

22. Prove the statement regarding T_e for a three-phase induction machine made in the first two statements of the last paragraph of Sec. 5.9.

*23. A four-pole 7.5-hp three-phase symmetrical induction motor has the following parameters: $r_s = 0.3\,\Omega$, $L_{ls} = 1.5\,\text{mH}$, $L_{ms} = 35\,\text{mH}$, $r_r' = 0.15\,\Omega$, $L_{lr}' = 0.7\,\text{mH}$. The machine is supplied from a 110-V line-to-neutral 60-Hz source.

   (a) Calculate the steady-state starting torque and current. Make valid approximations.

   (b) Calculate the no-load current. Neglect friction and windage losses.

24. Repeat Prob. 23 with the machine supplied from an 11-V line-to-neutral 6-Hz source.

CHAPTER

6

SYNCHRONOUS MACHINES

6.1 INTRODUCTION

Nearly all electric power is generated by synchronous machines driven either by hydroturbines or steam turbines or combustion engines. Just as the induction machine is the workhorse when it comes to converting energy from electric to mechanical, the synchronous machine is the principal means of converting energy from mechanical to electric. Although nearly all electric power is generated with three-phase synchronous machines, their electrical and electromechanical behavior can be predicted from the equations which describe the two-phase salient-pole synchronous machine. In particular, with only slight modifications, these equations can be used to predict the performance of large hydroturbine and steam turbine synchronous generators, synchronous motors, and reluctance motors used in low-power drive systems. It is for this reason that we will focus our attention on the two-phase machine since the work involved is less than with the three-phase device. However, those who wish to study the three-phase synchronous machine may do so since there is a section devoted to it near the end of the chapter.

The rotor of a synchronous machine is equipped with a field winding and one or more short-circuited windings which we will refer to as *damper windings*. In general, the rotor windings have different electrical characteristics. Moreover, the rotor of a salient-pole synchronous machine is magnetically asymmetrical. Owing to these rotor asymmetries, a change of variables offers no advantage in the case of the rotor variables. However, we will find it beneficial to define a change of variables or transformation for the voltages, currents, and flux linkages of the stator circuits. In effect, this transformation replaces these stator variables with fictitious variables associated with circuits fixed in the rotor.

In this chapter, the voltage and electromagnetic torque equations are first established for the synchronous machine in machine variables. Reference frame theory is then used to establish the machine equations with the stator variables transformed to a reference frame fixed in the rotor (Park's equations). The equations which describe the steady-state behavior are then derived from these equations. Attention is also given to the two-phase reluctance motor which finds wide use in control system applications.

6.2 TWO-PHASE SYNCHRONOUS MACHINE

A two-pole two-phase salient-pole synchronous machine is shown in Fig. 6.2-1. The stator windings are identical, sinusoidally distributed windings, as described in Chap. 4. The electrical characteristics of the rotor of a synchronous machine may be approximated by a field winding (fd winding) and short-circuited damper or amortisseur windings (kq and kd windings). Although the damper windings are shown with provisions to apply a voltage, they are, in fact, short-circuited windings which represent the paths for induced rotor currents. In particular, these short-circuited windings represent squirrel-cage-type windings (short-circuited copper bars) forged below the surface of the rotor or current paths in the iron of solid-iron rotors. Laminated salient-pole rotors with cage damper windings are used in machines with a large number of poles while solid-iron round rotors with or without cage-type damper windings are used in high-speed (two- or four-pole) machines. In any event, the electrical characteristics of the equivalent damper windings may be determined by test. We will assume that the damper windings are approximated by two sinusoidally distributed windings displaced 90°. The kd winding has the same magnetic axis as the fd winding, it has N_{kd} equivalent turns with resistance r_{kd}. The magnetic axis of kq winding is 90° ahead of the magnetic axis of the fd and kd windings. It has N_{kq} equivalent turns and r_{kq} resistance. It is important to mention that the rotor configuration shown in Fig. 6.2-1 for a two-phase machine is the same for any multiphase two-pole synchronous machine. In some cases, a more accurate representation of the electrical characteristics of the rotor is achieved by assuming that two or more damper windings exist in each axis (i.e., $kq1, kq2, \ldots$ and $kd1, kd2, \ldots$). We will consider only the kq and kd windings; the modifications and extensions necessary to accommodate any number of rotor windings are straightforward [1].

The quadrature axis (q axis) and direct axis (d axis) are introduced in Fig. 6.2-1. The q axis is the magnetic axis of the kq winding whereas the d axis is the magnetic axis for the fd and kd windings. The q and d axes are reserved to

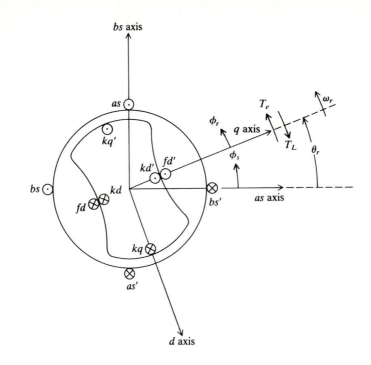

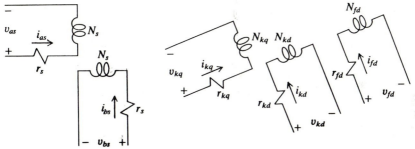

FIGURE 6.2-1
Two-pole two-phase salient-pole synchronous machine.

denote the rotor magnetic axes of a synchronous machine since, over the years, they have been associated with the physical structure of the rotor quite independent of any transformation. The angular displacement about the stator is denoted ϕ_s and it is referenced to the *as* axis. The angular displacement about the rotor is ϕ_r, which is referenced to the *q* axis. The electrical angular velocity of the rotor is ω_r and θ_r is the electrical angular displacement of the rotor measured from the *as* axis to the *q* axis. Thus, a given point on the rotor surface at the angular position ϕ_r may be related to an adjacent point on the inside stator surface with angular position ϕ_s as

$$\boxed{\phi_s = \phi_r + \theta_r} \tag{6.2-1}$$

The electromechanical torque T_e and the load torque T_L are also shown in Fig. 6.2-1. We are aware from Chap. 2 that T_e is assumed positive in the positive direction of θ_r. The load torque is positive in the opposite direction, opposing rotation.

The stator of a synchronous machine is symmetrical; however, the rotor is asymmetrical from two standpoints. The rotor windings are not identical since, in general, they do not have the same number of turns and the same value of resistance. Also, owing to the nonuniform air gap of the salient-pole synchronous machine, the magnetic characteristics of the q and d axes are not the same.

With balanced steady-state stator currents, an air gap mmf (mmf$_s$) is established which rotates about the air gap of a two-pole machine at ω_e, the angular velocity of the stator currents (4.4-11). Now, the damper windings are short-circuited and, for the machine to operate as a synchronous machine, a dc voltage is applied to the fd winding (field winding) by a brush and slip ring arrangement. The resulting field current i_{fd} establishes an air gap mmf (mmf$_r$) which is fixed with respect to the rotor. The air gap mmf (poles) established by the field winding must rotate at the same angular velocity as the rotating air gap mmf (poles) established by the stator windings in order to produce a nonzero average electromagnetic torque during steady-state operation. Therefore, the rotor must rotate in synchronism with the air gap mmf established by the stator windings ($\omega_r = \omega_e$); hence, the name synchronous machine. The main torque production mechanism is this interaction of the air gap mmf established by the stator currents (mmf$_s$) and the air gap mmf due to the direct current flowing in the field winding (mmf$_r$). However, electromagnetic torque (reluctance torque) is also developed at synchronous speed due to the nonuniform air gap (salient-pole rotor). The so-called salient-pole construction is common for slower speed machines (large number of poles) such as hydroturbine generators. In this type of rotor construction, the field winding is wound upon the rotor surface, as shown in Fig. 6.2-1, and the air gap is nonuniform to make room for the placement of the field winding. Therefore, the q-axis magnetic path has a higher reluctance than the d-axis magnetic path. Now, in Chap. 2 we learned that torque is produced in a reluctance machine to align the minimum-reluctance path of the rotor with the mmf produced by the stator. Let us apply this principle to the salient-pole synchronous machine. There is an electromagnetic torque developed to align the minimum-reluctance path (d axis) with the resultant air gap mmf (mmf$_s$ + mmf$_r$). We will find that in the case of salient-pole synchronous machines, the reluctance torque is a small part of the total torque developed. However, two facts warrant mentioning; first in high-speed synchronous machines (two-, four-, six-pole machines) the field windings are generally embedded in rotor slots and the air gap is, for the most part, uniform (round rotor). It is apparent that, in the case of a round-rotor synchronous machine, the reluctance torque is not present.

Second, if the field winding (fd winding) is removed from the salient-pole synchronous machine shown in Fig. 6.2-1, it would be a two-phase reluctance machine, which is used widely in low-power drive systems.

We have yet to discuss the damper windings. It was found early on that a synchronous machine with only a field winding on the rotor and without provisions for induced currents to circulate in the rotor iron would tend to oscillate about synchronous speed in a slowly damped manner following any slight disturbance. Adding damper windings (short-circuited rotor windings) provided the desired damping. To explain this damping action, let us compare the torque developed by the damper windings to that developed by an induction motor. The damper windings are short-circuited as are the rotor windings of an induction motor and, as we discussed in Chap. 4, currents are induced in these rotor (damper) windings whenever the speed of the rotor differs from the angular velocity of the rotating air gap mmf established by the stator currents (mmf$_s$). (It is assumed that you have not studied Chap. 5, but that you have read the material in Sec. 4.5 on induction machines.) Since the damper windings are not symmetrical and since the air gap is not uniform, the steady-state torque due to the interaction of the currents induced in the damper windings and mmf$_s$ will pulsate; however, an average torque will occur. Now, the main torque of a synchronous machine is developed at synchronous speed because of the interaction of mmf$_s$ and mmf$_r$. At synchronous speed, current is not induced in the damper windings and, hence, "induction motor" torque is not developed. However, if for any reason the speed of the rotor should vary around synchronous speed because of a disturbance, currents will be induced in the damper windings and the torque developed due to induction motor action, although small, will damp oscillations of the rotor speed. That is, a slight slowing down (speeding up) of the rotor will produce an induction motor torque to accelerate (decelerate) the rotor back to synchronous speed.

Although it is generally necessary to start large synchronous machines by auxiliary means, smaller-horsepower synchronous machines and reluctance motors develop sufficient induction motor torque because of the damper windings to accelerate the machine to near synchronous speed. During this starting period, the field winding of the synchronous machine is also short-circuited, hence, it too provides some induction motor torque. The synchronous machine will accelerate to near synchronous speed, whereupon it will operate as an induction machine developing the average torque necessary to satisfy the no-load losses. The field winding is then open-circuited and a dc voltage is applied to its terminals by means of a brush and slip ring combination. The machine then "pulls in" to synchronism with the rotating air gap mmf established by the stator currents and, thus, operates as a synchronous machine.

The damper windings of reluctance motors are often designed so that the device will develop sufficient induction motor torque to accelerate the rotor, sometimes under load, from stall to near synchronous speed. If the load is not

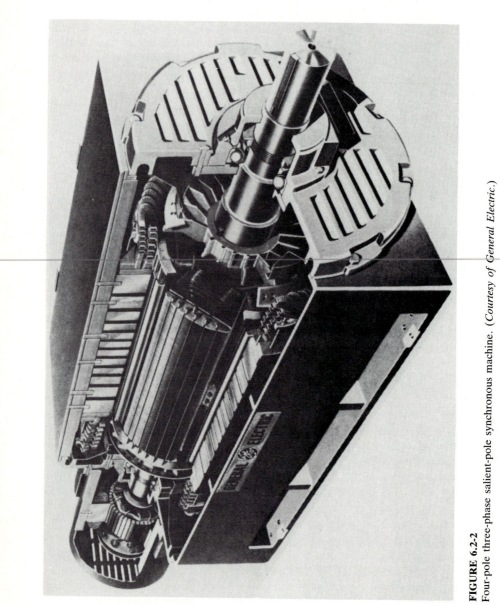

FIGURE 6.2-2
Four-pole three-phase salient-pole synchronous machine. (*Courtesy of General Electric.*)

too large, the reluctance torque will then pull the rotor in step with the rotating air gap mmf established by the stator currents, and the device will operate as a reluctance machine.

Torque is torque by whatever means it is developed and, perhaps, we should not emphasize the separation of torque into three types since the system is nonlinear and superposition cannot be applied. Nevertheless, this separation is helpful and, as we proceed, we will be able to identify what we have called the induction motor torque, the reluctance torque, and the torque due to the interaction of mmf$_s$ and mmf$_r$, all of which can occur in the machine shown in Fig. 6.2-1.

A four-pole three-phase salient-pole synchronous machine is shown in Fig. 6.2-2. Note the dc machine connected to the shaft for purposes of supplying voltage to the field winding of the synchronous machine. Note also, the squirrel-cage damper windings embedded in the pole faces. Figure 6.2-3 shows the stator and rotor of a miniature two-pole three-phase alternator with an alnico permanent-magnet rotor. This device produces 12 W at 4200 r/min to supply aircraft instruments. It mounts on an aircraft engine auxiliary drive pad where temperatures can be as high as 350°F.

SP6.2-1. Express mmf$_r$ for the two-pole two-phase synchronous machine shown in Fig. 6.2-1. [mmf$_r = -(N_{fd}/2)i_{fd} \sin(\phi_s - \theta_r)$]

SP6.2-2. A dc voltage is applied to the *fd* winding of the machine shown in Fig. 6.2-1.

FIGURE 6.2-3
Stator and rotor of a two-pole 12-W 4200-r/min permanent-magnet synchronous machine. (*Courtesy of Vickers ElectroMech.*)

The damper windings are short-circuited and the machine is driven at ω_r, counterclockwise. Assume that the stator currents are balanced 60-Hz currents with $\tilde{I}_{as} = -j\tilde{I}_{bs}$. Determine the frequency of the currents flowing in the damper windings. $[\omega_r + \omega_e]$

6.3 VOLTAGE EQUATIONS AND WINDING INDUCTANCES

The voltage equations for the two-pole two-phase salient-pole synchronous machine shown in Fig. 6.2-1 may be expressed as

$$v_{as} = r_s i_{as} + \frac{d\lambda_{as}}{dt} \tag{6.3-1}$$

$$v_{bs} = r_s i_{bs} + \frac{d\lambda_{bs}}{dt} \tag{6.3-2}$$

$$v_{kq} = r_{kq} i_{kq} + \frac{d\lambda_{kq}}{dt} \tag{6.3-3}$$

$$v_{fd} = r_{fd} i_{fd} + \frac{d\lambda_{fd}}{dt} \tag{6.3-4}$$

$$v_{kd} = r_{kd} i_{kd} + \frac{d\lambda_{kd}}{dt} \tag{6.3-5}$$

The above equations may be written in matrix form as

$$\boxed{\begin{aligned} \mathbf{v}_{abs} &= \mathbf{r}_s \mathbf{i}_{abs} + p\boldsymbol{\lambda}_{abs} \\ \mathbf{v}_{qdr} &= \mathbf{r}_r \mathbf{i}_{qdr} + p\boldsymbol{\lambda}_{qdr} \end{aligned}} \tag{6.3-6}$$
$$\tag{6.3-7}$$

where

$$(\mathbf{f}_{abs})^T = [f_{as} \quad f_{bs}] \tag{6.3-8}$$

$$(\mathbf{f}_{qdr})^T = [f_{kq} \quad f_{fd} \quad f_{kd}] \tag{6.3-9}$$

In the above equations, the s and r subscripts denote variables associated with the stator and rotor windings, respectively, and p is the operator d/dt. Also,

$$\mathbf{r}_s = \begin{bmatrix} r_s & 0 \\ 0 & r_s \end{bmatrix} \tag{6.3-10}$$

$$\mathbf{r}_r = \begin{bmatrix} r_{kq} & 0 & 0 \\ 0 & r_{fd} & 0 \\ 0 & 0 & r_{kd} \end{bmatrix} \tag{6.3-11}$$

A review of matrix algebra is given in Appendix D. The flux linkage equations may be expressed as

$$\lambda_{as} = L_{asas} i_{as} + L_{asbs} i_{bs} + L_{askq} i_{kq} + L_{asfd} i_{fd} + L_{askd} i_{kd} \tag{6.3-12}$$

$$\lambda_{bs} = L_{bsas} i_{as} + L_{bsbs} i_{bs} + L_{bskq} i_{kq} + L_{bsfd} i_{fd} + L_{bskd} i_{kd} \tag{6.3-13}$$

$$\lambda_{kq} = L_{kqas} i_{as} + L_{kqbs} i_{bs} + L_{kqkq} i_{kq} + L_{kqfd} i_{fd} + L_{kqkd} i_{kd} \tag{6.3-14}$$

$$\lambda_{fd} = L_{fdas}i_{as} + L_{fdbs}i_{bs} + L_{fdkq}i_{kq} + L_{fdfd}i_{fd} + L_{fdkd}i_{kd} \qquad (6.3\text{-}15)$$

$$\lambda_{kd} = L_{kdas}i_{as} + L_{kdbs}i_{bs} + L_{kdkq}i_{kq} + L_{kdfd}i_{fd} + L_{kdkd}i_{kd} \qquad (6.3\text{-}16)$$

In the case of a salient-pole device (nonuniform air gap), the self-inductances of the stator windings and the mutual inductances between stator windings are functions of θ_r. Although this is a review of the material in Chap. 1, let us consider L_{asas}. With $\theta_r = 0$, we see from Fig. 6.2-1 that the magnetizing inductance of L_{asas} is less than it would be when $\theta_r = \frac{1}{2}\pi$. Let the magnetizing inductance of the as winding be denoted L_{mq} when $\theta_r = 0$ since the q axis (high-reluctance path) is aligned with the magnetic axis of the as winding. Thus,

$$L_{asas} = L_{ls} + L_{mq} \qquad \theta_r = 0 \qquad (6.3\text{-}17)$$

where L_{ls} is the leakage inductance of the stator windings and

$$L_{mq} = \frac{N_s^2}{\mathcal{R}_{mq}} \qquad (6.3\text{-}18)$$

where $\mathcal{R}_{mq}$ is an equivalent reluctance of the magnetic path in the q axis. We called this $\mathcal{R}_m(0)$ in (1.7-24). Now, at $\theta_r = \frac{1}{2}\pi$ the d axis (low-reluctance path) is aligned with the magnetic axis of the as winding. Hence, denoting this magnetizing inductance as L_{md}, we can write

$$L_{asas} = L_{ls} + L_{md} \qquad \theta_r = \frac{1}{2}\pi \qquad (6.3\text{-}19)$$

where
$$L_{md} = \frac{N_s^2}{\mathcal{R}_{md}} \qquad (6.3\text{-}20)$$

where $\mathcal{R}_{md}$ is an equivalent reluctance of the magnetic path in the d axis. This is $\mathcal{R}_m(\frac{1}{2}\pi)$ in (1.7-25).

Since $\mathcal{R}_{mq} > \mathcal{R}_{md}$, $L_{mq} < L_{md}$, and we see that a minimum L_{asas} occurs at $\theta_r = 0$ and also again at $\theta_r = \pi$. Therefore, (6.3-17) is valid for $\theta_r = 0$ and π. Similarly, maximum L_{asas} occurs at $\theta_r = \frac{1}{2}\pi$ and again at $\theta_r = \frac{3}{2}\pi$; hence (6.3-19) applies for $\theta_r = \frac{1}{2}\pi$ and $\frac{3}{2}\pi$. The magnetizing inductance varies about an average value (which must be positive) and if we assume this variation to be sinusoidal, it would vary as a function of $2\theta_r$ (Fig. 1.7-3). Let L_A be the average value and L_B the amplitude of the sinusoidal variation about this average value. In this case,

$$L_{mq} = L_A - L_B \qquad (6.3\text{-}21)$$

$$L_{md} = L_A + L_B \qquad (6.3\text{-}22)$$

Substituting (6.3-18) and (6.3-20) for L_{mq} and L_{md}, respectively, into (6.3-21) and (6.3-22) and solving for L_A and L_B yields

$$L_A = \frac{N_s^2}{2}\left(\frac{1}{\mathcal{R}_{mq}} + \frac{1}{\mathcal{R}_{md}}\right) \qquad (6.3\text{-}23)$$

$$L_B = \frac{N_s^2}{2}\left(\frac{1}{\mathcal{R}_{mq}} - \frac{1}{\mathcal{R}_{md}}\right) \qquad (6.3\text{-}24)$$

Assuming a sinusoidal variation, we can write (Fig. 1.7-3)

$$L_{asas} = L_{ls} + L_A - L_B \cos 2\theta_r \qquad (6.3\text{-}25)$$

If the air gap were uniform as is the case in a round-rotor synchronous machine, $\mathcal{R}_{mq} = \mathcal{R}_{md}$ and, hence, from (6.3-24), $L_B = 0$.

By a similar procedure, it follows that, for the salient-pole device,

$$L_{bsbs} = L_{ls} + L_A + L_B \cos 2\theta_r \qquad (6.3\text{-}26)$$

Note when $\theta_r = 0$, L_{asas} is a minimum according to (6.3-25) and, according to (6.3-26), L_{bsbs} is a maximum. This, of course, corresponds to that which is portrayed in Fig. 6.2-1.

The mutual inductance $L_{asbs}(L_{bsas})$ is next. One would think that since the windings are orthogonal, the mutual coupling would always be zero. However, this is not the case due to the fact that the air gap is not uniform. Let us consider Fig. 6.3-1 where various rotor positions are shown with only the flux paths of the as winding illustrated. Coupling occurs when flux produced by one winding links the other winding; in particular, when the flux of the as winding links the bs winding. This will give us L_{bsas} and we know that $L_{asbs} = L_{bsas}$.

Note that, when $\theta_r = 0$, π, and 2π as shown in Fig. 6.3-1a or when $\theta_r = \frac{1}{2}\pi$ and $\frac{3}{2}\pi$ as shown in Fig. 6.3-1b, L_{bsas} is zero. In these positions, there is no channeling of the flux of one winding through the other. However, let the rotor start to turn counterclockwise from zero toward $\frac{1}{2}\pi$ and consider the flux produced by the as winding. As the rotor turns, the configuration of the rotor provides a low-reluctance path to the flux produced by the as winding and the flux is channeled through the bs winding with maximum coupling occurring at $\theta_r = \frac{1}{4}\pi$, as illustrated in Fig. 6.3-1c. We see that this same rotor position relative to the windings occurs also at $\theta_r = \frac{5}{4}\pi$. Maximum coupling will again occur at $\theta_r = \frac{3}{4}\pi$ and $\frac{7}{4}\pi$, as illustrated by Fig. 6.3-1d. Now, what is the sign of the mutual inductance? With the assumed direction of positive currents, the right-hand rule tells us that L_{bsas} (or L_{asbs}) is negative at $\theta_r = \frac{1}{4}\pi$, $\frac{5}{4}\pi$, ... (the fluxes of the windings oppose each other for positive currents) and positive for $\theta_r = \frac{3}{4}\pi$, $\frac{7}{4}\pi$, ... (the fluxes aid each other). If we sketch L_{bsas} versus θ_r using the above information, we see, from Fig. 6.3-1e, that, as a first approximation, L_{bsas} or L_{asbs} may be expressed as

$$L_{bsas} = L_{asbs} = -L_B \sin 2\theta_r \qquad (6.3\text{-}27)$$

In order for us to prove that the coefficient is L_B, it would be necessary to become quite involved [1]. We will accept this without proving it.

Let us now go back to the flux linkage equations, (6.3-12) through (6.3-16), and write these equations in matrix form as

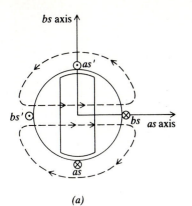

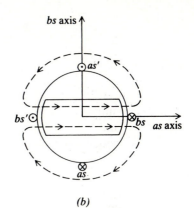

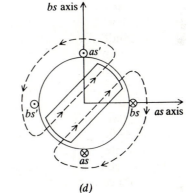

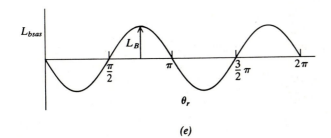

FIGURE 6.3-1
Flux path of *as* winding illustrating the mutual coupling between stator windings to determine L_{bsas} and L_{asbs}. (*a*) $\theta_r = 0$, π, and 2π; (*b*) $\theta_r = \frac{1}{2}\pi$ and $\frac{3}{2}\pi$; (*c*) $\theta_r = \frac{1}{4}\pi$ and $\frac{5}{4}\pi$; (*d*) $\theta_r = \frac{3}{4}\pi$ and $\frac{7}{4}\pi$; (*e*) approximation of L_{bsas} and L_{asbs}.

$$\begin{bmatrix} \boldsymbol{\lambda}_{abs} \\ \boldsymbol{\lambda}_{qdr} \end{bmatrix} = \begin{bmatrix} \mathbf{L}_s & \mathbf{L}_{sr} \\ (\mathbf{L}_{sr})^T & \mathbf{L}_r \end{bmatrix} \begin{bmatrix} \mathbf{i}_{abs} \\ \mathbf{i}_{qdr} \end{bmatrix} \qquad (6.3\text{-}28)$$

The matrix $\mathbf{L}_s$ can now be written as

$$\mathbf{L}_s = \begin{bmatrix} L_{asas} & L_{asbs} \\ L_{bsas} & L_{bsbs} \end{bmatrix}$$
$$= \begin{bmatrix} L_{ls} + L_A - L_B \cos 2\theta_r & -L_B \sin 2\theta_r \\ -L_B \sin 2\theta_r & L_{ls} + L_A + L_B \cos 2\theta_r \end{bmatrix} \qquad (6.3\text{-}29)$$

By inspection of Fig. 6.2-1, we can write

$$\mathbf{L}_{sr} = \begin{bmatrix} L_{askq} & L_{asfd} & L_{askd} \\ L_{bskq} & L_{bsfd} & L_{bskd} \end{bmatrix}$$
$$= \begin{bmatrix} L_{skq} \cos \theta_r & L_{sfd} \sin \theta_r & L_{skd} \sin \theta_r \\ L_{skq} \sin \theta_r & -L_{sfd} \cos \theta_r & -L_{skd} \cos \theta_r \end{bmatrix} \qquad (6.3\text{-}30)$$

$$\mathbf{L}_r = \begin{bmatrix} L_{kqkq} & L_{kqfd} & L_{kqkd} \\ L_{fdkq} & L_{fdfd} & L_{fdkd} \\ L_{kdkq} & L_{kdfd} & L_{kdkd} \end{bmatrix}$$
$$= \begin{bmatrix} L_{lkq} + L_{mkq} & 0 & 0 \\ 0 & L_{lfd} + L_{mfd} & L_{fdkd} \\ 0 & L_{fdkd} & L_{lkd} + L_{mkd} \end{bmatrix} \qquad (6.3\text{-}31)$$

In the above inductance matrices, the leakage inductances are denoted with an l in the subscript. The skq, sfd, and skd subscripts denote the peak mutual inductances between stator and rotor windings. The following equations define the inductances used in (6.3-30) and (6.3-31):

$$L_{skq} = \frac{N_{kq}}{N_s} L_{mq} \qquad (6.3\text{-}32)$$

$$L_{sfd} = \frac{N_{fd}}{N_s} L_{md} \qquad (6.3\text{-}33)$$

$$L_{skd} = \frac{N_{kd}}{N_s} L_{md} \qquad (6.3\text{-}34)$$

$$L_{mkq} = \left(\frac{N_{kq}}{N_s} \right)^2 L_{mq} \qquad (6.3\text{-}35)$$

$$L_{mfd} = \left(\frac{N_{fd}}{N_s} \right)^2 L_{md} \qquad (6.3\text{-}36)$$

$$L_{mkd} = \left(\frac{N_{kd}}{N_s} \right)^2 L_{md} \qquad (6.3\text{-}37)$$

$$L_{fdkd} = \frac{N_{kd}}{N_{fd}} L_{mfd} = \frac{N_{fd}}{N_{kd}} L_{mkd} \qquad (6.3\text{-}38)$$

As in the case of the induction machine, it is convenient to refer the rotor variables to a winding with N_s turns. Thus,

$$i'_j = \frac{N_j}{N_s} i_j \tag{6.3-39}$$

$$v'_j = \frac{N_s}{N_j} v_j \tag{6.3-40}$$

$$\lambda'_j = \frac{N_s}{N_j} \lambda_j \tag{6.3-41}$$

where j may be kq, fd, or kd. The flux linkage equations given by (6.3-28) may now be written as

$$\begin{bmatrix} \boldsymbol{\lambda}_{abs} \\ \boldsymbol{\lambda}'_{qdr} \end{bmatrix} = \begin{bmatrix} \mathbf{L}_s & \mathbf{L}'_{sr} \\ (\mathbf{L}'_{sr})^T & \mathbf{L}'_r \end{bmatrix} \begin{bmatrix} \mathbf{i}_{abs} \\ \mathbf{i}'_{qdr} \end{bmatrix} \tag{6.3-42}$$

where

$$\mathbf{L}'_{sr} = \begin{bmatrix} L_{mq} \cos \theta_r & L_{md} \sin \theta_r & L_{md} \sin \theta_r \\ L_{mq} \sin \theta_r & -L_{md} \cos \theta_r & -L_{md} \cos \theta_r \end{bmatrix} \tag{6.3-43}$$

$$\mathbf{L}'_r = \begin{bmatrix} L'_{lkq} + L_{mq} & 0 & 0 \\ 0 & L'_{lfd} + L_{md} & L_{md} \\ 0 & L_{md} & L'_{lkd} + L_{md} \end{bmatrix} \tag{6.3-44}$$

The voltage equations expressed in terms of machine variables referred by a turns-ratio to the stator windings are

$$\mathbf{v}_{abs} = \mathbf{r}_s \mathbf{i}_{abs} + p\boldsymbol{\lambda}_{abs} \tag{6.3-45}$$

$$\mathbf{v}'_{qdr} = \mathbf{r}'_r \mathbf{i}'_{qdr} + p\boldsymbol{\lambda}'_{qdr} \tag{6.3-46}$$

In terms of inductances, (6.3-45) and (6.3-46) become

$$\begin{bmatrix} \mathbf{v}_{abs} \\ \mathbf{v}'_{qdr} \end{bmatrix} = \begin{bmatrix} \mathbf{r}_s + p\mathbf{L}_s & p\mathbf{L}'_{sr} \\ p(\mathbf{L}'_{sr})^T & \mathbf{r}'_r + p\mathbf{L}'_r \end{bmatrix} \begin{bmatrix} \mathbf{i}_{abs} \\ \mathbf{i}'_{qdr} \end{bmatrix} \tag{6.3-47}$$

where in the matrices $\mathbf{r}'_r$ and $\mathbf{L}'_r$

$$r'_j = \left(\frac{N_s}{N_j}\right)^2 r_j \tag{6.3-48}$$

$$L'_{lj} = \left(\frac{N_s}{N_j}\right)^2 L_{lj} \tag{6.3-49}$$

where, as before, j may be kq, fd, or kd.

Since the synchronous machine is generally operated as a generator, it is often considered more convenient to assume positive current out of the machine. This may be done in the above equations by simply placing a negative sign preceding $\mathbf{i}_{abs}$.

SP6.3-1. Express L_{asbs} for positive θ_r in the clockwise direction in Fig. 6.2-1 with (a) positive direction of i_{as} reversed; (b) positive direction of i_{bs} reversed; and (c) positive direction of both i_{as} and i_{bs} reversed. [(a) and (b) $L_{asbs} = (6.3\text{-}27)$; (c) $L_{asbs} = -(6.3\text{-}27)$]

SP6.3-2. The current i_{fd} in Fig. 6.2-1 is 1 A, $L_{sfd} = 0.1$ H, and $\theta_r = 10t$. Determine the open-circuited steady-state voltages V_{as} and V_{bs}. [$V_{as} = \cos 10t$; $V_{bs} = \sin 10t$]

SP6.3-3. The current i'_{fd} in a round-rotor synchronous machine is 1 A, $L_{mq} = 0.1$ H, $L_{asfd} = \sin \theta_r$, and $\theta_r = 10t$. Determine the open-circuited steady-state voltages V_{as} and V_{bs}. [SP6.3-2]

6.4   TORQUE

The electromagnetic torque may be evaluated from Table 2.5-1:

$$T_e = \frac{P}{2} \frac{\partial W_c(\mathbf{i}, \theta_r)}{\partial \theta_r} \tag{6.4-1}$$

For a magnetically linear system, this yields

$$
\boxed{
\begin{aligned}
T_e = \frac{P}{2} \Big\{ & \frac{L_{md} - L_{mq}}{2} \left[(i_{as}^2 - i_{bs}^2) \sin 2\theta_r - 2 i_{as} i_{bs} \cos 2\theta_r \right] \\
& - L_{mq} i_{kq} (i_{as} \sin \theta_r - i_{bs} \cos \theta_r) \\
& + L_{md} (i'_{fd} + i'_{kd})(i_{as} \cos \theta_r + i_{bs} \sin \theta_r) \Big\}
\end{aligned}
}
\tag{6.4-2}
$$

The above expression for torque is positive for motor action. Obtaining (6.4-2) from (6.4-1) is a problem at the end of the chapter.

The torque and rotor speed are related by

$$\boxed{ T_e = J\left(\frac{2}{P}\right) \frac{d\omega_r}{dt} + B_m \left(\frac{2}{P}\right) \omega_r + T_L }\tag{6.4-3}$$

where J is the inertia expressed in kilogram $\cdot$ meter^2 (kg $\cdot$ m^2) or joule $\cdot$ second^2 (J $\cdot$ s^2). Often, the inertia is given as WR^2 in units of pound mass $\cdot$ feet^2 (lbm $\cdot$ ft^2). As indicated in Fig. 6.2-1, T_L is positive for a torque load when the machine is operated as a motor and negative when torque is supplied to the shaft of the machine by a prime mover (generator action). The constant B_m is a damping coefficient associated with the rotational system of the machine and mechanical load. It has the units N $\cdot$ m $\cdot$ s/rad of mechanical rotation, and it is generally small and often neglected in the case of the machine but may be considerable for the mechanical load.

SP6.4-1. Which of the terms on the right-hand side of (6.4-2) can be thought of as the reluctance torque? $\left\{\left(\dfrac{P}{2}\right)\left(\dfrac{L_{md} - L_{mq}}{2}\right)[\quad]\right\}$

SP6.4-2. Repeat SP6.4-1 for the damping (induction motor) torque. [Terms with i'_{kq} or i'_{kd}]

SP6.4-3. Repeat SP6.4-1 for the torque due to the interaction of mmf_s and the *fd* current. [Terms with i'_{fd}]

6.5 MACHINE EQUATIONS IN THE ROTOR REFERENCE FRAME

The mutual inductances between the stator and rotor windings vary sinusoidally with θ_r. Moreover, the self-inductances of the stator windings are sinusoidal functions of $2\theta_r$. Fortunately, a change of variables is helpful. It appears that R. H. Park was the first to incorporate a change of variables in the analysis of synchronous machines [2]. He transformed the stator variables to the rotor reference frame, thereby eliminating the time-varying inductances. For a two-phase machine, Park's transformation is

$$\begin{bmatrix} f^r_{qs} \\ f^r_{ds} \end{bmatrix} = \begin{bmatrix} \cos\theta_r & \sin\theta_r \\ \sin\theta_r & -\cos\theta_r \end{bmatrix} \begin{bmatrix} f_{as} \\ f_{bs} \end{bmatrix} \tag{6.5-1}$$

or

$$\boxed{\mathbf{f}^r_{qds} = \mathbf{K}^r_s \mathbf{f}_{abs}} \tag{6.5-2}$$

where f can represent either voltage, current, or flux linkage. It follows that

$$\mathbf{f}_{abs} = (\mathbf{K}^r_s)^{-1} \mathbf{f}^r_{qds} \tag{6.5-3}$$

where it can be shown that $(\mathbf{K}^r_s)^{-1} = \mathbf{K}^r_s$. Also,

$$\theta_r = \int_0^t \omega_r(\xi)\, d\xi + \theta_r(0) \tag{6.5-4}$$

where ξ is a dummy variable of integration and $\theta_r(0)$ is the time zero position of the rotor, which is generally selected to be zero. The *s* subscript denotes stator variables and the *r* superscript indicates that the transformation is to a reference frame fixed in the rotor. This same transformation is used in Chap. 7 to analyze the permanent-magnet synchronous machine (brushless dc motor).

Although the above change of variables does not require a physical connotation, it may be helpful to visualize this transformation as trigonometric relationships between variables with directions as shown in Fig. 6.5-1. The direction of f_{as} and f_{bs} variables shown in Fig. 6.5-1 happens to be the positive direction of the magnetic axes of the associated windings (*as* winding and *bs* winding). It follows then that the f^r_{qs} and f^r_{ds} variables can be thought of as

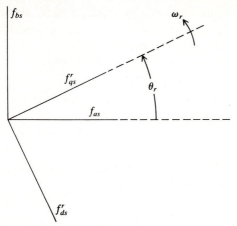

FIGURE 6.5-1
Trigonometric interpretation of the change of stator variables.

being associated with fictitious windings fixed on the rotor the positive magnetic axes of which are in the same direction as the direction of f^r_{qs} and f^r_{ds}. These fictitious windings result due to the change of variables for the stator variables. The positive magnetic axis of the winding associated with the f^r_{qs} variables coincides with the q axis of the rotor and the magnetic axis of the winding associated with the f^r_{ds} variables coincides with the d axis of the rotor.

The s (r) subscript denotes association with the stator (rotor) variables. As mentioned previously, the superscript r indicates that the transformation is to the rotor reference frame. Since the rotor reference frame is generally the only reference frame used in the analysis of synchronous machines, it could be omitted; however, we will carry it along for completeness.

Substituting (6.5-3) into (6.3-45) yields

$$(\mathbf{K}^r_s)^{-1}\mathbf{v}^r_{qds} = \mathbf{r}_s(\mathbf{K}^r_s)^{-1}\mathbf{i}^r_{qds} + p[(\mathbf{K}^r_s)^{-1}\boldsymbol{\lambda}^r_{qds}] \tag{6.5-5}$$

Multiplying (6.5-5) by $\mathbf{K}^r_s$ and rewriting (6.3-46) with the superscript r incorporated for notational completeness, we obtain Park's equations.

$$\mathbf{v}^r_{qds} = \mathbf{r}_s\mathbf{i}^r_{qds} + \omega_r\boldsymbol{\lambda}^r_{dqs} + p\boldsymbol{\lambda}^r_{qds} \tag{6.5-6}$$

$$\mathbf{v}'^r_{qdr} = \mathbf{r}'_r\mathbf{i}'^r_{qdr} + p\boldsymbol{\lambda}'^r_{qdr} \tag{6.5-7}$$

where

$$(\boldsymbol{\lambda}^r_{dqs})^T = [\lambda^r_{ds} \quad -\lambda^r_{qs}] \tag{6.5-8}$$

Equation (6.5-7) is (6.3-46) with the r superscript added to emphasize the fact that the voltage equations for the rotor circuits are written in the rotor frame of reference. We will not use a change of variables for the rotor variables. The last two terms of (6.5-6) come from the last term of (6.5-5) multiplied by $\mathbf{K}^r_s$. That is,

$$\mathbf{K}^r_s p[(\mathbf{K}^r_s)^{-1}\boldsymbol{\lambda}^r_{qds}] = \mathbf{K}^r_s[p(\mathbf{K}^r_s)^{-1}]\boldsymbol{\lambda}^r_{qds} + \mathbf{K}^r_s(\mathbf{K}^r_s)^{-1}p\boldsymbol{\lambda}^r_{qds} \tag{6.5-9}$$

It is left to the reader to show that the right-hand side of (6.5-9) reduces to the last two terms of (6.5-6). As a guide, one may wish to refer to (4.8-16) where a similar procedure was carried out for the induction machine.

Equations (6.5-6) and (6.5-7) are valid for a linear or nonlinear magnetic system. For a linear magnetic system, we can express $\boldsymbol{\lambda}_{abs}$ and $\boldsymbol{\lambda}'_{qdr}$ from (6.3-42):

$$\boldsymbol{\lambda}_{abs} = \mathbf{L}_s \mathbf{i}_{abs} + \mathbf{L}'_{sr} \mathbf{i}'_{qdr} \tag{6.5-10}$$

$$\boldsymbol{\lambda}'_{qdr} = (\mathbf{L}'_{sr})^T \mathbf{i}_{abs} + \mathbf{L}'_r \mathbf{i}'_{qdr} \tag{6.5-11}$$

where $\mathbf{L}_s$, $\mathbf{L}'_{sr}$, and $\mathbf{L}'_r$ are given by (6.3-29), (6.3-43), and (6.3-44), respectively. Substituting (6.5-3) for $\boldsymbol{\lambda}_{abs}$ and $\mathbf{i}_{abs}$ and adding the r superscript to $\boldsymbol{\lambda}'_{qdr}$ and $\mathbf{i}'_{qdr}$ and then solving (6.5-10) for $\boldsymbol{\lambda}'_{qds}$ yields

$$\boldsymbol{\lambda}^r_{qds} = \mathbf{K}^r_s \mathbf{L}_s (\mathbf{K}^r_s)^{-1} \mathbf{i}^r_{qds} + \mathbf{K}^r_s \mathbf{L}'_{sr} \mathbf{i}''_{qdr} \tag{6.5-12}$$

$$\boldsymbol{\lambda}''^r_{qdr} = (\mathbf{L}'_{sr})^T (\mathbf{K}^r_s)^{-1} \mathbf{i}^r_{qds} + \mathbf{L}'_r \mathbf{i}''^r_{qdr} \tag{6.5-13}$$

We can show that

$$\mathbf{K}^r_s \mathbf{L}_s (\mathbf{K}^r_s)^{-1} = \begin{bmatrix} L_{ls} + L_{mq} & 0 \\ 0 & L_{ls} + L_{md} \end{bmatrix} \tag{6.5-14}$$

$$\mathbf{K}^r_s \mathbf{L}'_{sr} = \begin{bmatrix} L_{mq} & 0 & 0 \\ 0 & L_{md} & L_{md} \end{bmatrix} \tag{6.5-15}$$

$$(\mathbf{L}'_{sr})^T (\mathbf{K}^r_s)^{-1} = \begin{bmatrix} L_{mq} & 0 \\ 0 & L_{md} \\ 0 & L_{md} \end{bmatrix} \tag{6.5-16}$$

where L_{mq} and L_{md} are defined by (6.3-21) and (6.3-22), respectively. In a problem at the end of the chapter, you are asked to obtain (6.5-14) through (6.5-16). The flux linkage equations may now be written as

$$\begin{bmatrix} \lambda^r_{qs} \\ \lambda^r_{ds} \\ \lambda''^r_{kq} \\ \lambda''^r_{fd} \\ \lambda''^r_{kd} \end{bmatrix} = \begin{bmatrix} L_{ls} + L_{mq} & 0 & L_{mq} & 0 & 0 \\ 0 & L_{ls} + L_{md} & 0 & L_{md} & L_{md} \\ L_{mq} & 0 & L'_{lkq} + L_{mq} & 0 & 0 \\ 0 & L_{md} & 0 & L'_{lfd} + L_{md} & L_{md} \\ 0 & L_{md} & 0 & L_{md} & L'_{lkd} + L_{md} \end{bmatrix} \begin{bmatrix} i^r_{qs} \\ i^r_{ds} \\ i''^r_{kq} \\ i''^r_{fd} \\ i''^r_{kd} \end{bmatrix} \tag{6.5-17}$$

We have accomplished our goal; the self- and mutual inductances in (6.5-17) are constant. Moreover, all q circuits are magnetically decoupled from d circuits. We now see that the fictitious windings are indeed fixed in the rotor reference frame. Since the mutual inductances between the fictitious windings ($\overset{r}{qs}$ and $\overset{r}{ds}$ windings) and the rotor windings are all constant, the fictitious windings and the rotor windings are not in relative motion. Hence, the $\overset{r}{qs}$ and $\overset{r}{ds}$ windings are fixed on the rotor.

The inductance $L_{ls} + L_{mq}$ is commonly called the q-axis inductance and denoted L_q. Similarly, $L_{ls} + L_{md}$ is called the d-axis inductance and denoted L_d. That is,

$$L_q = L_{ls} + L_{mq} \tag{6.5-18}$$

$$L_d = L_{ls} + L_{md} \tag{6.5-19}$$

If the air gap is uniform, $L_q = L_d$. Otherwise, $L_q < L_d$.

Park's equations are often written in expanded form. From (6.5-6) and (6.5-7),

$$v_{qs}^r = r_s i_{qs}^r + \omega_r \lambda_{ds}^r + p\lambda_{qs}^r \tag{6.5-20}$$

$$v_{ds}^r = r_s i_{ds}^r - \omega_r \lambda_{qs}^r + p\lambda_{ds}^r \tag{6.5-21}$$

$$v_{kq}''^r = r_{kq}' i_{kq}''^r + p\lambda_{kq}''^r \tag{6.5-22}$$

$$v_{fd}''^r = r_{fd}' i_{fd}''^r + p\lambda_{fd}''^r \tag{6.5-23}$$

$$v_{kd}''^r = r_{kd}' i_{kd}''^r + p\lambda_{kd}''^r \tag{6.5-24}$$

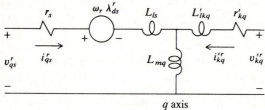

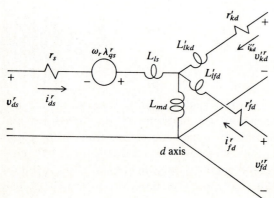

q axis

d axis

FIGURE 6.5-2
Equivalent circuits of a two-phase synchronous machine with reference frame fixed in rotor—Park's equations.

Although we will carry the s subscript, the r superscript, and the primes, Park's equations are generally written without these notations. Also, we realize that the damper windings are always short-circuited, hence, v''^r_{kq} and v''^r_{kd} are zero.

For a linear magnetic system, the flux linkage equations may be written from (6.5-17) as

$$\lambda^r_{qs} = L_{ls}i^r_{qs} + L_{mq}(i^r_{qs} + i''^r_{kq})$$
$$= L_q i^r_{qs} + L_{mq}i''^r_{kq} \tag{6.5-25}$$

$$\lambda^r_{ds} = L_{ls}i^r_{ds} + L_{md}(i^r_{ds} + i''^r_{fd} + i''^r_{kd})$$
$$= L_d i^r_{ds} + L_{md}(i''^r_{fd} + i''^r_{kd}) \tag{6.5-26}$$

$$\lambda''^r_{kq} = L'_{lkq}i''^r_{kq} + L_{mq}(i^r_{qs} + i''^r_{kq})$$
$$= L'_{kq}i''^r_{kq} + L_{mq}i^r_{qs} \tag{6.5-27}$$

$$\lambda''^r_{fd} = L'_{lfd}i''^r_{fd} + L_{md}(i^r_{ds} + i''^r_{fd} + i''^r_{kd})$$
$$= L'_{fd}i''^r_{fd} + L_{md}(i^r_{ds} + i''^r_{kd}) \tag{6.5-28}$$

$$\lambda''^r_{kd} = L'_{lkd}i''^r_{kd} + L_{md}(i^r_{ds} + i''^r_{fd} + i''^r_{kd})$$
$$= L'_{kd}i''^r_{kd} + L_{md}(i^r_{ds} + i''^r_{fd}) \tag{6.5-29}$$

where L_q and L_d are defined by (6.5-18) and (6.5-19), respectively, and

$$L'_{kq} = L'_{lkq} + L_{mq} \tag{6.5-30}$$

$$L'_{fd} = L'_{lfd} + L_{md} \tag{6.5-31}$$

$$L'_{kd} = L'_{lkd} + L_{md} \tag{6.5-32}$$

The voltage and flux linkage equations suggest the equivalent circuits shown in Fig. 6.5-2. Substituting (6.5-25) through (6.5-29) into (6.5-20) through (6.5-24) yields the voltage equations in terms of currents.

$$
\begin{bmatrix} v^r_{qs} \\ v^r_{ds} \\ v''^r_{kq} \\ v''^r_{fd} \\ v''^r_{kd} \end{bmatrix} =
\begin{bmatrix}
r_s + pL_q & \omega_r L_d & pL_{mq} & \omega_r L_{md} & \omega_r L_{md} \\
-\omega_r L_q & r_s + pL_d & -\omega_r L_{mq} & pL_{md} & pL_{md} \\
pL_{mq} & 0 & r'_{kq} + pL'_{kq} & 0 & 0 \\
0 & pL_{md} & 0 & r'_{fd} + pL'_{fd} & pL_{md} \\
0 & pL_{md} & 0 & pL_{md} & r'_{kd} + pL'_{kd}
\end{bmatrix}
\begin{bmatrix} i^r_{qs} \\ i^r_{ds} \\ i''^r_{kq} \\ i''^r_{fd} \\ i''^r_{kd} \end{bmatrix}
$$

$$\tag{6.5-33}$$

The expression for the electromagnetic torque in rotor reference frame variables may be obtained by substituting the equation of transformation into (6.4-2). After considerable work,

$$T_e = \frac{P}{2}\left[L_{md}(i_{ds}^r + i_{fd}^{'r} + i_{kd}^{'r})i_{qs}^r - L_{mq}(i_{qs}^r + i_{kq}^{'r})i_{ds}^r\right] \qquad (6.5\text{-}34)$$

which may also be written as

$$T_e = \frac{P}{2}(\lambda_{ds}^r i_{qs}^r - \lambda_{qs}^r i_{ds}^r) \qquad (6.5\text{-}35)$$

It is helpful to review what has been done thus far in this chapter. Since the stator and rotor windings of a synchronous machine are in relative motion, it is necessary to implement a change of variables which, in effect, eliminates the relative motion between circuits. However, the synchronous machine is not too cooperative. Not only are there circuits in relative motion, but also the rotor of the salient-pole synchronous machine gives rise to sinusoidal variations of $2\theta_r$ in the self-inductances of the stator windings. To make matters worse, the rotor windings are not symmetrical.

As we think about this situation, we conclude that Park really had no choice but to devise a change of variables which would transform the stator variables to fictitious circuits fixed on the rotor. First, the air gap of a salient-pole synchronous machine is not uniform, hence, only circuits fixed on the rotor could experience a constant self-inductance. Yes, that is right, but what about a round-rotor synchronous machine? Why, in the case of a round-rotor machine, would it be necessary to transform the stator variables to fictitious rotor circuits? The windings must be symmetrical to benefit from a change of variables which is a function of an angular displacement. (You may wish to refer to Example 5A if you have questions about this statement.) The rotor windings of a synchronous, salient-pole, or round-rotor machine, are, in general, asymmetrical. We are unaware of a change of variables which provides an advantage in transforming the variables of asymmetrical windings to a reference frame other than that where the windings physically exist. Perhaps another Park will come along and help us out. Until then, we shall have to be content to use the rotor reference frame to analyze synchronous machines.

Although we are probably getting too involved in reference frame theory, you might have thought by this time that, if the rotor were round and if the rotor windings were made symmetrical, we could use any reference frame. This is indeed the case. If the air gap were uniform (round rotor) and if the kd and kq windings were identical, and if an fq winding identical to the fd winding were added in the q axis, the rotor windings would form two symmetrical sets. The machine would then have the same configuration as an induction machine with two sets of symmetrical rotor windings and we would not be restricted by the configuration of the machine to any reference frame.

Example 6A. A two-pole two-phase reluctance machine is identical in configuration to the synchronous machine shown in Fig. 6.2-1 with the field winding (fd

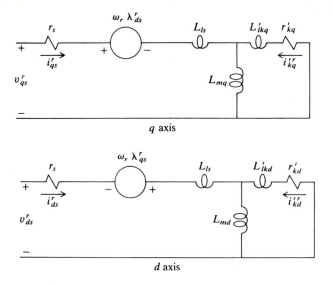

FIGURE 6A-1
Rotor reference frame equivalent circuits for reluctance machine.

winding) absent. It is our job to derive the equivalent circuits for the reluctance machine in the rotor reference frame.

Actually we need not do any derivation; it has already been done. We need only to eliminate the field winding from the equivalent circuits shown in Fig. 6.5-2. In particular, Fig. 6A-1 shows equivalent circuits in the rotor reference frame for a two-phase reluctance machine wherein the damper windings are shown short-circuited as they always are in real life.

SP6.5-1. $f_{as} = \cos \omega_e t$ and $f_{bs} = \sin \omega_e t$. Determine f'_{qs} and f'_{ds} if $\theta_r = \omega_r t$ and $\omega_r = \omega_e$. $[f'_{qs} = 1; f'_{ds} = 0]$

SP6.5-2. Determine $\theta_r(0)$ so that $f'_{qs} = f'_{ds}$ in SP6.5-1, where here $\theta_r = \omega_r t + \theta_r(0)$. $[\theta_r(0) = \frac{1}{4}\pi$ and $-\frac{3}{4}\pi]$

SP6.5-3. Determine $\theta_{ef}(0)$ so that $f'_{qs} = f'_{ds}$ in SP6.5-1, where $\theta_{ef} = \omega_e t + \theta_{ef}(0)$. $[\theta_{ef}(0) = -\frac{1}{4}\pi$ and $\frac{3}{4}\pi]$

SP6.5-4. Repeat SP6.5-1 for $\omega_r = -\omega_e$. $[f'_{qs} = \cos 2\omega_e t; f'_{ds} = -\sin 2\omega_e t]$

6.6 ROTOR ANGLE

It would seem that we have already done our share of defining concepts and terms which we have not seen or used before; the sinusoidally distributed winding, self- and mutual inductances that vary, changes of variables that give rise to fictitious windings, and a machine that develops "three types of torque." We cannot help but wonder when these "new concepts" are going to stop. Unfortunately, we are now faced with another definition which has evolved over the years to become deeply ingrained in synchronous-machine theory. The rotor angle is the case in point.

In its broadest definition, the rotor angle δ is

$$\delta = \theta_r - \theta_{esv}$$

$$= \int_0^t [\omega_r(\xi) - \omega_e(\xi)]\, d\xi + \theta_r(0) - \theta_{esv}(0) \qquad (6.6\text{-}1)$$

where θ_{esv} is the angular displacement of a stator phase voltage, generally v_{as}. In (6.6-1) ξ is a dummy variable of integration and $\omega_r(\xi)$ and $\omega_e(\xi)$ are the electrical angular velocity of the rotor and the terminal voltages, respectively. The time zero position is generally selected so that the fundamental component of v_{as} is maximum at $t = 0$, for example, a cosine with $\theta_{esv}(0) = 0$. Although the above definition of δ is valid regardless of the mode of operation (either or both ω_r and ω_e may vary), a physical interpretation is most easily visualized during balanced steady-state operation. It is, however, important to mention in passing that here we see the mixing of a variable associated with the electric system, θ_e (ω_e), and a variable associated with the mechanical system, θ_r (ω_r). Fortunately, good will come from this, even though it can be somewhat confusing when, in the next section, we superimpose a phasor diagram which rotates at ω_e upon the rotor, which also rotates at ω_e during steady-state operation, and then note that the angle between $\tilde{V}_{as}$ and the q axis is δ, the rotor angle.

SP6.6-1. Calculate δ for SP6.5-1, SP6.5-2, and SP6.5-4. Assume f_{as} and f_{bs} are v_{as} and v_{bs}, respectively. $[\delta = 0; \; \delta = \tfrac{1}{4}\pi$ and $-\tfrac{3}{4}\pi; \; \delta = -2\omega_e t]$
SP6.6-2. Why did we not ask you to calculate δ for SP6.5-3? [We could have, but $\theta_{esv}(0)$ is generally selected to be zero.]

6.7 ANALYSIS OF STEADY-STATE OPERATION

In the case of the synchronous machine, we have found it necessary to refer the stator variables to the rotor reference frame. What will be the frequency of the variables in the rotor reference frame during balanced steady-state operation? Actually, we have our answer from SP6.5-1, but let us think more about this before doing anything analytical. First, we know that, during steady-state operation, the electrical angular velocity of the rotor, ω_r, is equal to ω_e. Hence, the circuits that physically exist on the rotor (kq, fd, and kd windings) or fictitious windings put there because of a change of variables ($\overset{r}{q}s$ and $\overset{r}{d}s$ windings) do not experience a change of flux linkages. How do we know this? Well, the air gap mmf established by the constant (dc) field current is, of course, constant relative to the windings on the rotor. Now what about mmf_s, the rotating air gap mmf established by the balanced, sinusoidal stator currents? It rotates at ω_e which is also the speed ω_r of the rotor. Hence, since neither mmf_s nor mmf_r is changing relative to the rotor, the windings on the rotor (physical or fictitiously) will not experience a change in flux linkages. At this point, we can forget about the short-circuited damper windings since, without a change of flux linkages, there can be no induced voltage and, thus, $i_{kq}^{\prime r}$ and $i_{kd}^{\prime r}$ must be zero for balanced steady-state operation where $\omega_r = \omega_e$. Actually, if we accept the fact that there is not a change of flux linkages relative to the rotor circuits, then there can be no induced voltage due to transformer action in any of the circuits on the rotor. One would then guess that the currents and voltages associated with all rotor windings, actual or fictitious, would have to be constant ($i_{kq}^{\prime r}$ and $i_{kd}^{\prime r}$ are constant at zero). This seems logical, but what has happened to the balanced, sinusoidal stator variables? Remember the balanced, steady-state sinusoidal stator currents give rise to a constant mmf$_s$ rotating at ω_e. But, if mmf$_s$, which rotates at ω_e, is now to be produced by currents flowing in fictitious windings (i_{qs}^r and i_{ds}^r), which are mathematically fixed on the rotor which is rotating at ω_e during steady-state operation, what must be the frequency of these currents flowing in the fictitious windings? They must be constant (dc).

Now that we know what to expect, let us proceed. During balanced steady-state operation, the stator variables may be expressed as

$$F_{as} = \sqrt{2} F_s \cos\left[\omega_e t + \theta_{esf}(0)\right] \tag{6.7-1}$$

$$F_{bs} = \sqrt{2} F_s \sin\left[\omega_e t + \theta_{esf}(0)\right] \tag{6.7-2}$$

Substituting F_{as} and F_{bs} into the equations of transformation, (6.4-2), with $\omega_r = \omega_e$, yields

$$F_{qs}^r = \sqrt{2} F_s \cos\left[\theta_{esf}(0) - \theta_r(0)\right] \tag{6.7-3}$$

$$F_{ds}^r = -\sqrt{2} F_s \sin\left[\theta_{esf}(0) - \theta_r(0)\right] \tag{6.7-4}$$

Clearly, $\theta_{esf}(0)$ and $\theta_r(0)$ are constants and, thus, F_{qs}^r and F_{ds}^r are constants. In other words, a balanced set of sinusoidal stator variables become constants in the rotor reference frame during steady-state conditions where $\omega_r = \omega_e$.

Let us now go back to the voltage equations in the rotor reference frame, (6.5-20) through (6.5-24). As we have agreed, for balanced steady-state operation we can forget about the damper windings. Hence, (6.5-22) and (6.5-24) play no role in the analysis of steady-state operation of a synchronous machine. Moreover, since the voltages and currents in the rotor reference frame are constants during balanced steady-state operation, we can apply dc circuit theory to the $\overset{r}{qs}$, $\overset{r}{ds}$, and $\overset{r}{fd}$ voltage equations. In particular, since all variables are constant during steady-state operation, all variables multiplied by the operator p (d/dt) are zero. Hence, (6.5-20), (6.5-21), and (6.5-23) may be written as

$$V_{qs}^r = r_s I_{qs}^r + \omega_r \lambda_{ds}^r \tag{6.7-5}$$

$$V_{ds}^r = r_s I_{ds}^r - \omega_r \lambda_{qs}^r \tag{6.7-6}$$

$$V_{fd}^{\prime r} = r_{fd}^\prime I_{fd}^{\prime r} \tag{6.7-7}$$

where capital letters have been used to denote steady-state quantities. Now λ^r_{qs} and λ^r_{ds} are (6.5-25) and (6.5-26), respectively, wherein the damper winding currents are set equal to zero. Appropriate substitution of (6.5-25) and (6.5-26) into (6.7-5) and (6.7-6), wherein ω_r is set equal to ω_e, yields

$$V^r_{qs} = r_s I^r_{qs} + X_d I^r_{ds} + X_{md} I''_{fd} \qquad (6.7\text{-}8)$$

$$V^r_{ds} = r_s I^r_{ds} - X_q I^r_{qs} \qquad (6.7\text{-}9)$$

Note these are dc voltage equations and, yet, we have reactances X_d, X_q, and X_{md} without j's. We are not dealing with phasors, yet X times I is a voltage, regardless. Reactances in dc voltage equations? Add another "new" concept to the list at the beginning of Sec. 6.6.

The above voltage equations can be used in their present form to analyze the synchronous machine; however, it is convenient and customary to relate the F^r_{qs} and F^r_{ds} quantities, which are constants, to $\tilde{F}_{as}$, which is a phasor representing a sinusoidal voltage. Once we have done this, we will have another one of these "new" concepts. They seem to be coming one right after another. Sorry! To accomplish this goal, let us first look at δ for steady-state operation. In particular, if we arbitrarily select or "call" time zero with $\omega_r = \omega_e$, the steady-state rotor angle from (6.6-1) becomes

$$\delta = \theta_r(0) - \theta_{esv}(0) \qquad (6.7\text{-}10)$$

Later, we will set $\theta_{esv}(0) = 0$, but for now we shall let it be. If (6.7-10) is solved for $\theta_r(0)$ and the result substituted into (6.7-3) and (6.7-4), we obtain

$$F^r_{qs} = \sqrt{2} F_s \cos\left[\theta_{esf}(0) - \theta_{esv}(0) - \delta\right] \qquad (6.7\text{-}11)$$

$$F^r_{ds} = -\sqrt{2} F_s \sin\left[\theta_{esf}(0) - \theta_{esv}(0) - \delta\right] \qquad (6.7\text{-}12)$$

Now let us leave these equations for just a moment. From (6.7-1) and (6.7-2), F_{as} and F_{bs} may be written as

$$F_{as} = \text{Re}\left[\sqrt{2} \tilde{F}_{as} e^{j\omega_e t}\right] \qquad (6.7\text{-}13)$$

$$F_{bs} = \text{Re}\left[\sqrt{2} \tilde{F}_{bs} e^{j\omega_e t}\right] \qquad (6.7\text{-}14)$$

where

$$\tilde{F}_{as} = F_s e^{j\theta_{esf}(0)} \qquad (6.7\text{-}15)$$

and $\tilde{F}_{bs} = -j\tilde{F}_{as}$. If each side of (6.7-15) is multiplied by $\sqrt{2} e^{-j\delta}$, we will obtain

$$\sqrt{2}\tilde{F}_{as} e^{-j\delta} = \sqrt{2} F_s \cos\left[\theta_{esf}(0) - \delta\right] + j\sqrt{2} F_s \sin\left[\theta_{esf}(0) - \delta\right] \qquad (6.7\text{-}16)$$

We will now do what we promised. We will select or call time zero at the maximum positive value of V_{as}. That is, $\theta_{esv}(0) = 0$, whereupon

$$V_{as} = \sqrt{2} V_s \cos \omega_e t \qquad (6.7\text{-}17)$$

$$V_{bs} = \sqrt{2} V_s \sin \omega_e t \qquad (6.7\text{-}18)$$

and $\tilde{V}_{as}$ is at zero degrees. Let us remember that from now on whenever we conduct an analysis of steady-state operation of synchronous machines, $\theta_{esv}(0) = 0$. With this restriction, compare the right-hand terms of (6.7-16) with (6.7-11) and (6.7-12). Here is the "new" concept coming at us head on. From this comparison we can write

$$\sqrt{2}\tilde{F}_{as}e^{-j\delta} = F^r_{qs} - jF^r_{ds} \qquad (6.7-19)$$

Can this be correct? Here, we are equating a phasor which represents a sinusoidal quantity to F^r_{qs} and F^r_{ds}, which are constants. Yes, but, in its naked form, a phasor is nothing more than a complex number. The $e^{j\omega_e t}$ gives it rotation. (See Appendix B.) So if we forget about what a phasor is used to represent and think only that we are equating complex numbers, we can go along with it, at least for now.

We have only a few more steps. From (6.7-19) we can write

$$\sqrt{2}\tilde{V}_{as}e^{-j\delta} = V^r_{qs} - jV^r_{ds} \qquad (6.7-20)$$

Substituting (6.7-8) and (6.7-9) into (6.7-20) yields

$$\sqrt{2}\tilde{V}_{as}e^{-j\delta} = r_s I^r_{qs} + X_d I^r_{ds} + X_{md}I''_{fd} + j(-r_s I^r_{ds} + X_q I^r_{qs}) \qquad (6.7-21)$$

If $X_q I^r_{ds}$ is added to and subtracted from the right-hand side of (6.7-21) and if it is noted from (6.7-19) that

$$j\sqrt{2}\tilde{I}_{as}e^{-j\delta} = I^r_{ds} + jI^r_{qs} \qquad (6.7-22)$$

then (6.7-21) may be written as

$$\tilde{V}_{as} = (r_s + jX_q)\tilde{I}_{as} + \frac{1}{\sqrt{2}}[(X_d - X_q)I^r_{ds} + X_{md}I''_{fd}]e^{j\delta} \qquad (6.7-23)$$

It is convenient to define the last term of (6.7-23) as

$$\tilde{E}_a = \frac{1}{\sqrt{2}}[(X_d - X_q)I^r_{ds} + X_{md}I''_{fd}]e^{j\delta} \qquad (6.7-24)$$

which is often referred to as the *excitation voltage*. Thus, (6.7-23) becomes

$$\tilde{V}_{as} = (r_s + jX_q)\tilde{I}_{as} + \tilde{E}_a \qquad (6.7-25)$$

Equation (6.7-25) is used widely in the analysis of steady-state operation of synchronous machines. It is very compact and easy to use, far more reduced than one would have expected at the outset of this development. However, there is something more involved here than first meets the eye. Note that the angle of the phasor $\tilde{E}_a$ is δ, but δ has to do with the rotor. In particular, from (6.7-10) with $\theta_{esv}(0) = 0$ ($\tilde{V}_{as}$ is at zero degrees) the steady-state value of δ is $\theta_r(0)$. In other words, if we look back to Fig. 6.2-1, we see that δ or $\theta_r(0)$ is the position of the q axis at the instant we called time zero, which was at the positive maximum value of V_{as}. (We will not mention the fact that another one of these "new" concepts is appearing on the horizon.) In the end, we will have superimposed the phasor diagram upon the rotor. To start, consider Fig. 6.7-1. In the upper left-hand corner of Fig. 6.7-1a we see the position of the rotor at $t = 0$. However, we know that the rotor is rotating at ω_r (ω_e). Now, because at time zero $\theta_{esv}(0) = 0$, $\tilde{V}_{as}$ is at zero degrees, and the phase angle of $\tilde{E}_a$ is equal to δ, as shown in the upper right-hand corner of Fig. 6.7-1b, where we have assumed $\tilde{I}_{as}$ leading $\tilde{V}_{as}$ to illustrate generator operation. How can a and b of Fig. 6.7-1 be superimposed? Well, the rotor is rotating counterclockwise at ω_e and, although we generally do not think of it in this way, the phasors are also rotating counterclockwise at ω_e (Appendix B). Perhaps, this is enough of an explanation to allow us to superimpose the two as shown in c of Fig. 6.7-1. If not, consider this. With your eyes closed, keep your left eye ready to look at (a) and your right eye ready to look at (b). Blink them open at time zero and you see what is shown in Fig. 6.7-1a and b. Now close them immediately and

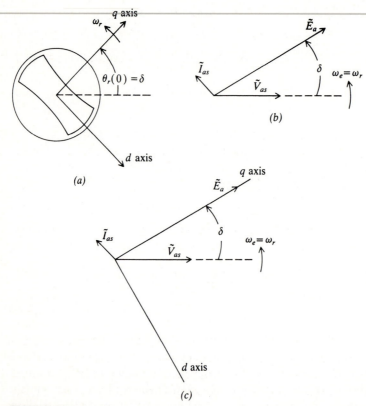

FIGURE 6.7-1
Superimposing the phasor diagram on the rotor of a synchronous machine $[\theta_{esv}(0) = 0]$. (a) Rotor at $t = 0$; (b) phasor at $t = 0$; (c) together at $t = 0$.

keep them closed until the V_{as} voltage is again a positive maximum. At that instant, blink them open again. What do you see? Exactly what you saw at time zero. Keep doing this and each time let the "focus" of each eye come closer and closer. Did you notice that we could superimpose these two and consider them to be stationary as in Fig. 6.7-1c?

We can now express the electromagnetic torque in terms of the rotor angle. If (6.7-8) and (6.7-9) are solved for I_{qs}^r and I_{ds}^r and the results substituted into (6.5-34), we obtain

$$T_e = -\frac{P}{2}\frac{1}{\omega_e}\left\{\frac{r_s X_{md} I_{fd}''}{r_s^2 + X_q X_d}\left(V_{qs}^r - X_{md}I_{fd}'' - \frac{X_d}{r_s}V_{ds}^r\right)\right.$$

$$+\frac{X_d - X_q}{(r_s^2 - X_q X_d)^2}[r_s X_q(V_{qs}^r - X_{md}I_{fd}'')^2$$

$$+\left.(r_s^2 - X_q X_d)V_{ds}^r(V_{qs}^r - X_{md}I_{fd}'') - r_s X_d(V_{ds}^r)^2]\right\} \qquad (6.7\text{-}26)$$

Note that, if we want to use (6.7-11) and (6.7-12), respectively, to express V_{qs}^r and V_{ds}^r, then $\theta_{esv} = \theta_{esf}$ and

$$V_{qs}^r = \sqrt{2}V_s \cos[\theta_{esv}(0) - \theta_{esv}(0) - \delta]$$

$$= \sqrt{2}V_s \cos\delta \qquad (6.7\text{-}27)$$

$$V_{ds}^r = -\sqrt{2}V_s \sin[\theta_{esv}(0) - \theta_{esv}(0) - \delta]$$

$$= \sqrt{2}V_s \sin\delta \qquad (6.7\text{-}28)$$

We should mention in passing that (6.7-27) and (6.7-28) are valid for transient as well as steady-state operation. Although this fact is interesting, we will use the equations only for steady-state operation in this section. For compactness, we will define

$$E_{xfd}'' = X_{md}I_{fd}'' \qquad (6.7\text{-}29)$$

If (6.7-27) through (6.7-29) are substituted into (6.7-26) and if the resistance r_s of the stator windings is neglected, the steady-state electromagnetic torque may be written as

$$T_e = -\frac{P}{2}\frac{1}{\omega_e}\left[\frac{E_{xfd}''\sqrt{2}V_s}{X_d}\sin\delta + \frac{1}{2}\left(\frac{1}{X_q} - \frac{1}{X_d}\right)(\sqrt{2}V_s)^2\sin2\delta\right] \qquad (6.7\text{-}30)$$

Neglecting r_s is justified if its ohmic value is small relative to the magnetizing reactances (X_{mq} and X_{md}) of the machine. In variable-frequency drive systems, this is not the case at low frequencies, whereupon (6.7-26) must be used to calculate the steady-state torque rather than (6.7-30). The electromagnetic torque evaluated by (6.7-26) or (6.7-30) is positive for motor action (torque load) and negative for generator action (torque input).

Although (6.7-30) is a valid expression for the electromagnetic torque during balanced steady-state operation only if the stator winding resistance is small relative to the magnetizing reactances, it permits a quantitative description of two of the three torques produced by a salient-pole synchronous machine. Since $\omega_r = \omega_e$, the damper winding currents are zero and, hence, induction motor torque is not present. The first term on the right-hand side of (6.7-30) is due to the interaction of the mmf produced by the stator currents and the mmf produced by the field current. The second term is the reluctance torque which occurs owing to the forces set up to attempt to align the minimum-reluctance path of the rotor with the resultant air gap mmf.

A word of caution seems appropriate. With the advent of controlled electronic switching devices, electric machines are often operated in systems where the frequency and amplitude of applied stator voltages can be varied. In the above steady-state voltage, (6.7-24), (6.7-25), and (6.7-29), and torque equations, (6.7-26) and (6.7-30), inductive reactances are used. There reactances are calculated by using the frequency of the applied stator voltages. Therefore, the ω_e in the reactances as well as the ω_e which appears in the torque equations must be changed as frequency changes.

Let us return to the expression for steady-state electromagnetic torque given by (6.7-30). Remember that it is valid only if r_s is small relative to X_{mq} and X_{md}. The first term on the right side of (6.7-30) is plotted in Fig. 6.7-2a, the second in Fig 6.7-2b, and the total or sum of the two components is plotted in Fig. 6.7-2c. (We shall talk about the points 1 and 1' appearing in Fig. 6.7-2c a little later.) It is noted that, for a given frequency of the applied stator voltages and for a given machine design, the amplitude of the first term (denoted as A in Fig. 6.7-2a) is proportional to the product of the amplitude of the stator voltages ($\sqrt{2}V_s$) and the field voltage V_{fd}'' since, from (6.7-29), during steady-state operation

$$E_{xfd}'' = \frac{X_{md}}{r_{fd}'} V_{fd}'' \tag{6.7-31}$$

When synchronous machines are used as motors in variable-frequency drive systems, both V_s and E_{xfd}'' can be varied. However, in applications where the synchronous machine is used as a generator to produce electric power such as in a power system, the stator voltage (V_s) is generally regulated and, consequently, it is not allowed to vary more than, say, 1 to 3 percent during normal steady-state operation. In this case, the amplitude of the first term, which is the main torque component in generators, is changed by changing the field voltage V_{fd}''.

The amplitude of the second term, the reluctance torque, is denoted as B in Fig. 6.7-2b. For a fixed frequency of operation and a given machine design, the reluctance torque varies as the square of the amplitude of the applied stator voltages ($\sqrt{2}V_s$). In variable-frequency drive systems, the reluctance torque may be changed by changing V_s for a given frequency of operation. However, in a power system where both the frequency and amplitude of the stator

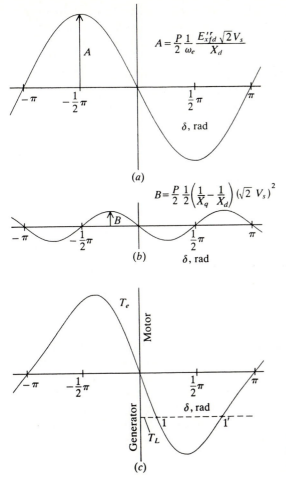

$$A = \frac{P}{2} \frac{1}{\omega_e} \frac{E''_{xfd} \sqrt{2} V_s}{X_d}$$

(a)

$$B = \frac{P}{2} \frac{1}{2} \left(\frac{1}{X_q} - \frac{1}{X_d} \right) (\sqrt{2} \, V_s)^2$$

(b)

(c)

FIGURE 6.7-2
Steady-state electromagnetic torque
of synchronous machine.

voltages are essentially constant, the amplitude of the reluctance torque is also essentially constant, depending upon the design of the machine.

Recall that the rotor angle δ is the angle between the phasor $\tilde{E}_a$ which lays along the q axis and the phasor $\tilde{V}_{as}$. Also, for a given load or input torque, the rotor angle is constant during steady-state operation and, since T_e versus δ is periodic, we need only consider the plot for $-\pi < \delta < \pi$. Now, the load torque and the electromagnetic torque are related by (6.4-3) from which it is clear that during steady-state operation $T_e = T_L$. For illustrative purposes, let the synchronous machine be connected to an electric system and let the load torque be negative. That is, torque is applied to the shaft by some external means—a steam or hydroturbine or a combustion engine, or, perhaps a wind turbine. Regardless of how the applied torque is developed, it is an input torque to the generator shaft and, if friction and windage losses are neglected, the steady-state T_e must equal T_L, thus T_e is negative also. Let us think about this for a minute. The machine is connected to an electric system. If a torque is applied to the shaft and since torque times rotor speed is equal to power, the synchronous machine must deliver an equal amount of power (neglecting friction and windage losses and ohmic losses) to the electric system. Otherwise, there would be a torque or power imbalance and, if T_e is not equal to T_L, the synchronous machine would accelerate for $T_e - T_L > 0$ and decelerate for $T_e - T_L < 0$.

In the next section, we will discuss the dynamic performance of the synchronous machine. It is interesting, however, to continue our example of steady-state operation to determine the stable operating region. We can do as we did in the case of the relay. With T_L negative, there are two possible operating points between $-\pi < \delta < \pi$. They are denoted 1 and 1' in Fig. 6.7-2c. Now we must remember that T_L is constant. First, assume that steady-state operation occurs at point 1. This is a valid operating point if the system will return to this point when distributed from it. To test this, let δ decrease ever so slightly. In this case $T_e - T_L > 0$ and the rotor will accelerate, thereby increasing δ, (6.6-1); hence, a torque is developed to move the system back to operating point 1. If now δ increases ever so slightly, $T_e - T_L < 0$, and the system will again move back to point 1. Hence, point 1 is a stable operating point.

Although we all suspect what is going to happen at point 1', let us go through the exercise anyway. If we assume that the system is operating stably at point 1', then a displacement in δ from this operating point should cause a torque to restore the system to this operating point. If δ decreases slightly from point 1', $T_e - T_L < 0$ and the machine would decelerate which would further decrease δ. The system would move away from point 1' and, after all transients have subsided, the system would operate stably at point 1. If, instead, δ increases from point 1', $T_e - T_L > 0$, whereupon the machine will accelerate, moving away from point 1'. We can conclude that, although point 1' satisfies the "torque balance equation" (6.4-3), it is not a stable operating condition. Let us go back a step. If δ increases from point 1' and the rotor accelerates, where will it end up? At point $1 + 2\pi$ if it does not go unstable dynamically. We really cannot appreciate the meaning of "unstable dynamically" until we can study the dynamic characteristics in the next section. It is sufficient here to say that we are dealing with steady-state characteristics, and one can get into trouble using steady-state characteristics to explain the large-excursion, dynamic (transient) characteristics of a system.

It is apparent that the only way that the steady-state electric power output or input (or T_e) can be changed is to change the torque T_L. However, the electrical characteristics of a synchronous machine connected to a system can be changed by changing the field voltage $V_{fd}''^r$. Although the following is, perhaps, of most interest to power system engineers, it is worth a passing consideration by all. To explain the influence of $E_{xfd}''^r$ ($V_{fd}''^r$), we will assume that the machine is connected to a large system so that, regardless what we do with the synchronous machine, it will not change the magnitude or phase of the system voltage, i.e., $\tilde{V}_{as}$. This is commonly called an *infinite bus* in power system language. If we now assume that the load torque T_L is zero and if we neglect friction and windage losses along with the stator resistance, then T_e and δ are also zero and the machine will run at synchronous speed without absorbing energy from either the electric or mechanical systems.

Although this mode of operation is not feasible in practice since the machine will actually absorb a small amount of power to satisfy the ohmic, friction, and windage losses and thus a small δ would exist at no load, it is convenient for purposes of explanation. With the machine "floating on the line" the field voltage can be adjusted to establish the desired terminal conditions. Three situations may exist:

1. $|\tilde{E}_a| = |\tilde{V}_{as}|$, whereupon $\tilde{I}_{as} = 0$.
2. $|\tilde{E}_a| > |\tilde{V}_{as}|$, whereupon $\tilde{I}_{as}$ leads $\tilde{V}_{as}$ and the synchronous machine appears as a capacitor supplying reactive power to the system.
3. $|\tilde{E}_a| < |\tilde{V}_{as}|$ with $\tilde{I}_{as}$ lagging $\tilde{V}_{as}$, whereupon the machine is absorbing reactive power appearing as an inductor to the system.

We should define reactive power which is generally denoted as Q. In particular, the reactive power per phase is (Appendix B)

$$Q = |\tilde{V}_{as}||\tilde{I}_{as}| \sin [\theta_{esv}(0) - \theta_{esi}(0)]$$
$$= |\tilde{V}_{as}||\tilde{I}_{as}| \sin \phi_{pf} \qquad (6.7\text{-}32)$$

where ϕ_{pf} is the power factor angle and the units of Q are var (voltampere reactive). An inductance is said to absorb reactive power and thus, by definition, Q is positive for an inductor and negative for a capacitor. Actually, Q is a measure of the exchange of energy stored in the electric (capacitor) and magnetic (inductance) fields; however, there is no average power interchanged between these energy storage devices.

Now, to maintain the voltage in a power system at rated value, the synchronous generators are normally operated in the overexcited mode, $|\tilde{E}_a| > |\tilde{V}_{as}|$, since the generators are the main source of reactive power for the inductive loads throughout the system. In the past, synchronous machines have been placed in the power system for the sole purpose of supplying reactive power without any provision to absorb or provide real power. During peak load conditions when the system voltage is depressed, these so-called *synchronous condensers* are brought on line and the field voltage is adjusted to help increase the system voltage. In this mode of operation, the synchronous machine behaves like an adjustable capacitor. On the other hand, it may be necessary for a generator to absorb reactive power in order to regulate voltage in a high-voltage transmission system during light load conditions. This mode of operation is, however, not desirable and should be avoided since machine oscillations become less damped as the reactive power required is decreased.

The influence of the field voltage during motor operation is illustrated in Example 6B.

As a finale to the analysis of steady-state operation of synchronous machines, let us consider the procedure by which generator action is established and then look at the phasor diagram for this mode of operation. A prime mover is mechanically connected to the shaft of the synchronous generator. As mentioned, this prime mover can be either a steam turbine, a hydroturbine, or a combustion engine. If initially the torque input to the shaft due to the prime mover is zero, the synchronous machine is essentially floating on the line. If now the input torque is increased to some value (T_L negative), for example, by supplying steam to the turbine blades, a torque imbalance occurs since T_e must remain at its original value (zero) until δ changes. Hence, the rotor will temporarily accelerate slightly above synchronous speed, whereupon δ will increase in accordance with (6.6-1). Thus, T_e increases negatively and a new operating point will be established with a positive δ where T_L is equal to T_e. The rotor will again rotate at synchronous speed. The actual dynamic response of the electric and mechanical systems during this loading process is illustrated by computer traces in the following section. If, during generator operation, the torque input from the prime mover is increased (T_L negative) to a value greater than the maximum possible value of T_e, the machine will be unable to maintain steady-state operation since it cannot electrically transmit the power supplied to the shaft. In this case, the device will accelerate above synchronous speed theoretically without bound. However, protection is normally provided in power systems which disconnects the machine from the system and reduces the input torque to zero by closing the steam valves of the steam turbine, for example, when it exceeds synchronous speed by 3 to 5 percent.

Normally, steady-state generator operation is depicted by the phasor diagram shown in Fig. 6.7-3. Here $\theta_{esi}(0)$ is the angle between the voltage and

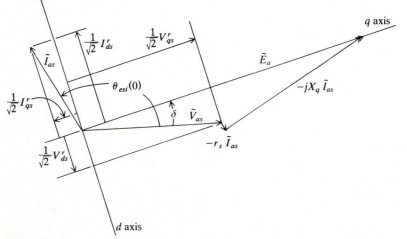

FIGURE 6.7-3
Phasor diagram for generator operation.

the current since the time zero position is $\theta_{esv}(0) = 0$. Since the phasor diagram and the q and d axes of the machine may be superimposed, the rotor reference frame voltages and currents are also shown. If we wish to show each component of V_{qs}^r and V_{ds}^r, they can be broken up according to (6.7-8) and (6.7-9), and each term added algebraically along the appropriate axes. However, care must be taken when interpreting this diagram. $\tilde{V}_{as}$, $\tilde{I}_{as}$, and $\tilde{E}_a$ are phasors representing sinusoidal quantities. On the other hand, all rotor reference frame quantities are constants. They do not represent phasors in the rotor reference frame even though we have displayed them on a phasor diagram.

Example 6B. A six-pole two-phase salient-pole synchronous machine is supplied from a 440-V 60-Hz source. The machine is operated as a motor with a total power input of 40 kW at the terminals. The parameters of the machine are $r_s = 0.3\,\Omega$, $L_{ls} = 0.001$ H, $L_{md} = 0.015$ H, $L_{mq} = 0.008$ H, $r'_{fd} = 0.03\,\Omega$, $L'_{lfd} = 0.001$ H. (a) The excitation is adjusted so that $\tilde{I}_{as}$ lags $\tilde{V}_{as}$ by 30°. Calculate $\tilde{E}_a$. (b) Repeat (a) with the excitation adjusted so that $\tilde{I}_{as}$ is in phase with $\tilde{V}_{as}$. (c) Repeat (a) with the excitation adjusted so that $\tilde{I}_{as}$ leads $\tilde{V}_{as}$ by 30°.

(a) The phase current may be calculated from the power as

$$|\tilde{I}_{as}| = \frac{\frac{1}{2}(40 \times 10^3)}{440 \cos 30°} = 52.5 \text{ A} \tag{6B-1}$$

Note that the 40×10^3 W is the total power which is the sum of the two phases. With $\tilde{V}_{as} = 440\underline{/0°}$,

$$\tilde{I}_{as} = 52.5\underline{/-30°} \text{ A} \tag{6B-2}$$

From (6.7-25),

$$\tilde{E}_a = \tilde{V}_{as} - (r_s + jX_q)\tilde{I}_{as}$$
$$= 440\underline{/0°} - [0.3 + j377(0.001 + 0.008)]52.5\underline{/-30°}$$
$$= 368\underline{/-23.4°} \text{ V} \tag{6B-3}$$

(b) The phase current is

$$|\tilde{I}_{as}| = \frac{20 \times 10^3}{440} = 45.4 \text{ A} \tag{6B-4}$$

From (6.7-25),

$$\tilde{E}_a = 440\underline{/0°} - (0.3 + j3.39)45.4\underline{/0°}$$
$$= 453\underline{/-19.9°} \text{ V} \tag{6B-5}$$

(c) The phase current is

$$|\tilde{I}_{as}| = \frac{20 \times 10^3}{440 \cos 30°} = 52.5 \text{ A} \tag{6B-6}$$

From (6.7-25),

$$\tilde{E}_a = 440\underline{/0°} - (0.3 + j3.39)52.5\underline{/30°}$$
$$= 540\underline{/-17.4°} \text{ V} \tag{6B-7}$$

It is important to note that the characteristics of the reactive component of the input power of the machine may be changed by changing the magnitude of $\tilde{E}_a$. If r_s is negligibly small, the power output is determined entirely by the input torque and, therefore, it is the same in (a), (b), and (c). It is left to the reader to construct the phasor diagram for each case.

Example 6C. A two-pole 60-Hz 110-V $\frac{3}{4}$-hp two-phase reluctance machine has the following parameters: $r_s = 1 \, \Omega$, $L_{ls} = 0.005$ H, $L_{md} = 0.10$ H, $L_{mq} = 0.02$ H. The machine is operating at rated torque output. Calculate δ and $\tilde{I}_{as}$.

With the machine operating at rated conditions, the power output is

$$P_{out} = (0.75)(746) = 559.5 \text{ W} \tag{6C-1}$$

Therefore, the electromagnetic torque is

$$T_e = \frac{P_{out}}{(2/P)\omega_r} = \frac{559.5}{(\frac{2}{2})377} = 1.484 \text{ N} \cdot \text{m} \tag{6C-2}$$

Substituting into (6.7-30) and noting that I_{fd}'' is zero, we can solve for δ. In particular,

$$
\begin{aligned}
\sin 2\delta &= \frac{-(2/P)\omega_e T_e (2)(1/X_q - 1/X_d)^{-1}}{(\sqrt{2}V_s)^2} \\
&= \frac{-(\frac{2}{2})(377)(1.484)(2)[1/(377)(0.025) - 1/(377)(0.105)]^{-1}}{(2)(110)^2} \\
&= \frac{-(377)(1.484)(0.0808)^{-1}}{(110)^2} = -0.572
\end{aligned} \tag{6C-3}
$$

Therefore $\delta = -17.4°$.

Although we could use (6.7-25) to obtain $\tilde{I}_{as}$, it is more straightforward to use (6.7-8) and (6.7-9). We know, from (6.7-27) and (6.7-28) that

$$
\begin{aligned}
V_{qs}^r &= \sqrt{2}V_s \cos \delta \\
&= \sqrt{2} \, 110 \cos (-17.4°) = 148.4 \text{ V} \tag{6C-4} \\
V_{ds}^r &= \sqrt{2} \, 110 \sin (-17.4°) = -46.5 \text{ V} \tag{6C-5}
\end{aligned}
$$

Therefore, we can write (6.7-8) and (6.7-9) as

$$
\begin{bmatrix} V_{qs}^r \\ V_{ds}^r \end{bmatrix} = \begin{bmatrix} r_s & X_d \\ -X_q & r_s \end{bmatrix} \begin{bmatrix} I_{qs}^r \\ I_{ds}^r \end{bmatrix} \tag{6C-6}
$$

which may be written as

$$
\begin{bmatrix} 148.4 \\ -46.5 \end{bmatrix} = \begin{bmatrix} 1 & (377)(0.105) \\ -(377)(0.025) & 1 \end{bmatrix} \begin{bmatrix} I_{qs}^r \\ I_{ds}^r \end{bmatrix} \tag{6C-7}
$$

Solving for I_{qs}^r and I_{ds}^r yields

$$I_{qs}^r = 5.32 \text{ A} \tag{6C-8}$$

$$I_{ds}^r = 3.61 \text{ A} \tag{6C-9}$$

From (6.7-19),

$$\tilde{I}_{as} = \frac{1}{\sqrt{2}} (I'_{qs} - jI'_{ds})e^{j\delta}$$

$$= \frac{1}{\sqrt{2}} (5.32 - j3.61)e^{-j17.4°} = 4.55\underline{/-51.6°} \text{ A} \qquad (6C\text{-}10)$$

If we calculate the input power from the voltage and current, we obtain approximately 620 W. If we add the output power to the ohmic losses, we obtain approximately 601 W. Why the discrepancy? [*Hint:* What are the restrictions on (6.7-30)?]

SP6.7-1. A two-pole two-phase synchronous machine is operated as a generator with $\tilde{V}_{as} = 110\underline{/0°}$ and $\tilde{I}_{as} = 5\underline{/150°}$. Calculate (*a*) total real power and (*b*) total reactive power. [(*a*) $P = -952.6$ W; (*b*) $Q = -550$ var]

SP6.7-2. The machine in SP6.7-1 is a round-rotor device with $\omega_r = 377$ rad/s. $L_{ls} = 4$ mH, $L_{md} = 50$ mH, and $r_s \cong 0$. Calculate δ. [$\delta = 28.7°$]

SP6.7-3. Calculate I''^r_{fd} for SP6.7-2. [$I''^r_{fd} = 13.76$ A]

SP6.7-4. The reluctance machine in Example 6C is operating as a motor with $\delta = -30°$. Calculate T_e. Neglect r_s. [$T_e = -2.25$ N·m]

SP6.7-5. Determine the approach in SP6.7-4 if we are to take the stator resistance into account. [(6.7-26) with $I''^r_{fd} = 0$]

6.8 DYNAMIC AND STEADY-STATE PERFORMANCE

It is instructive to observe the variables of the synchronous machine during dynamic and steady-state operation. In this section, generator operation of a synchronous machine is illustrated by computer traces as well as motor operation of a reluctance machine. Although a two-phase reluctance machine is often used in practice, a two-phase synchronous generator would not normally be used. Instead, the three-phase synchronous machine is the device normally used for generating electric power. Nevertheless, our purpose is to understand the theory and principles of operation of a synchronous machine. A two-phase machine is just as applicable in this regard as a three-phase machine. The following section on the three-phase synchronous machine provides the information necessary for the power system engineer to make the straightforward "transformation" from a two-phase to a three-phase machine.

Two-Phase Synchronous Machine

The two-phase synchronous machine which we will consider is a four-pole 150-hp 440-V (rms) 60-Hz machine with the following parameters: $r_s = 0.26$ Ω, $L_{ls} = 1.14$ mH, $r'_{kq} = 0.02$ Ω, $L'_{lkq} = 1$ mH, $L_{mq} = 11$ mH, $L_{md} = 13.7$ mH, $r'_{fd} = 0.13$ Ω, $L'_{lfd} = 2.1$ mH, $r'_{kd} = 0.0224$ Ω, $L'_{lkd} = 1.4$ mH. The inertia of the rotor and connected mechanical load is $J = 16.6$ kg·m² and B_m is assumed to be zero.

The dynamic performance of this synchronous machine during a step decrease in load torque from zero to -400 N·m is illustrated in Fig. 6.8-1.

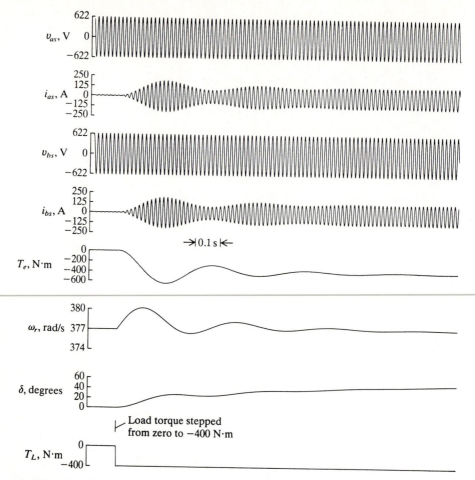

FIGURE 6.8-1
Dynamic performance of a two-phase synchronous generator during a step increase in input torque.

Since we are talking about generator operation, perhaps it is more appropriate to consider this as a step increase in input torque from zero to 400 N · m. In any event, the machine is initially operating at synchronous speed with the field voltage adjusted so that the open-circuit voltage of the stator windings is equal to the rated voltage of the machine (440 V). Therefore, the stator currents are very small since $T_L = 0$. Plotted are v_{as}, i_{as}, v_{bs}, i_{bs}, T_e, ω_r (electrical angular velocity), δ, and T_L.

Immediately upon the application of the input torque $(-T_L)$, the machine accelerates above synchronous speed as predicted by (6.4-3) and the rotor angle increases in accordance with (6.6-1). The rotor continues to speed up until the accelerating torque on the rotor is zero. This occurs when T_e is equal in magnitude to the input torque. As noted in Fig. 6.8-1, the speed increases to approximately 380 rad/s (electrical angular velocity). Even though the accelerating torque on the rotor is zero at this time, the rotor is still running above synchronous speed. Hence, δ will continue to increase and, consequently, T_e will continue to decrease (increase negatively). The decrease in T_e causes the rotor to decelerate and the speed of the rotor decreases toward synchronous speed. Note at the first synchronous speed crossing of ω_r after the torque disturbance, the rotor angle is approximately 28 electrical degrees and T_e is approximately -600 N·m. The rotor speed decreases below synchronous speed, whereupon the integrand of (6.6-1) becomes negative and the rotor

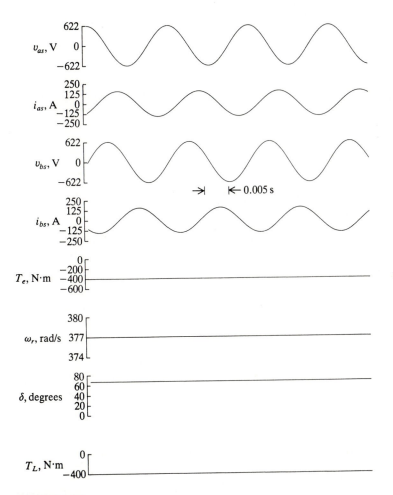

FIGURE 6.8-2
Steady-state operation of the two-phase synchronous generator with an input torque of 400 N·m ($T_L = 400$ N·m).

angle will begin to decrease. Damped oscillations of the rotor about synchronous speed continue until the new steady-state operating point is attained. We might wish to think of the instantaneous electromagnetic torque during this disturbance resulting from the interaction between (1) the stator and field currents, (2) the stator currents and saliency of the rotor, and (3) the stator and damper winding currents. Although this line of thinking may be helpful in visualizing what is going on, we must be careful since we cannot actually separate the expression for T_e given by (6.4-2) or by (6.5-35) into these three different torques during this transient period.

Steady-state operation at the new operating point is depicted in Fig. 6.8-2 (p. 253). Note from the phase relationship between v_{as} and i_{as} or v_{bs} and i_{bs} that the synchronous machine "looks like" a negative resistance (generator action) in series with an inductor. [$\pi < \theta_{esi}(0) < \frac{3}{2}\pi$; $\tilde{I}_{as}$ lags $\tilde{V}_{as}$ by more than 90° but less than 180°.]

The dynamic torque versus rotor angle characteristics during and following the step change in input torque is shown in Fig. 6.8-3. It is interesting to note that it requires considerable time before the machine establishes steady-state operation at $T_L = -400\,\text{N}\cdot\text{m}$. The steady-state torque-angle curve which is also shown, in part, in Fig. 6.8-3 will pass through $T_e = 0$ at $\delta \cong 0$ and $T_L = -400\,\text{N}\cdot\text{m}$ at $\delta = 68°$; however, it is much different from the T_e versus δ during the transient period.

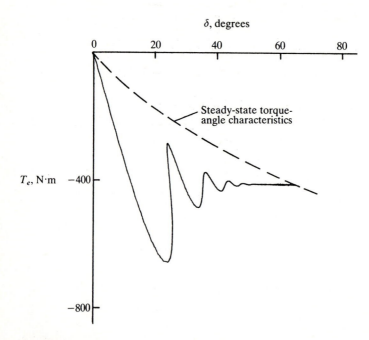

FIGURE 6.8-3
Dynamic torque versus rotor angle characteristic.

Recall that, if we slowly increase the input torque in small increments, theoretically we could reach the maximum value of T_e shown in Fig. 6.7-2c before the machine would fall out of synchronism. The machine is generally rated at 50 to 70 percent of maximum torque capability. It is interesting to mention in passing that the maximum value of input torque (or load torque) that can be applied, with T_L initially zero and with the machine still being able to return to synchronous speed, is referred to as the *transient stability limit*.

It is necessary to employ a computer to predict the dynamic torque-angle characteristics as shown in Fig. 6.8-3 and to determine the transient stability limit. However, before computers, the dynamic torque-angle characteristics were approximated for the "first swing" of the rotor by replacing X_d in (6.7-30) with X_d', the so-called *transient reactance*, and E_{xfd}'' with a voltage behind this

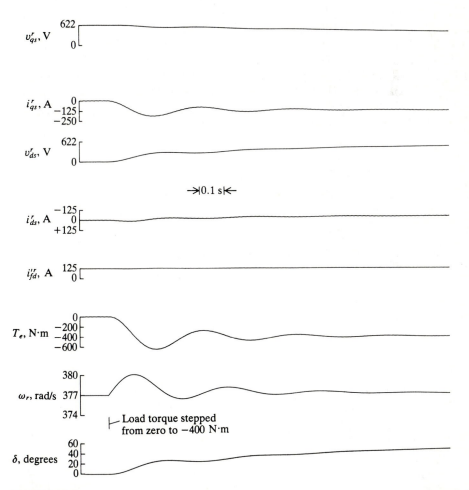

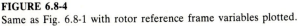

FIGURE 6.8-4
Same as Fig. 6.8-1 with rotor reference frame variables plotted.

transient reactance [1]. The transient reactance X'_d is always smaller than X_d and approximately equal to the sum of X_{ls} and X'_{lfd}. Also, the voltage behind this reactance which replaces E''_{xfd} is always larger than E''_{xfd}. It is shown in [1] that the resulting approximate, dynamic torque-angle characteristic is quite accurate during the first swing of the rotor. We shall leave this all to the power system engineer because it is, indeed, a topic which should be studied by one working in the area of power system stability. However, our purpose here is to make the first-time reader aware that the steady-state and dynamic torque versus rotor angle characteristics are different, sometimes markedly different as illustrated here.

Figures 6.8-4 and 6.8-5 are repeats of Figs. 6.8-1 and 6.8-2, respectively, with the rotor reference frame variables plotted rather than the stator or

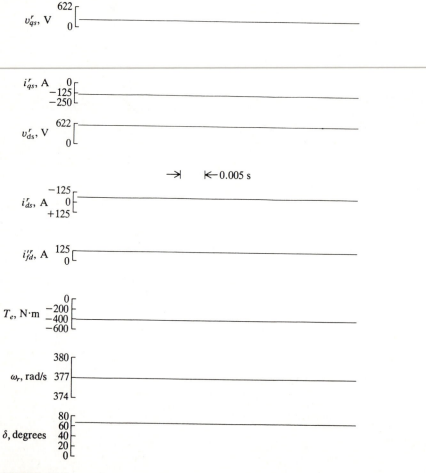

FIGURE 6.8-5
Same as Fig. 6.8-2 with rotor reference frame variables plotted.

machine variables. Also plotted is the field current $i_{fd}^{\prime\prime}$. Although for this machine the field current changed only slightly owing to a change in flux linkages, this is not typical of all machines. In some cases, depending upon the parameters and the type of the disturbance, a considerable voltage may be induced in the field winding resulting in a marked change in field current during the transient period [1].

Two-Phase Reluctance Machine

As we have mentioned, the two-phase reluctance machine is a two-phase salient-pole synchronous machine without a field winding. It is instructive to observe the steady-state and dynamic performance of a low-power two-phase reluctance motor. The parameters of a 115-V (rms) 60-Hz two-pole two-phase

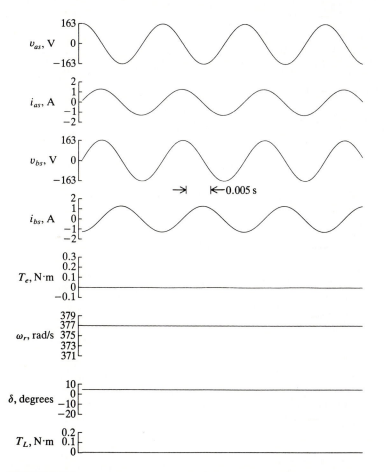

FIGURE 6.8-6
Steady-state operation of a two-pole two-phase $\frac{1}{10}$-hp reluctance motor with $T_L = 0$.

$\frac{1}{10}$-hp reluctance motor are $r_s = 10\,\Omega$, $L_{ls} = 26.5\,\text{mH}$, $r'_{kq} = 2\,\Omega$, $L'_{lkq} = 26.5\,\text{mH}$, $J = 1 \times 10^{-3}\,\text{kg} \cdot \text{m}^2$, $L_{mq} = 132.6\,\text{mH}$, $L_{md} = 318.3\,\text{mH}$, $r'_{kd} = 4\,\Omega$, $L'_{lkd} = 26.5\,\text{mH}$, $B_m = 0$.

Steady-state operation with rated stator voltages and no-load torque ($T_L = 0$) is shown in Fig. 6.8-6 (p. 257). The following variables are plotted: v_{as}, i_{as}, v_{bs}, i_{bs}, T_e, ω_r, δ, and T_L. The dynamic performance when T_L is stepped from zero to $0.2\,\text{N} \cdot \text{m}$ is shown in Fig. 6.8-7. Steady-state operation with $T_L = 0.2\,\text{N} \cdot \text{m}$ is depicted in Fig. 6.8-8. Figure 6.8-9 illustrates the dynamic performance when T_L is stepped back to zero from $0.2\,\text{N} \cdot \text{m}$. The

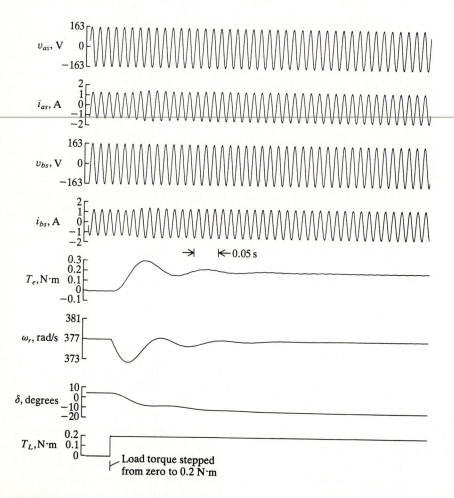

FIGURE 6.8-7
Dynamic performance of a two-pole two-phase $\frac{1}{10}$-hp reluctance motor when T_L is stepped from zero to $0.2\,\text{N} \cdot \text{m}$.

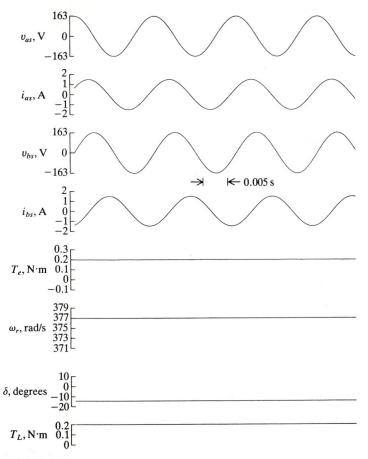

FIGURE 6.8-8
Same as Fig. 6.8-6 with $T_L = 0.2\,\text{N}\cdot\text{m}$.

steady-state torque versus rotor angle characteristic is shown in Fig. 6.8-10. It is interesting to note that this characteristic does not pass through the origin as does the reluctance component of the steady-state torque portrayed in Fig. 6.7-2b. Recall that the characteristics plotted in Fig. 6.7-2 were calculated by using (6.7-30) wherein the stator resistance is neglected. The stator resistance of this small reluctance motor is relatively large. The characteristics shown in Fig. 6.8-10 are calculated without neglecting r_s.

We understand that a reluctance or synchronous machine is equipped with short-circuited rotor windings for the purpose of damping rotor oscillations about synchronous speed and that is why we call them damper windings. Also, we understand that this damping torque occurs because of currents induced in the rotor circuits whenever $\omega_r \neq \omega_e$. As we have mentioned, this is

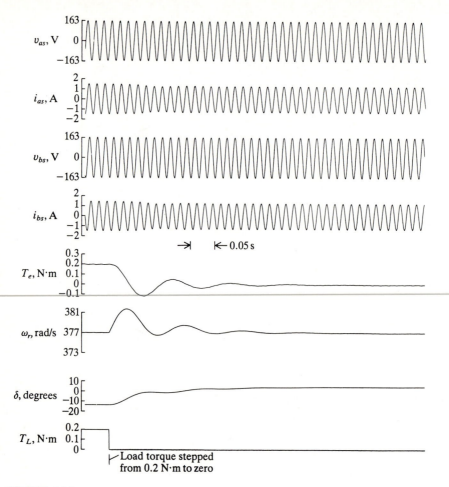

FIGURE 6.8-9
Same as Fig. 6.8-7 with T_L stepped from 0.2 N · m back to zero.

often called induction motor action since, with the short-circuited rotor windings, the reluctance machine produces an average torque-speed characteristic
similar to an induction machine. If the rotor should slow down ever so slightly
from synchronous speed, the induction motor torque would become positive
which would tend to accelerate the rotor back to synchronous speed. Likewise,
if the rotor should speed up from synchronous speed, the induction motor
torque would be negative, which would tend to slow the rotor speed. One
wonders then if it is possible for the reluctance machine to produce a starting
torque by induction motor action. Yes; in most cases the machine is designed
so that it can accelerate from stall and "pull in" to synchronous-speed
operation often under 50 to 70 percent of rated load. Actually, we are not too
surprised at this since we expected that somehow the reluctance motors of the

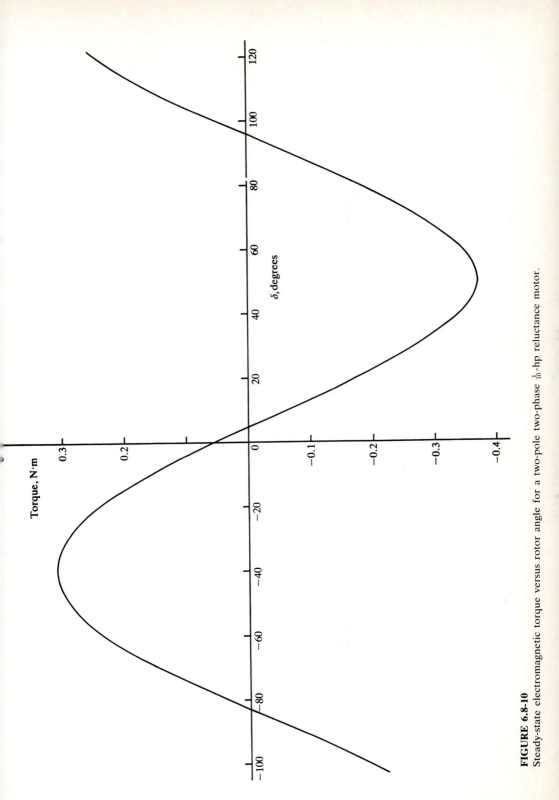

FIGURE 6.8-10

Steady-state electromagnetic torque versus rotor angle for a two-pole two-phase $\frac{1}{10}$-hp reluctance motor.

world would be started by some means other than physically twisting each by its shaft in order to get it up to synchronous speed.

The acceleration characteristics of the reluctance motor from stall to synchronous speed are shown in Fig. 6.8-11 and the torque versus speed characteristics are shown in Fig. 6.8-12. The load torque during acceleration is $T_L = K\omega_{rm}^2$, where $K = 0.2(377)^{-2}$ N·m·s^2/rad^2. The electromagnetic torque pulsates until synchronous speed is reached, whereupon the device operates as a reluctance motor producing a constant torque. But what causes the pulsating torque? Well, immediately following the application of the stator voltages, the pulsation in torque appears to be 60-Hz. This is caused by the transient dc offset in the stator currents; however, as this 60-Hz pulsation decays in

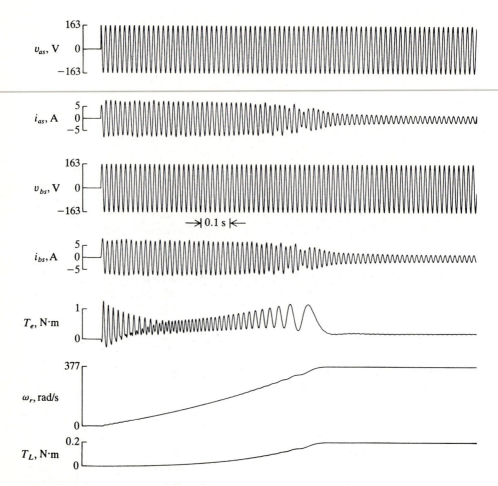

FIGURE 6.8-11

Acceleration from stall of the $\frac{1}{10}$-hp reluctance machine with $T_L = K\omega_{rm}^2$, where $K = 0.2(377)^{-2}$ N·m·s^2/rad^2.

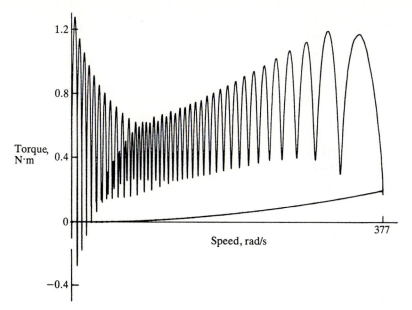

FIGURE 6.8-12
Torque versus speed during acceleration shown in Fig. 6.8-11.

amplitude, a higher-frequency pulsation starts to appear. As the rotor accelerates, this pulsation in torque increases in amplitude while decreasing in frequency. Actually, the frequency of this pulsating component is $2(\omega_e - \omega_r)$ and it is due primarily to the saliency of the rotor (instantaneous reluctance torque) and, to a much less degree, due to the fact that the kq and kd windings have different resistances. The "pulsating reluctance torque" is understandable; however, the fact that a pulsating torque with a frequency of $2(\omega_e - \omega_r)$ occurs because of unequal r'_{kq} and r'_{kd} is not at all apparent. It requires considerable work to prove that it exists; however, the interested reader will find this proof in [1].

The reluctance motor is often used in variable-speed drive applications. In this type of operation, the reluctance motor is supplied from an inverter which can be made to vary rapidly the frequency and/or the amplitude of the fundamental component of the stator voltages. The dynamic behavior of the reluctance motor following a step decrease in frequency from 60 to 50 Hz with an accompanying step decrease in the amplitude of the applied voltages from 110 V (rms) to $(\frac{5}{6})110$ V (rms) is shown in Fig. 6.8-13. Once the machine reaches steady-state operation, the frequency and voltage amplitude are stepped back to their original values. Although the stator voltages supplied from an inverter would contain harmonics, the assumption of a sinusoidal variable-frequency source allows us a first look at variable-frequency operation without getting involved with the operation of the inverter.

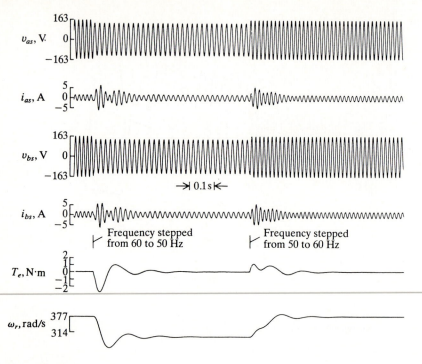

FIGURE 6.8-13
Step changes in frequency of the stator voltages of a $\frac{1}{10}$-hp reluctance motor.

SP6.8-1. From the steady-state voltage and current waveforms given in Fig. 6.8-2 approximate the power input from the electric system and compare this value to the shaft power ($T_L \omega_{rm}$). [$\tilde{V}_{as} = 440\underline{/0°}$, $\tilde{I}_{as} \cong 111\underline{/-135°}$, $P_{in} \cong -69$ kW, $P_{shaft} \cong -75$ kW]

SP6.8-2. Why is there a difference between P_{in} and P_{shaft} in SP6.8-1? Check your answer. [Ohmic loss]

SP6.8-3. Repeat SP6.8-1 by using the voltage and current waveforms shown in Fig. 6.8-5

SP6.8-4. Use either Fig. 6.8-1 or Fig. 6.8-4 to determine each term of (6.4-3) at the first time $\omega_r = \omega_e$ after the step in T_L from zero to -400 N · m. [-400 N · m, -57 N · m, -170 N · m]

SP6.8-5. From Fig. 6.8-8, approximate the per-phase input impedance of the reluctance motor with $T_L = 0.2$ N · m. Compare the imaginary part of this impedance to X_q and X_d. Show that the real part is r_s plus a resistance which represents the power output at the shaft. [$Z \cong 42 + j100 \ \Omega$]

6.9 THREE-PHASE SYNCHRONOUS MACHINE

A two-pole three-phase salient-pole synchronous machine is shown in Fig. 6.9-1. The stator windings are identical and sinusoidally distributed with their magnetic axes displaced 120° from each other. The extension from the analysis

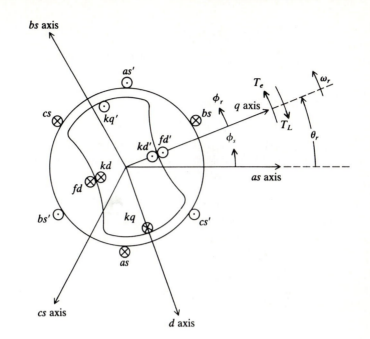

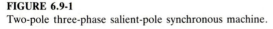

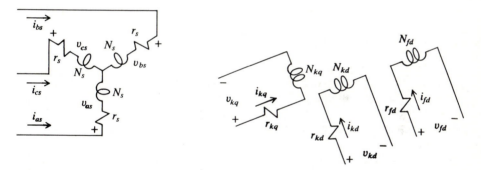

FIGURE 6.9-1
Two-pole three-phase salient-pole synchronous machine.

of a two-phase machine to a three-phase machine is straightforward. However, it is worthwhile to note the expressions of the mutual inductances of the stator windings. Also, the addition of a third substitute variable, the zero variable, is necessary since we have three stator variables.

Voltage Equations and Winding Inductances

The voltage equations for the three-phase synchronous machine are those given by (6.3-1) through (6.3-5) for the two-phase machine, with the voltage equation for the c phase added. In matrix form,

$$\mathbf{v}_{abcs} = \mathbf{r}_s \mathbf{i}_{abcs} + p\boldsymbol{\lambda}_{abcs} \qquad (6.9\text{-}1)$$

$$\mathbf{v}_{qdr} = \mathbf{r}_r \mathbf{i}_{qdr} + p\boldsymbol{\lambda}_{qdr} \qquad (6.9\text{-}2)$$

where

$$(\mathbf{f}_{abcs})^T = [f_{as} \quad f_{bs} \quad f_{cs}] \qquad (6.9\text{-}3)$$

$$(\mathbf{f}_{qdr})^T = [f_{kq} \quad f_{fd} \quad f_{kd}] \qquad (6.9\text{-}4)$$

The matrix $\mathbf{r}_s$ is an equal-element diagonal matrix and $\mathbf{r}_r$ is defined by (6.3-11). The flux linkage equations may be written as

$$\begin{bmatrix} \boldsymbol{\lambda}_{abcs} \\ \boldsymbol{\lambda}_{qdr} \end{bmatrix} = \begin{bmatrix} \mathbf{L}_s & \mathbf{L}_{sr} \\ (\mathbf{L}_{sr})^T & \mathbf{L}_r \end{bmatrix} \begin{bmatrix} \mathbf{i}_{abcs} \\ \mathbf{i}_{qdr} \end{bmatrix} \qquad (6.9\text{-}5)$$

where

$$\mathbf{L}_s = \begin{bmatrix} L_{ls} + L_A - L_B \cos 2\theta_r & -\tfrac{1}{2}L_A - L_B \cos 2(\theta_r - \tfrac{1}{3}\pi) & -\tfrac{1}{2}L_A - L_B \cos 2(\theta_r + \tfrac{1}{3}\pi) \\ -\tfrac{1}{2}L_A - L_B \cos 2(\theta_r - \tfrac{1}{3}\pi) & L_{ls} + L_A - L_B \cos 2(\theta_r - \tfrac{2}{3}\pi) & -\tfrac{1}{2}L_A - L_B \cos 2(\theta_r + \pi) \\ -\tfrac{1}{2}L_A - L_B \cos 2(\theta_r + \tfrac{1}{3}\pi) & -\tfrac{1}{2}L_A - L_B \cos 2(\theta_r + \pi) & L_{ls} + L_A - L_B \cos 2(\theta_r + \tfrac{2}{3}\pi) \end{bmatrix}$$

$$(6.9\text{-}6)$$

where L_{ls} is the leakage inductance, and L_A and L_B are defined by (6.3-23) and (6.3-24), respectively. The matrix $\mathbf{L}_{sr}$ is an extension of (6.3-30) to account for a three-phase stator.

$$\mathbf{L}_{sr} = \begin{bmatrix} L_{skq} \cos \theta_r & L_{sfd} \sin \theta_r & L_{skd} \sin \theta_r \\ L_{skq} \cos (\theta_r - \tfrac{2}{3}\pi) & L_{sfd} \sin (\theta_r - \tfrac{2}{3}\pi) & L_{skd} \sin (\theta_r - \tfrac{2}{3}\pi) \\ L_{skq} \cos (\theta_r + \tfrac{2}{3}\pi) & L_{sfd} \sin (\theta_r + \tfrac{2}{3}\pi) & L_{skd} \sin (\theta_r + \tfrac{2}{3}\pi) \end{bmatrix}$$

$$(6.9\text{-}7)$$

The matrix $\mathbf{L}_r$ is (6.3-31).

The $\mathbf{L}_s$ given by (6.9-6) requires some discussion. The expressions for the self-inductances, which are the diagonal terms in (6.9-6), are apparent from our explanation for the self-inductances of a two-phase machine. Also the $-\tfrac{1}{2}L_A$ factor in the off-diagonal terms of (6.9-6) seems logical since the mutual inductance between two sinusoidally distributed windings of the stator can be adequately portrayed by the cosine of the angle between their magnetic axes (120°). Therefore, if the air gap were uniform, as in the case of the round-rotor synchronous machine, the off-diagonal terms (mutual inductances) would be $-\tfrac{1}{2}L_A$. However, as in the case of the two-phase synchronous machine, it is not clear that the variation of the mutual inductance would be L_B, nor is it immediately obvious that maximum coupling between the as and bs phase, for example, would occur at $\theta_r = \tfrac{1}{3}\pi$ and $\tfrac{4}{3}\pi$ and minimum coupling at $\theta_r = \tfrac{5}{6}\pi$ and $\tfrac{11}{6}\pi$. We will accept (6.9-6) without proof. A derivation is given in [1] for those who wish additional information.

In the case of the three-phase synchronous machine, the stator magnetizing inductances are defined as one and one-half times the magnetizing inductances of a two-phase machine. In particular,

$$L_{mq} = \tfrac{3}{2}(L_A - L_B) \qquad (6.9\text{-}8)$$

$$L_{md} = \tfrac{3}{2}(L_A + L_B) \qquad (6.9\text{-}9)$$

With the above definition of L_{mq} and L_{md}, the right-hand sides of (6.3-32) through (6.3-37) must be multiplied by $\tfrac{2}{3}$ in order to define the amplitudes of the stator-to-rotor mutual inductances used in $\mathbf{L}_{sr}$ and the rotor magnetizing inductances used in $\mathbf{L}_r$. With the selection of (6.9-8) and (6.9-9) as the stator magnetizing inductances, the turns-ratio for the rotor currents is two-thirds that given by (6.3-39); however, the turns-ratio for the rotor voltages and flux linkages are unchanged from (6.3-40) and (6.3-41). The flux linkage equations can now be written as

$$\begin{bmatrix} \boldsymbol{\lambda}_{abcs} \\ \boldsymbol{\lambda}'_{qdr} \end{bmatrix} = \begin{bmatrix} \mathbf{L}_s & \mathbf{L}'_{sr} \\ \tfrac{2}{3}(\mathbf{L}'_{sr})^T & \mathbf{L}'_r \end{bmatrix} \begin{bmatrix} \mathbf{i}_{abcs} \\ \mathbf{i}'_{qdr} \end{bmatrix} \qquad (6.9\text{-}10)$$

where

$$\mathbf{L}'_{sr} = \begin{bmatrix} L_{mq} \cos\theta_r & L_{md} \sin\theta_r & L_{md} \sin\theta_r \\ L_{mq} \cos(\theta_r - \tfrac{2}{3}\pi) & L_{md} \sin(\theta_r - \tfrac{2}{3}\pi) & L_{md} \sin(\theta_r - \tfrac{2}{3}\pi) \\ L_{mq} \cos(\theta_r + \tfrac{2}{3}\pi) & L_{md} \sin(\theta_r + \tfrac{2}{3}\pi) & L_{md} \sin(\theta_r + \tfrac{2}{3}\pi) \end{bmatrix} \qquad (6.9\text{-}11)$$

The matrix $\mathbf{L}'_r$ is (6.3-44) with our new definition of L_{mq} and L_{md}.

The voltage equations become

$$\mathbf{v}_{abcs} = \mathbf{r}_s \mathbf{i}_{abcs} + p\boldsymbol{\lambda}_{abcs} \qquad (6.9\text{-}12)$$

$$\mathbf{v}'_{qdr} = \mathbf{r}'_r \mathbf{i}'_{qdr} + p\boldsymbol{\lambda}'_{qdr} \qquad (6.9\text{-}13)$$

In terms of inductances,

$$\begin{bmatrix} \mathbf{v}_{abcs} \\ \mathbf{v}'_{qdr} \end{bmatrix} = \begin{bmatrix} \mathbf{r}_s + p\mathbf{L}_s & p\mathbf{L}'_{sr} \\ \tfrac{2}{3}p(\mathbf{L}'_{sr})^T & \mathbf{r}'_r + p\mathbf{L}'_r \end{bmatrix} \begin{bmatrix} \mathbf{i}_{abcs} \\ \mathbf{i}'_{qdr} \end{bmatrix} \qquad (6.9\text{-}14)$$

where r'_j and L'_{lj} are one and one-half times (6.3-48) and (6.3-49), respectively.

Torque

The electromagnetic torque, positive for motor action, may be expressed by using the second entry in Table 2.5-1. In particular,

$$T_e = \frac{P}{2} \left\{ \frac{L_{md} - L_{mq}}{3} \left[(i_{as}^2 - \tfrac{1}{2}i_{bs}^2 - \tfrac{1}{2}i_{cs}^2 - i_{as}i_{bs} - i_{as}i_{cs} + 2i_{bs}i_{cs}) \sin 2\theta_r \right. \right.$$

$$\left. + \frac{\sqrt{3}}{2} (i_{bs}^2 + i_{cs}^2 - 2i_{as}i_{bs} + 2i_{as}i_{cs}) \cos 2\theta_i \right]$$

$$- L_{mq}i_{kq} \left[(i_{as} - \tfrac{1}{2}i_{bs} - \tfrac{1}{2}i_{cs}) \sin \theta_r - \frac{\sqrt{3}}{2} (i_{bs} - i_{cs}) \cos \theta_r \right]$$

$$\left. + L_{md}(i'_{fd} + i'_{kd}) \left[(i_{as} - \tfrac{1}{2}i_{bs} - \tfrac{1}{2}i_{cs}) \cos \theta_r + \frac{\sqrt{3}}{2} (i_{bs} - i_{cs}) \sin \theta_r \right] \right\}$$

(6.9-15)

The torque and rotor speed are related by (6.4-3), which is repeated here for convenience:

$$T_e = J \left(\frac{2}{P} \right) \frac{d\omega_r}{dt} + B_m \left(\frac{2}{P} \right) \omega_r + T_L$$

(6.9-16)

where J is the inertia and B_m is the damping coefficient the units of which are discussed following (6.4-3). The load torque T_L is positive for a torque load (motor action) and negative for a torque input (generator action), as shown in Fig. 6.9-1.

Machine Equations in the Rotor Reference Frame

Since there are three stator variables (f_{as}, f_{bs}, f_{cs}), we must use three substitute variables in the transformation of the stator variables to the rotor reference frame. In particular,

$$\mathbf{f}_{qd0s}^r = \mathbf{K}_s^r \mathbf{f}_{abcs}$$

(6.9-17)

$$(\mathbf{f}_{qd0s}^r)^T = [f_{qs}^r \quad f_{ds}^r \quad f_{0s}]$$

(6.9-18)

$$\mathbf{K}_s^r = \frac{2}{3} \begin{bmatrix} \cos \theta_r & \cos (\theta_r - \tfrac{2}{3}\pi) & \cos (\theta_r + \tfrac{2}{3}\pi) \\ \sin \theta_r & \sin (\theta_r - \tfrac{2}{3}\pi) & \sin (\theta_r + \tfrac{2}{3}\pi) \\ \tfrac{1}{2} & \tfrac{1}{2} & \tfrac{1}{2} \end{bmatrix}$$

(6.9-19)

where the rotor displacement θ_r is defined by (6.5-4). The inverse of $\mathbf{K}_s^r$ is

$$(\mathbf{K}_s^r)^{-1} = \begin{bmatrix} \cos \theta_r & \sin \theta_r & 1 \\ \cos (\theta_r - \tfrac{2}{3}\pi) & \sin (\theta_r - \tfrac{2}{3}\pi) & 1 \\ \cos (\theta_r + \tfrac{2}{3}\pi) & \sin (\theta_r + \tfrac{2}{3}\pi) & 1 \end{bmatrix}$$

(6.9-20)

It is important to note that the same notation ($\mathbf{K}_s^r$) is used for the transformation for both the two- and three-phase change of variables. A trigonometric interpretation of the above change of variables is shown in Fig. 6.9-2.

As in the case of the induction machine, the zero variable ($0s$ variable) is the third substitute variable. We note that f_{0s} is zero for balanced conditions and that f_{0s} is not a function of θ_r and, therefore, the $0s$ quantities (voltage, current, and flux linkage) are associated with stationary circuits. For this reason, a raised index is not incorporated with the zero variables.

If the above change of variables is substituted into the stator voltage equations and the flux linkage equations, the resulting q and d voltage equations are identical to (6.5-20) through (6.5-24). The voltage equation for the $0s$ variables must be added. In particular,

$$v_{0s} = r_s i_{0s} + p\lambda_{0s} \tag{6.9-21}$$

For a linear magnetic system, the q and d flux linkage equations for a three-phase synchronous machine are identical to those given by (6.5-25) through (6.5-29) for the two-phase machine. One must remember, however, that for a three-phase machine L_{mq} and L_{md} are defined by (6.9-8) and (6.9-9), respectively. The expression for λ_{0s} is

$$\lambda_{0s} = L_{ls} i_{0s} \tag{6.9-22}$$

Hence, the voltage equation in the rotor reference frame may be written in terms of inductances as

$$
\begin{bmatrix} v^r_{qs} \\ v^r_{ds} \\ v_{0s} \\ v'^r_{kq} \\ v'^r_{fd} \\ v'^r_{kd} \end{bmatrix} =
\begin{bmatrix}
r_s + pL_q & \omega_r L_d & 0 & pL_{mq} & \omega_r L_{md} & \omega_r L_{md} \\
-\omega_r L_q & r_s + pL_d & 0 & -\omega_r L_{mq} & pL_{md} & pL_{md} \\
0 & 0 & r_s + pL_{ls} & 0 & 0 & 0 \\
pL_{mq} & 0 & 0 & r'_{kq} + pL'_{kq} & 0 & 0 \\
0 & pL_{md} & 0 & 0 & r'_{fd} + pL'_{fd} & pL_{md} \\
0 & pL_{md} & 0 & 0 & pL_{md} & r'_{kd} + pL'_{kd}
\end{bmatrix}
\begin{bmatrix} i^r_{qs} \\ i^r_{ds} \\ i_{0s} \\ i'^r_{kq} \\ i'^r_{fd} \\ i'^r_{kd} \end{bmatrix}
$$

$$\tag{6.9-23}$$

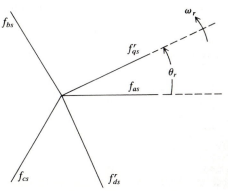

FIGURE 6.9-2
Trigonometric interpretation of the change of stator variables for a three-phase synchronous machine.

where L_q and L_d are defined by (6.5-18) and (6.5-19) while L'_{kq}, L'_{fd}, and L'_{kd} are defined by (6.5-30) through (6.5-32), respectively. We must realize that L_{mq} and L_{md} in these equations are defined by (6.9-8) and (6.9-9), respectively, for the three-phase machine.

The q and d equivalent circuits given in Fig. 6.5-2 for a two-phase machine are valid for the three-phase machine if L_{mq} and L_{md} are defined by (6.9-8) and (6.9-9), respectively, and the appropriate turns-ratio is used for the rotor currents [one and one-half times (6.3-39)]. The equivalent circuit for the $0s$ quantities is a series rL circuit.

The expression for the electromagnetic torque for a three-phase synchronous machine in terms of q and d variables is identical to (6.5-34) and (6.5-35) if each expression is multiplied by $\frac{3}{2}$ and, of course, with the appropriate expression for L_{mq} and L_{md}. It follows that the steady-state voltage and torque equations given for the two-phase machine are also valid for the three-phase machine with the $\frac{3}{2}$ factor properly taken into account in the torque equation and in L_{mq} and L_{md}.

SP6.9-1. The parameters of a three-phase synchronous machine are identical to those for the two-phase synchronous machine given in Sec. 6.8. What must J and T_L be to make the dynamic and steady-state response of v^r_{qs}, i^r_{qs}, v^r_{ds}, and i^r_{ds} shown in Figs. 6.8-4 and 6.8-5 the same for the two- and three-phase machines? $[J_3 = \frac{3}{2}J_2; T_{L3} = \frac{3}{2}T_{L2}]$

SP6.9-2. The values of L_A and L_B are identical for two machines. One is a two-phase reluctance motor, the other a three-phase reluctance motor; otherwise, the parameters are identical. Will v^r_{qs}, i^r_{qs}, v^r_{ds}, and i^r_{ds} be identical for steady-state operation with $T_{L3} = \frac{3}{2}T_{L2}$? Why? $[No; L_{mq3} = \frac{3}{2}L_{mq2}, L_{md3} = \frac{3}{2}L_{md2}]$

6.10   RECAPPING

To conduct a rigorous analysis of a synchronous machine, it was necessary to incorporate a change of variables. In effect, this change of variables replaces the stator variables (voltages, currents, and flux linkages) with variables that are associated with fictitious windings fixed in the rotor. In this way, the time-varying stator self-inductances as well as the time-varying stator-to-rotor mutual inductances are eliminated and all inductances are constant. Although this analysis is rather involved, the resulting equations form the basis for the analysis and the computer simulation of synchronous machines. Also, we found that, for steady-state operation, the voltage equations reduce to a single phasor equation, making steady-state motor or generator operation readily analyzable.

Since the equations which describe the dynamic behavior of synchronous machines are nonlinear, it is necessary to use a computer to solve these equations. Computer traces are given to illustrate the dynamic and steady-state performance of a synchronous generator and a reluctance motor. Although implementation of a computer simulation which can be used for these calculations is beyond the goals of this text, the advantage of portraying machine variables by computer simulation is vividly illustrated by the dynamic response

The angular displacement about the rotor is ϕ_r, referenced to the q axis. The angular velocity of the rotor is ω_r and θ_r is the angular displacement of the rotor measured from the as axis to the q axis. Thus, a given point on the rotor surface at the angular position ϕ_r may be related to an adjacent point on the inside stator surface with angular position ϕ_s as

$$\phi_s = \phi_r + \theta_r \qquad\qquad (7.2\text{-}1)$$

The d axis is fixed at the center of the north pole of the permanent-magnet rotor and the q axis is displaced 90° counterclockwise from the d axis. The electromechanical torque T_e and the load torque T_L are also indicated in Fig. 7.2-1. As defined in Chap. 2, T_e is assumed positive in the direction of increasing θ_r; T_L is positive in the opposite direction.

In the following analysis, it is assumed that:

1. The magnetic system is linear.
2. The open-circuit stator voltages induced by rotating the permanent-magnet rotor at a constant speed are sinusoidal.
3. Large stator currents can be tolerated without significant demagnetization of the permanent magnet.
4. Damper windings (short-circuited rotor windings) are not considered.

Neglecting damper windings, in effect, neglects currents circulating in the surface of the rotor (eddy currents), which are induced by harmonics in the applied voltages and/or oscillations in rotor speed (induction motor action). For permanent-magnet machines with surface-mounted magnets of low permeability, neglecting circulating rotor currents appears to be justified. However, in buried-magnet type of rotor construction, eddy current effects occur and it may be necessary to include short-circuited rotor windings in the analysis. If so, this may be readily achieved by application of the material presented in Chap. 6.

The two sensors shown in Fig. 7.2-1 may be Hall-effect devices. When the north pole is under a sensor, its output is nonzero; with a south pole under the sensor, its output is zero. In brushless dc motor applications, the stator of the permanent-magnet synchronous machine is supplied from a dc-to-ac inverter the frequency of which corresponds to the rotor speed. The states of the sensors are used to determine the switching logic for the inverter which, in turn, determines the output frequency of the inverter. In the actual machine, the sensors are not positioned over the rotor, as shown in Fig. 7.2-1. Instead, they are placed over a ring which is mounted on the shaft external to stator windings and which is magnetized by the rotor. We will return to these sensors and the role they play later. It is first necessary to establish the voltage and torque equations which can be used to describe the behavior of the permanent-magnet synchronous machine when used as a brushless dc motor. This chapter is written with the assumption that the reader is not familiar with synchronous machine theory. Hence, those who have studied the material in Chap. 6 may find most of the material in the next two sections a review, offering little challenge.

In this chapter, as in previous chapters, we will consider the two-phase machine before the three-phase counterpart for convenience of analysis. From an analytical standpoint, it is to our advantage to consider the two-phase device and then extend our work to the three-phase system.

In Fig. 7.2-1, the magnetic axes of the stator windings are denoted as the *as* and *bs* axis. The *d* axis (direct axis) is used to denote the magnetic axis of the permanent-magnet rotor and *q* axis (quadrature axis) is used to denote an axis 90° ahead of the *d* axis. The concept of the *q* axis and *d* axis is reserved for association with the rotor magnetic axes of synchronous machines since, over the years, this association has become convention.

(a)

(b)

FIGURE 7.2-2
Four-pole three-phase 28-V $\frac{1}{3}$-hp brushless dc motor. (*Courtesy of EG and G Rotron.*)

angular position corresponding to the time zero value of the instantaneous current

↗ p.137

FIGURE 7.2-3
Ten-pole three-phase 28-V 0.63-hp 4500-r/min brushless dc motor. (*Courtesy of Vickers Elec-troMech.*)

$$(4.4\text{-}11) \quad mmf_s = \frac{N_s}{2}\sqrt{2}\,I_s \cos\left[\omega_e t + \boxed{\theta_{esi}(0)} - \phi_s\right]$$

Electromagnetic torque is produced by the interaction of the poles of the permanent-magnet rotor and the poles resulting from the rotating air gap mmf established by currents flowing in the stator windings. The rotating mmf (mmf_s) established by symmetrical two-phase stator windings carrying balanced two-phase currents is given by (4.4-11).

A four-pole three-phase 28-V $\frac{1}{3}$-hp brushless dc motor is shown in Fig. 7.2-2. The disassembled motor is shown in Fig. 7.2-2a wherein the stator windings are visible. The opposite end of the stator housing is shown in Fig. 7.2-2b. Housed therein are the Hall-effect sensors, the drive inverter, the filter capacitor, and the logic circuitry. The stator and rotor of a ten-pole three-phase 28-V 0.63-hp 4500-r/min brushless dc motor is shown in Fig. 7.2-3. The magnets are samarium cobalt and the drive inverter is supplied from a 28-V dc source. The magnetic end cap is used in conjunction with Hall-effect sensors mounted in the stator housing (not shown) to determine the rotor position.

SP7.2-1. Express mmf_r for the two-pole two-phase permanent-magnet synchronous machine shown in Fig. 7.2-1. Let F_p denote the peak value. [$mmf_r = -F_p \sin(\phi_s - \theta_r)$]

7.3 VOLTAGE EQUATIONS AND WINDING INDUCTANCES

The voltage equations for the two-pole two-phase permanent-magnet synchronous machine shown in Fig. 7.2-1 may be expressed as

$$v_{as} = r_s i_{as} + \frac{d\lambda_{as}}{dt} \tag{7.3-1}$$

$$v_{bs} = r_s i_{bs} + \frac{d\lambda_{bs}}{dt} \tag{7.3-2}$$

In matrix form,

$$\mathbf{v}_{abs} = \mathbf{r}_s \mathbf{i}_{abs} + p\boldsymbol{\lambda}_{abs} \tag{7.3-3}$$

where p is the operator d/dt, and for voltages, currents, and flux linkages,

$$(\mathbf{f}_{abs})^T = [f_{as} \quad f_{bs}] \tag{7.3-4}$$

with

$$\mathbf{r}_s = \begin{bmatrix} r_s & 0 \\ 0 & r_s \end{bmatrix} \tag{7.3-5}$$

A review of matrix algebra is given in Appendix D. The flux linkage equations may be expressed as

$$\lambda_{as} = L_{asas} i_{as} + L_{asbs} i_{bs} + \lambda_{asm} \tag{7.3-6}$$

$$\lambda_{bs} = L_{bsas} i_{as} + L_{bsbs} i_{bs} + \lambda_{bsm} \tag{7.3-7}$$

In matrix form,

$$\boldsymbol{\lambda}_{abs} = \mathbf{L}_s \mathbf{i}_{abs} + \boldsymbol{\lambda}'_m \tag{7.3-8}$$

where $\boldsymbol{\lambda}'_m$ is the column vector

$$\boldsymbol{\lambda}'_m = \begin{bmatrix} \lambda_{asm} \\ \lambda_{bsm} \end{bmatrix} = \lambda'_m \begin{bmatrix} \sin\theta_r \\ -\cos\theta_r \end{bmatrix} \tag{7.3-9}$$

In (7.3-9), λ'_m is the amplitude of the flux linkages established by the permanent magnet as viewed from the stator phase windings. In other words, the magnitude of λ'_m is proportional to the magnitude of the open-circuit sinusoidal voltage induced in each stator phase winding. It may be helpful to visualize the permanent-magnet rotor as a rotor with a winding carrying a constant current and in such a position to cause the north and south poles to appear as shown in Fig. 7.2-1. The rotor displacement θ_r is expressed as

$$\theta_r = \int_0^t \omega_r(\xi) \, d\xi + \theta_r(0) \tag{7.3-10}$$

where ξ is a dummy variable of integration.

We will assume that the air gap of the permanent-magnet synchronous machine is uniform. Actually, this assumption may be an oversimplification in some cases. However, it markedly reduces our work and allows us to establish directly the basic principles of a brushless dc motor without significant error. Saliency (nonuniform air gap) is taken into account in [1].

With the assumption of a uniform air gap, the mutual inductance between the *as* and *bs* windings is zero. Since the windings are identical, the self-inductances L_{asas} and L_{bsbs} are equal and denoted as L_{ss}. As in the case of a transformer, the self-inductance is made up of a leakage L_{ls} and a magnetizing inductance L_{ms}. Thus,

$$L_{ss} = L_{ls} + L_{ms} \tag{7.3-11}$$

The machine is designed to minimize the leakage inductance; it generally makes up approximately 10 percent of L_{ss}. The magnetizing inductances may be expressed in terms of turns and reluctance. In particular,

$$L_{ms} = \frac{N_s^2}{\mathcal{R}_m} \tag{7.3-12}$$

The magnetizing reluctance $\mathcal{R}_m$ is an equivalent reluctance due to the stator steel, the permanent magnet, and the air gap. We will assume that $\mathcal{R}_m$ is independent of rotor position θ_r. The self-inductance matrix $\mathbf{L}_s$ may be expressed as

$$\mathbf{L}_s = \begin{bmatrix} L_{ss} & 0 \\ 0 & L_{ss} \end{bmatrix} \tag{7.3-13}$$

Example 7A. The parameters of a four-pole two-phase permanent-magnet machine used as a brushless dc motor are $r_s = 3.4\,\Omega$, $L_{ls} = 1.1\,\text{mH}$, and $L_{ms} = 11\,\text{mH}$. When the device is driven at $1000\,\text{r/min}$, the open-circuit phase voltage is sinusoidal with a peak-to-peak value of $34.6\,\text{V}$. Determine λ'_m.

The actual rotor speed at which the measurement was taken is

$$\omega_{rm} = \frac{(\text{r/min})(\text{rad/r})}{\text{s/min}}$$

$$= \frac{(1000)(2\pi)}{60} = \tfrac{100}{3}\pi \text{ rad/s} \tag{7A-1}$$

From (4.6-10), the electrical angular velocity is

$$\omega_r = \frac{P}{2}\,\omega_{rm}$$

$$= \frac{4}{2}\frac{100\pi}{3} = \tfrac{200}{3}\pi \text{ rad/s} \tag{7A-2}$$

With the phases open-circuited, $i_{as} = i_{bs} = 0$. Thus, from (7.3-1) and (7.3-9),

$$v_{as} = \frac{d(\lambda'_m \sin\theta_r)}{dt} = \lambda'_m \omega_r \cos\theta_r \tag{7A-3}$$

Now the peak-to-peak voltage is $34.6\,\text{V}$; hence, from (7A-3), with the peak-to-peak voltage divided by 2, we have

$$\frac{34.6}{2} = \lambda'_m \left(\tfrac{200}{3}\pi\right) \tag{7A-4}$$

Solving for λ'_m yields

$$\lambda'_m = \frac{(34.6)(3)}{(2)(200\pi)} = 0.0826 \text{ V} \cdot \text{s/rad} \tag{7A-5}$$

SP7.3-1. The stator windings of the permanent-magnet synchronous machine shown in Fig. 7.2-1 are open-circuited. The rotor is driven clockwise and $V_{as} = -10 \sin 100t$. Determine V_{bs}. [$V_{bs} = -10 \cos 100t$]

SP7.3-2. Determine λ'_m for SP7.3-1. [$\lambda'_m = 0.1 \, \text{V} \cdot \text{s/rad}$]

SP7.3-3. What is $\theta_r(0)$ in SP7.3-1? [$\theta_r(0) = -\frac{1}{2}\pi$]

7.4   TORQUE

An expression for the electromagnetic torque may be obtained by using the second entry in Table 2.5-1. The coenergy may be expressed as

$$W_c = \tfrac{1}{2}L_{ss}(i_{as}^2 + i_{bs}^2) + \lambda'_m i_{as} \sin \theta_r - \lambda'_m i_{bs} \cos \theta_r + W_{pm} \qquad (7.4\text{-}1)$$

where W_{pm} relates to the energy associated with the permanent magnet, which is constant for the device shown in Fig. 7.2-1. Taking the partial derivative with respect to θ_r yields

$$T_e = \frac{P}{2} \lambda'_m (i_{as} \cos \theta_r + i_{bs} \sin \theta_r) \qquad (7.4\text{-}2)$$

The above expression is positive for motor action. The torque and speed may be related as

$$T_e = J\left(\frac{2}{P}\right)\frac{d\omega_r}{dt} + B_m \left(\frac{2}{P}\right)\omega_r + T_L \qquad (7.4\text{-}3)$$

where J is in $\text{kg} \cdot \text{m}^2$; it is the inertia of the rotor and sometimes the connected load. Since we will be concerned primarily with motor action, the load torque T_L is assumed positive, as indicated in Fig. 7.2-1. The constant B_m is a damping coefficient associated with the rotational system of the machine and mechanical load. It has the units $\text{N} \cdot \text{m} \cdot \text{s/rad}$ of mechanical rotation, and it is generally small and often neglected in the case of the machine but may be considerable for the mechanical load.

SP7.4-1. Calculate T_e for a two-pole permanent-magnet synchronous motor if $i_{as} = \cos \theta_r$ and $i_{bs} = \sin \theta_r$. $\lambda'_m = 0.1 \, \text{V} \cdot \text{s/rad}$. [$T_e = 0.1 \, \text{N} \cdot \text{m}$]

SP7.4-2. Repeat SP7.4-1 with $\theta_r = \omega_r t$ and $i_{as} = \cos(\omega_r t + \frac{1}{2}\pi)$ and $i_{bs} = \sin(\omega_r t + \frac{1}{2}\pi)$. [$T_e = 0$]

7.5   MACHINE EQUATIONS IN THE ROTOR REFERENCE FRAME

A change of variables is helpful in the analysis of the permanent-magnet synchronous machine; however, we are restricted as to the reference frame we can use. For the moment, consider the permanent magnet as a winding with a constant current. We are unaware of a change of variables which can be used for a single winding or, in general, an asymmetrical set of windings. The objective of a change of variables is to transform all machine variables to a common reference frame, thereby eliminating θ_r from the inductance equations.

R. H. Park was the first to incorporate a change of variables in the analysis of synchronous machines [2]. He replaced the stator variables with fictitious variables, thereby eliminating θ_r from the expressions for the inductances. The question may arise as to the reason for using a transformation since the inductances in $\mathbf{L}_s$, (7.3-13), are not functions of θ_r. However, the flux linkage $\boldsymbol{\lambda}'_m$, (7.3-9), is a function of θ_r. In other words, the magnetic system of the permanent magnet is viewed as a time-varying flux linkage by the stator windings.

For a two-phase machine, Park's transformation is

$$\begin{bmatrix} f^r_{qs} \\ f^r_{ds} \end{bmatrix} = \begin{bmatrix} \cos\theta_r & \sin\theta_r \\ \sin\theta_r & -\cos\theta_r \end{bmatrix} \begin{bmatrix} f_{as} \\ f_{bs} \end{bmatrix} \tag{7.5-1}$$

or

$$\mathbf{f}^r_{qds} = \mathbf{K}^r_s \mathbf{f}_{abs} \tag{7.5-2}$$

where f can represent either voltage, current, or flux linkage and θ_r is defined by (7.3-10). It follows that

$$\mathbf{f}_{abs} = (\mathbf{K}^r_s)^{-1} \mathbf{f}^r_{qds} \tag{7.5-3}$$

where $(\mathbf{K}^r_s)^{-1} = \mathbf{K}^r_s$. The s subscript denotes stator variables and the r superscript indicates that the transformation is to a reference frame fixed in the rotor.

Although the above change of variables does not require a physical connotation, it may be convenient to visualize this transformation as trigonometric relationships among variables with directions as shown in Fig. 7.5-1. The direction of f_{as} and f_{bs} variables shown in Fig. 7.5-1 is the positive direction of the magnetic axes of the associated windings (*as* winding and *bs* winding). The f^r_{qs} and f^r_{ds} variables are associated with fictitious windings the positive magnetic axes of which are in the same direction as the direction of f^r_{qs} and f^r_{ds}. The s subscript denotes association with the stator variables. The superscript r indicates that the transformation is to the rotor reference frame. Since the rotor reference frame is generally the only reference frame used in the analysis of synchronous machines, it is often omitted; however, we will carry it along for completeness.

Substituting (7.5-3) into (7.3-3) yields

$$(\mathbf{K}^r_s)^{-1} \mathbf{v}^r_{qds} = \mathbf{r}_s (\mathbf{K}^r_s)^{-1} \mathbf{i}^r_{qds} + p[(\mathbf{K}^r_s)^{-1} \boldsymbol{\lambda}^r_{qds}] \tag{7.5-4}$$

Multiplying (7.5-4) by $\mathbf{K}^r_s$ yields

$(7.3\text{-}3)$ $\mathbf{V}_{abs} = \mathbf{r}_s\, \mathbf{i}_{abs} + p\,\boldsymbol{\lambda}_{abs}$

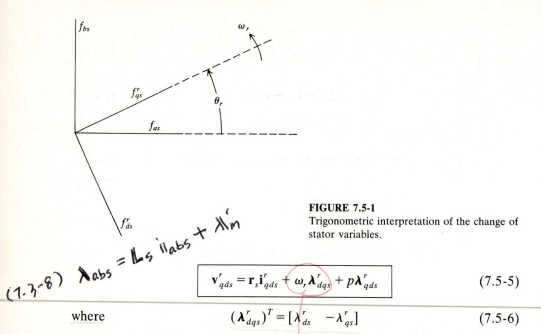

FIGURE 7.5-1
Trigonometric interpretation of the change of stator variables.

(handwritten) $(7.3-8)$ $\lambda_{abs} = L_s i_{abs} + \lambda_m$

$$\boxed{\mathbf{v}^r_{qds} = \mathbf{r}_s \mathbf{i}^r_{qds} + \omega_r \boldsymbol{\lambda}^r_{dqs} + p\boldsymbol{\lambda}^r_{qds}} \qquad (7.5\text{-}5)$$

where $\qquad\qquad (\boldsymbol{\lambda}^r_{dqs})^T = [\lambda^r_{ds} \quad -\lambda^r_{qs}] \qquad\qquad (7.5\text{-}6)$

The last two terms of (7.5-5) come from the last term of (7.5-4). That is,

$$\mathbf{K}^r_s p[(\mathbf{K}^r_s)^{-1}\boldsymbol{\lambda}^r_{qds}] = \mathbf{K}^r_s [p(\mathbf{K}^r_s)^{-1}]\boldsymbol{\lambda}^r_{qds} + \mathbf{K}^r_s(\mathbf{K}^r_s)^{-1}p\boldsymbol{\lambda}^r_{qds} \qquad (7.5\text{-}7)$$

It is left to the reader to show that the right-hand side of (7.5-7) reduces to the last two terms of (7.5-5).

For a magnetically linear system, the stator flux linkages are expressed by (7.3-8). Substituting the change of variables into (7.3-8) yields

$$(\mathbf{K}^r_s)^{-1}\boldsymbol{\lambda}^r_{qds} = \mathbf{L}_s(\mathbf{K}^r_s)^{-1}\mathbf{i}^r_{qds} + \boldsymbol{\lambda}'_m \qquad (7.5\text{-}8)$$

Premultiplying by $\mathbf{K}^r_s$ and substituting (7.3-13) and (7.3-11) for $\mathbf{L}_s$ yields

$$\boldsymbol{\lambda}^r_{qds} = \begin{bmatrix} L_{ls} + L_{ms} & 0 \\ 0 & L_{ls} + L_{ms} \end{bmatrix}\begin{bmatrix} i^r_{qs} \\ i^r_{ds} \end{bmatrix} + \lambda''_m\begin{bmatrix} 0 \\ 1 \end{bmatrix} \qquad (7.5\text{-}9)$$

To be consistent with our previous notation, we have added the superscript r to λ'_m. We see from (7.5-9) that, in our new system of variables, the flux linkage created by the permanent magnet appears constant. Hence, our fictitious circuits are fixed relative to the permanent magnet and, therefore, fixed in the rotor. We have accomplished the goal of eliminating flux linkages which vary with θ_r.

In expanded form, the voltage equations are

$$\boxed{v^r_{qs} = r_s i^r_{qs} + \omega_r \lambda^r_{ds} + p\lambda^r_{qs}} \qquad (7.5\text{-}10)$$

$$\boxed{v^r_{ds} = r_s i^r_{ds} - \omega_r \lambda^r_{qs} + p\lambda^r_{ds}} \qquad (7.5\text{-}11)$$

where
$$\lambda^r_{qs} = L_{ss}i^r_{qs} \tag{7.5-12}$$

$$\lambda^r_{ds} = L_{ss}i^r_{ds} + \lambda''_m \tag{7.5-13}$$

and, as before,

$$L_{ss} = L_{ls} + L_{ms} \tag{7.5-14}$$

Substituting (7.5-12) and (7.5-13) into (7.5-10) and (7.5-11) and, since λ''_m is constant, $p\lambda''_m = 0$.

$$v^r_{qs} = (r_s + pL_{ss})i^r_{qs} + \omega_r L_{ss}i^r_{ds} + \omega_r \lambda''_m \tag{7.5-15}$$

$$v^r_{ds} = (r_s + pL_{ss})i^r_{ds} - \omega_r L_{ss}i^r_{qs} \tag{7.5-16}$$

The above equations may be rewritten in matrix form as

$$\begin{bmatrix} v^r_{qs} \\ v^r_{ds} \end{bmatrix} = \begin{bmatrix} r_s + pL_{ss} & \omega_r L_{ss} \\ -\omega_r L_{ss} & r_s + pL_{ss} \end{bmatrix} \begin{bmatrix} i^r_{qs} \\ i^r_{ds} \end{bmatrix} + \begin{bmatrix} \omega_r \lambda''_m \\ 0 \end{bmatrix} \tag{7.5-17}$$

The expression for electromagnetic torque is obtained by expressing i_{as} and i_{bs} in (7.4-2) in terms of i^r_{qs} and i^r_{ds}. In particular,

$$T_e = \frac{P}{2} \lambda''_m i^r_{qs} \tag{7.5-18}$$

which is positive for motor action.

SP7.5-1. Let $f_{as} = -\cos \theta_r$ and $f_{bs} = -\sin \theta_r$. Determine f^r_{qs} and f^r_{ds}. $[f^r_{qs} = -1; f^r_{ds} = 0]$
SP7.5-2. Let $f_{as} = \cos \theta_{esf}$, $f_{bs} = \sin \theta_{esf}$, and $\theta_r = \omega_r t$. Determine θ_{esf} for $f^r_{qs} = 0$ and $f^r_{ds} = 1$. $[\theta_{esf} = \omega_r t - \frac{1}{2}\pi]$
SP7.5-3. If the speed ω_r is constant, $V_{as} = \cos \theta_r$, and $V_{bs} = \sin \theta_r$, what will be the waveform of I^r_{qs} and I^r_{ds} (steady-state currents)? [dc]

7.6 ANALYSIS OF STEADY-STATE OPERATION

As pointed out earlier, the state of the sensors shown in Fig. 7.2-1 provides information regarding the position of the rotor poles and, thus, the position of the q and d axes. This information can be used to determine the switching of the source inverter relative to the position of the rotor. For our purposes, the source inverter may be thought of as a voltage source or a current source; however, we will work primarily with a voltage source. When the machine is supplied from a voltage source inverter, the fundamental component of the stator applied voltages may be expressed as

$$v_{as} = \sqrt{2}v_s \cos \theta_{esv} \tag{7.6-1}$$

$$v_{bs} = \sqrt{2}v_s \sin \theta_{esv} \tag{7.6-2}$$

Equations (7.6-1) and (7.6-2) form a balanced two-phase set, where the amplitude v_s may be a function of time. In (7.6-1) and (7.6-2),

$$\theta_{esv} = \int_0^t \omega_e(\xi)\, d\xi + \theta_{esv}(0) \qquad (7.6\text{-}3)$$

where $\theta_{esv}(0)$ is the time zero position of the applied voltages, which will play an important role in the operation of the brushless dc motor since we will find that it can be controlled by delaying or advancing the switching of the source inverter relative to the measured rotor position. Now, the brushless dc machine is, by definition, a device where the frequency of the fundamental component of the applied stator voltages corresponds to the speed of the rotor. At this stage, we do not know the details of how this is done; however, we have been told that this is accomplished by appropriately switching the inverter supplying (driving) the permanent-magnet synchronous machine. We will go along with this and set ω_e in (7.6-3) equal to ω_r for the brushless dc machine.

Summarizing, the permanent-magnet synchronous machine is equipped with shaft or rotor position sensors. If this machine is supplied from a dc-to-ac source inverter which is switched so that the fundamental angular velocity of the stator applied voltages is always the same as the rotor speed ($\omega_e = \omega_r$), the inverter machine combination is a brushless dc machine. In addition, the switching of the inverter may be shifted relative to the rotor position which, in effect, changes $\theta_{esv}(0)$; hence, $\theta_{esv}(0)$ is a variable which can be controlled. Moreover, the fundamental amplitude of the voltage applied to the stator windings, v_s, may be changed by what is called pulse width modulation (PWM) of the inverter. As we said before, we do not really know how all of this is accomplished by this thing called an inverter; however, this is not important at this stage. Let us accept the fact that the amplitude of the phase voltages and $\theta_{esv}(0)$ are controllable and the frequency of the stator voltages is always ω_r. We will become more aware as to how this is all done as we go along.

Assuming a voltage source, the fundamental component of each of the applied stator voltages is given by (7.6-1) and (7.6-2). If these voltages are substituted into the equation of transformation, (7.5-1), with $\omega_e = \omega_r$, we obtain

$$v_{qs}^r = \sqrt{2}\, v_s \cos \phi_v \qquad (7.6\text{-}4)$$

$$v_{ds}^r = -\sqrt{2}\, v_s \sin \phi_v \qquad (7.6\text{-}5)$$

For compactness,

$$\boxed{\phi_v = \theta_{esv}(0) - \theta_r(0)} \qquad (7.6\text{-}6)$$

From our earlier discussion, we know that $\theta_{esv}(0)$ is a controllable variable. Since time zero and, thus, $\theta_r(0)$ may be selected arbitrarily, it is generally set equal to zero, which is what we will do. Hence, with $\theta_r(0)$ selected as zero, ϕ_v is $\theta_{esv}(0)$.

At this point in the analysis of the brushless dc machine, it is important to be aware that v'_{qs} and v'_{ds} may be changed in two ways. In particular, consider v'_{qs} and v'_{ds} given by (7.6-4) and (7.6-5), respectively. The amplitude of both v'_{qs} and v'_{ds} may be changed by changing v_s and the relative magnitude of v'_{qs} and v'_{ds} may be changed by changing ϕ_v. Provisions to control both the amplitude and the relative magnitude are possible. Let us look at this again. Equations (7.6-1) and (7.6-2) form a balanced two-phase set of voltages varying at a frequency corresponding to ω_r. Did you notice something very interesting? When we transform these voltages to the rotor reference frame which is rotating at an electrical angular velocity of ω_r, the substitute voltages (v'_{qs} and v'_{ds}) do not vary at a frequency corresponding to ω_r. In fact, if the amplitude (peak value) of v_{as} and v_{bs} ($\sqrt{2}v_s$) is constant and if $\theta_{esv}(0)$ is also constant, then v'_{qs} and v'_{ds} are constant (dc voltages). That is interesting, even though SP7.5-1 through SP7.5-3 may have forewarned us of this coming event. We have transformed a balanced set of two-phase voltages which vary at ω_r to the rotor reference frame and, to our surprise, we find that, regardless of the rotor speed, this balanced set appears in the rotor reference frame as constant voltages if v_s and $\theta_{esv}(0)$ are constants. If this device is going to look like a dc motor, then one would like to believe that there must be a reference frame (substitute variables) wherein the applied voltages will always appear as dc voltages during steady-state operation. Perhaps we are on the right track; let us hope so.

If the applied stator voltages are sinusoidal, then v'_{qs} and v'_{ds} are given by (7.6-4) and (7.6-5). If v_s and ϕ_v are constant, as would be the case for steady-state operation, then v'_{qs} and v'_{ds} will be constant and the voltage equations given by (7.5-15) and (7.5-16), with $p = 0$, become

$$V'_{qs} = r_s I'_{qs} + \omega_r L_{ss} I'_{ds} + \omega_r \lambda''_m \qquad (7.6\text{-}7)$$

$$V'_{ds} = r_s I'_{ds} - \omega_r L_{ss} I'_{qs} \qquad (7.6\text{-}8)$$

where capital letters denote steady-state (constant) quantities, except for λ''_m which is assumed to be always constant. If (7.6-7) and (7.6-8) are solved for I'_{qs} and the result substituted into the expression for electromagnetic torque given by (7.5-18), we obtain

$$T_e = \frac{P}{2} \frac{\lambda''_m r_s}{r_s^2 + \omega_r^2 L_{ss}^2} \left(V'_{qs} - \frac{\omega_r L_{ss} V'_{ds}}{r_s} - \omega_r \lambda''_m \right) \qquad (7.6\text{-}9)$$

We will find that the voltage equations in the q and d axes are, perhaps, the most convenient for analyzing the steady-state operation of the brushless dc motor. However, it is instructive to take a moment to derive a steady-state voltage equation in phasor form. To accomplish this, consider the stator variables during steady-state balanced operation. In particular, with $\omega_e = \omega_r$,

$$F_{as} = \sqrt{2}F_s \cos\left[\omega_r t + \theta_{esf}(0)\right] \tag{7.6-10}$$

$$F_{bs} = \sqrt{2}F_s \sin\left[\omega_r t + \theta_{esf}(0)\right] \tag{7.6-11}$$

Substituting F_{as} and F_{bs} into the equation of transformation, (7.5-1), yields

$$F_{qs}^r = \sqrt{2}F_s \cos\left[\theta_{esf}(0) - \theta_r(0)\right] \tag{7.6-12}$$

$$F_{ds}^r = -\sqrt{2}F_s \sin\left[\theta_{esf}(0) - \theta_r(0)\right] \tag{7.6-13}$$

The phasor that represents F_{as} given by (7.6-10) is

$$\tilde{F}_{as} = F_s e^{j\theta_{esf}(0)} \tag{7.6-14}$$

which may be written as

$$\tilde{F}_{as} = F_s \cos\theta_{esf}(0) + jF_s \sin\theta_{esf}(0) \tag{7.6-15}$$

Now, since we can choose our time zero as we wish, let us select it so that $\theta_r(0) = 0$. In other words, our time zero will always be selected at the time the q axis in Fig. 7.2-1, is horizontal to the right. For those of you who have read Chap. 6 on synchronous machines, you will note a difference; there we selected time zero so that $\theta_{esv}(0) = 0$.

If $\theta_r(0) = 0$, then F_{qs}^r, (7.6-12), and $-F_{ds}^r$, (7.6-13), are, respectively, the same as $\sqrt{2}$ times the real and imaginary parts of (7.6-15). Thus, for $\theta_r(0) = 0$,

$$\boxed{\sqrt{2}\tilde{F}_{as} = F_{qs}^r - jF_{ds}^r} \tag{7.6-16}$$

Here, we have equated a phasor, which represents a sinusoidal quantity, to F_{qs}^r and F_{ds}^r which are constants. However, without the $e^{j\omega_r t}$ which gives rotation (Appendix B), the phasor is simply a complex number. Hence, we are doing nothing illegal.

Substituting (7.6-7) and (7.6-8) into (7.6-16) yields

$$\sqrt{2}\tilde{V}_{as} = V_{qs}^r - jV_{ds}^r$$
$$= r_s I_{qs}^r + \omega_r L_{ss} I_{ds}^r + \omega_r \lambda_m^{\prime r} - j(r_s I_{ds}^r - \omega_r L_{ss} I_{qs}^r) \tag{7.6-17}$$

From (7.6-16),

$$j\sqrt{2}\tilde{F}_{as} = F_{ds}^r + jF_{qs}^r \tag{7.6-18}$$

Hence, (7.6-17) may be written as

$$\tilde{V}_{as} = (r_s + j\omega_r L_{ss})\tilde{I}_{as} + \frac{1}{\sqrt{2}}\omega_r \lambda_m^{\prime r} e^{j0} \tag{7.6-19}$$

We can define

$$\boxed{\tilde{E}_a = \frac{1}{\sqrt{2}}\omega_r \lambda_m^{\prime r} e^{j0}} \tag{7.6-20}$$

and write (7.6-19) as

$$\tilde{V}_{as} = (r_s + j\omega_r L_{ss})\tilde{I}_{as} + \tilde{E}_a \qquad (7.6\text{-}21)$$

One should not confuse $\tilde{E}_a$ given by (7.6-20) with $\tilde{E}_a$ defined for a synchronous machine in Chap. 6.

Since $\theta_r(0) = 0$, we know from (7.6-12) that $\phi_v = \theta_{esv}(0)$. Thus, the phase angle of $\tilde{V}_{as}$ is $\theta_{esv}(0)$ or ϕ_v, which can be controlled. Also, we know that the phase angle of $\tilde{E}_a$ is zero. If we wish, we can superimpose the phasor diagram and the rotor. We have selected $\theta_r(0) = 0$, as shown in Fig. 7.6-1a. At $t = 0$ the phasor diagram for a given mode of operation would be as shown in Fig. 7.6-1b. During steady-state operation, the rotor rotates at ω_r in the counterclockwise direction and the phasor diagram also rotates in the counterclockwise direction at ω_e (Appendix B). However, since $\omega_e = \omega_r$, it follows that Fig. 7.6-1a and Fig. 7.6-1b may be superimposed, as shown in Fig. 7.6-1c. To clarify this, you may wish to read the "blinking eye" explanation given with Fig. 6.7-1 in the case of the synchronous machine. If so, be careful. Although $\tilde{E}_a$ coincides with the q axis, we made $\theta_{esv}(0) = 0$ in the case of the synchronous machine rather than $\theta_r(0) = 0$. Hence, $\tilde{E}_a$ (the q axis) is at the phase angle of δ in Fig. 6.7-1; otherwise, the explanation is valid.

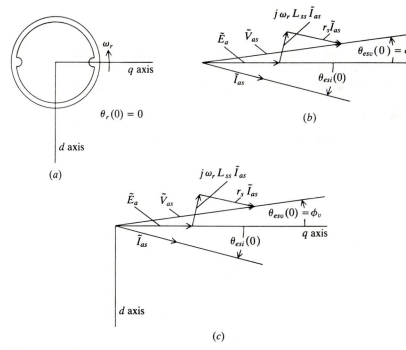

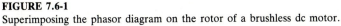

FIGURE 7.6-1
Superimposing the phasor diagram on the rotor of a brushless dc motor.

Operation Without Phase Shift

Operation without phase shift of the stator voltages is common. Here, $\phi_v = 0$ and, from (7.6-4) and (7.6-5), $V_{qs}^r = \sqrt{2}V_s$ and $V_{ds}^r = 0$, where the capital letters are used to denote steady-state operation. In this case, (7.6-8) may be solved for I_{ds}^r in terms of I_{qs}^r.

$$I_{ds}^r = \frac{\omega_r L_{ss}}{r_s} I_{qs}^r \tag{7.6-22}$$

Substituting (7.6-22) into (7.6-7) yields

$$V_{qs}^r = \frac{r_s^2 + \omega_r^2 L_{ss}^2}{r_s} I_{qs}^r + \omega_r \lambda_m'^r \tag{7.6-23}$$

where $V_{qs}^r = \sqrt{2}V_s$, and we know that V_s is the rms value of the applied stator phase voltages. We now start to see a similarity between the voltage equation for the brushless dc machine with $\phi_v = 0$ and the shunt-connected dc machine discussed in Chap. 3. The steady-state armature voltage equation of a dc shunt machine is given by (3.4-2) and repeated here.

$$V_a = r_a I_a + \omega_r L_{AF} I_f \tag{7.6-24}$$

If we neglect the $\omega_r^2 L_{ss}^2$ term in (7.6-23), which is the same as neglecting I_{ds}^r in (7.6-7), and if we assume the field current is constant in (7.6-24), the two equations are identical in form. Yes, but we would have not been able to show this similarity analytically without incorporating the change of variables. Thank you R. H. Park.

There is another similarity. The expression for torque given by (7.5-18) is identical in form to that of a dc shunt machine with a constant field current given by (3.3-5); $T_e = L_{AF} i_f i_a$. We now see why the brushless dc motor is so called, not because it has the same physical configuration as a dc machine but its operating characteristics may be made to resemble those of a dc machine. We must be careful, however, since in order for (7.6-23) and (7.6-24) to be identical in form, ϕ_v must be zero and the term $\omega_r^2 L_{ss}^2$ must be significantly less than r_s^2. There is another very interesting point. In the voltage equation of the dc shunt machine, the current is the armature current I_a. In the similar voltage equation for the brushless dc motor, the current is I_{qs}^r. It is not the stator phase current of the permanent-magnet machine when $\phi_v = 0$; that is, $\sqrt{2}\tilde{I}_{as} = I_{qs}^r - jI_{ds}^r$. With $\phi_v = 0$, $\tilde{V}_{as}$ and $\tilde{E}_a$ in (7.6-21) are at e^{j0} with $\theta_r(0) = 0$. In other words, $\tilde{V}_{as}$ and $\tilde{E}_a$ are in phase and they will lie along the q axis if we superimpose the phasor diagram upon the q and d axes as in Fig. 7.6-1c.

Let us consider the influence of the $\omega_r^2 L_{ss}^2$ term upon the torque versus speed characteristics. If, in (7.6-9), we set $V_{qs}^r = \sqrt{2}V_s$ and $V_{ds}^r = 0$ ($\phi_v = 0$), we obtain

$$T_e = \frac{P}{2} \frac{r_s \lambda_m''}{r_s^2 + \omega_r^2 L_{ss}^2} (\sqrt{2}V_s - \omega_r \lambda_m'') \qquad (7.6\text{-}25)$$

which can also be written as

$$T_e = \frac{P}{2} \left[\frac{2V_s^2 \tau_v}{r_s(1 + \tau_s^2 \omega_r^2)} (1 - \tau_v \omega_r) \right] \qquad (7.6\text{-}26)$$

In (7.6-26), time constants are used in an equation for steady-state operation. Although it is not a common practice to include time constants in a steady-state equation, it is convenient in this instance. The time constants are defined as

$$\tau_s = \frac{L_{ss}}{r_s} \qquad (7.6\text{-}27)$$

$$\tau_v = \frac{\lambda_m''}{\sqrt{2}V_s} \qquad (7.6\text{-}28)$$

The time constant τ_s is analogous to the armature circuit time constant τ_a of a dc machine. Although τ_v has the units of seconds, it is not a constant and will change with V_s. It is, in fact, the inverse of the no-load rotor speed with $\phi_v = 0$. [How did we know this? Set $T_e = 0$ in (7.6-26) and solve for τ_v.] In speed control systems, pulse width modulation of the inverter can change the value of V_s and, thus, τ_v. Regardless of this, it is still convenient for purposes of analysis to use τ_v as defined by (7.6-28).

The steady-state torque-speed characteristics for a typical brushless dc servomotor are shown in Fig. 7.6-2 which is a plot of (7.6-25). If $\omega_r^2 L_{ss}^2$ is

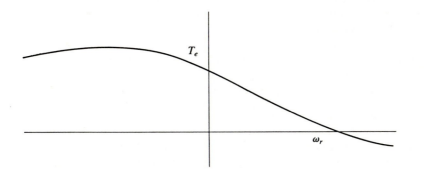

FIGURE 7.6-2
Torque-speed characteristics of a brushless dc motor with $\phi_v = 0$; $V_{qs}' = \sqrt{2}V_s$ and $V_{ds}' = 0$.

neglected, then (7.6-25) yields a straight-line T_e versus ω_r characteristic for a constant V_{qs}^r. Thus, if $\omega_r^2 L_{ss}^2$ could be neglected, the plot shown in Fig. 7.6-2 would be a straight line just as a dc shunt machine (Fig. 3.4-3). Although the T_e versus ω_r plot is approximately linear over the region where $T_e > 0$ and $\omega_r > 0$, it is not linear over the complete speed range. In fact, we see from Fig. 7.6-2 that there is a maximum and a minimum torque. Let us take the derivative of (7.6-25) with respect to ω_r and set the result equal to zero and solve for ω_r. Thus, zero slope of the torque versus speed characteristics (maximum or minimum torque) for $V_{qs}^r = \sqrt{2}V_s$ and $V_{ds}^r = 0$ ($\phi_v = 0$) occurs at

$$\omega_{rMT} = \frac{V_{qs}^r}{\lambda_m''} \pm \sqrt{\left(\frac{V_{qs}^r}{\lambda_m''}\right)^2 + \left(\frac{r_s}{L_{ss}}\right)^2}$$

$$= \frac{1}{\tau_v} \pm \frac{1}{\tau_s \tau_v} \sqrt{\tau_s^2 + \tau_v^2} \qquad \text{for } \phi_v = 0 \qquad (7.6\text{-}29)$$

Example 7B. We are aware that the operating characteristics of a brushless dc motor supplied from a voltage source without phase shift ($\phi_v = 0$) are similar to those of a dc shunt motor with a constant field current. It is instructive, however, to set the steady-state equations side by side and look at the similarities.

Brushless DC Motor	DC Shunt Motor
$V_{qs}^r = \dfrac{r_s^2 + \omega_r^2 L_{ss}^2}{r_s} I_{qs}^r + \omega_r \lambda_m''$	$V_a = r_a I_a + \omega_r L_{AF} I_f$
$T_e = \dfrac{P}{2} \lambda_m'' I_{qs}^r$	$T_e = L_{AF} I_f I_a$
$T_e = \dfrac{P}{2} \dfrac{r_s \lambda_m''}{r_s^2 + \omega_r^2 L_{ss}^2} (V_{qs}^r - \omega_r \lambda_m'')$	$T_e = \dfrac{L_{AF} V_a}{r_a R_f} \left(V_a - \dfrac{\omega_r L_{AF} V_a}{R_f}\right)$

We must realize that the voltage equations are of the same form only if the $\omega_r^2 L_{ss}^2$ term is neglected and I_f, which is V_a/R_f, is constant. We see that V_{qs}^r is analogous to V_a, I_{qs}^r to I_a, λ_m'' to $L_{AF} I_f$, or $L_{AF}(V_a/R_f)$, or k_v, (3.3-4), which are one and the same.

Example 7C. Consider the permanent-magnet machine the parameters of which are given in Example 7A. Assume that the applied stator voltages form a balanced two-phase set with $V_{as} = \sqrt{2}\,11.25 \cos \omega_r t$ and with $\phi_v = 0$. The machine is operating as a brushless dc motor, and the steady-state rotor speed is 600 r/min. Calculate T_e and $\tilde{I}_{as}$.

Comparing the given V_{as} with (7.6-10), we know that $\theta_{esv}(0)$ is zero and, since $\phi_v = 0$, $\theta_r(0) = 0$. Also, $V_s = 11.25$ V. We can calculate the electromagnetic torque by direct substitution into (7.6-25) or (7.6-26). However, we should first calculate ω_r. From (4.6-10),

$$\omega_r = \frac{P}{2} \omega_{rm} = \frac{4}{2} \frac{(600)(2\pi)}{60} = 40\pi \text{ rad/s} \qquad (7C\text{-}1)$$

We can calculate the torque from (7.6-25):

$$T_e = \frac{P}{2} \frac{r_s \lambda_m''^r}{r_s^2 + \omega_r^2 L_{ss}^2} (\sqrt{2}V_s - \omega_r \lambda_m'')$$

$$= \frac{4}{2} \frac{(3.4)(0.0826)}{(3.4)^2 + (40\pi)^2 (12.1 \times 10^{-3})^2} [\sqrt{2}(11.25) - (40\pi)(0.0826)]$$

$$= 0.224 \, \text{N} \cdot \text{m} \tag{7C-2}$$

If B_m is small, which it often is, then from (7.4-3) $T_e = T_L$; thus T_L is $0.224 \, \text{N} \cdot \text{m}$.

There are several ways to calculate $\tilde{I}_{as}$. We could use (7.5-18) to calculate I_{qs}', then (7.6-22) to obtain I_{ds}', and, finally, (7.6-16) to arrive at $\tilde{I}_{as}$. You are asked to follow this procedure in a problem at the end of the chapter; however, we shall use another method here. From (7.6-20),

$$\tilde{E}_a = \frac{1}{\sqrt{2}} \omega_r \lambda_m''^r \underline{/0^\circ}$$

$$= \frac{1}{\sqrt{2}} (40\pi)(0.0826) \underline{/0^\circ} = 7.34 \underline{/0^\circ} \tag{7C-3}$$

From (7.6-21),

$$\tilde{I}_{as} = \frac{1}{r_s + j\omega_r L_{ss}} (\tilde{V}_{as} - \tilde{E}_a)$$

$$= \frac{1}{3.4 + j(40\pi)(12.1 \times 10^{-3})} (11.25 \underline{/0^\circ} - 7.34 \underline{/0^\circ})$$

$$= \frac{3.91 \underline{/0^\circ}}{3.72 \underline{/24.1^\circ}} = 1.05 \underline{/-24.1^\circ} \, \text{A} \tag{7C-4}$$

From (7.6-16),

$$\sqrt{2}\tilde{I}_{as} = I_{qs}' - jI_{ds}'$$

$$\sqrt{2} \, 1.05 \underline{/-24.1^\circ} = 1.36 - j0.606 \tag{7C-5}$$

Thus $I_{qs}' = 1.36 \, \text{A}$ and $I_{ds}' = 0.606 \, \text{A}$. It is left to the reader to draw the phasor diagram. Recall when $\phi_v = 0$, $\tilde{V}_{as}$ and $\tilde{E}_a$ are both at zero degrees.

Operating Modes Achievable by Phase Shifting

We found that, with $\phi_v = 0$, the permanent-magnet synchronous machine has operating characteristics which resemble those of a shunt dc machine if the frequency of the stator applied voltages corresponds to the speed of the rotor. What if we do not restrict ϕ_v to zero? In this case, V_{ds}' is not restricted to zero, whereupon solving (7.6-8) for I_{ds}' and substituting the result into (7.6-7) yields

$$\boxed{V_{qs}' = \frac{r_s^2 + \omega_r^2 L_{ss}^2}{r_s} I_{qs}' + \frac{\omega_r L_{ss}}{r_s} V_{ds}' + \omega_r \lambda_m''} \tag{7.6-30}$$

If now we solve (7.6-30) for I_{qs}' and substitute it into (7.5-18) as we did earlier in this section, we will obtain the expression for torque given by (7.6-9). If now we substitute the steady-state version of (7.6-4) and (7.6-5) for V_{qs}^r and V_{ds}^r, respectively, into (7.6-9), the electromagnetic torque may be expressed as

$$T_e = \frac{P}{2} \frac{r_s \lambda_m''}{r_s^2 + \omega_r^2 L_{ss}^2} \left(\sqrt{2} V_s \cos \phi_v + \frac{\omega_r L_{ss}}{r_s} \sqrt{2} V_s \sin \phi_v - \omega_r \lambda_m'' \right)$$

(7.6-31)

which may also be expressed as

$$T_e = \frac{P}{2} \left\{ \frac{2 V_s^2 \tau_v}{r_s (1 + \tau_s^2 \omega_r^2)} \left[\cos \phi_v + (\tau_s \sin \phi_v - \tau_v) \omega_r \right] \right\} \qquad (7.6\text{-}32)$$

where τ_s and τ_v are defined by (7.6-27) and (7.6-28), respectively.

Although (7.6-31) is valid for all rotor speeds, positive values of ω_r and T_e are of primary interest. From (7.6-32), we see that for the electromagnetic torque to be positive for positive values of rotor speed,

$$\cos \phi_v + \tau_s \omega_r \sin \phi_v > \tau_v \omega_r \qquad (7.6\text{-}33)$$

The torque is zero when (7.6-33) is an equality. Therefore, if $\phi_v = 0$, T_e is positive if $\tau_v \omega_r < 1$ and T_e is zero when $\omega_r = \tau_v^{-1}$; however, we already knew this latter relationship from (7.6-26). When $\phi_v = \frac{1}{2}\pi$, T_e is positive for all positive values of ω_r, provided that $\tau_s > \tau_v$. Also, with $\phi_v = \frac{1}{2}\pi$, T_e is zero for all values of ω_r if $\tau_s = \tau_v$.

What can we do with ϕ_v in regard to the effect it might have on T_e? Can it be used to maximize T_e? Yes! The value of ϕ_v at which a maximum or minimum electromagnetic torque occurs for a given rotor speed may be obtained by taking the derivative of T_e, given by (7.6-32), with respect to ϕ_v, setting the result equal to zero, and then solving for ϕ_v. In particular,

$$\phi_{vMT} = \tan^{-1} (\tau_s \omega_r) \qquad (7.6\text{-}34)$$

where ϕ_{vMT} is the shift in phase of the voltage v_{as} relative to the q axis, which will yield maximum or minimum steady-state torque at a given rotor speed. It is interesting to note that this phase shift angle is the angle of the impedance of the machine. That is, since $\omega_e = \omega_r$, $\tau_s \omega_r = \omega_e L_{ss}/r_s$. That is something one would not have expected at first glance or even at second glance.

The maximum or minimum steady-state electromagnetic torque may be determined by substituting (7.6-34) into (7.6-32). This gives

$$T_{eM} = \frac{P}{2} \frac{2 V_s^2 \tau_v}{r_s (1 + \tau_s^2 \omega_r^2)} \left(\sqrt{1 + \tau_s^2 \omega_r^2} - \tau_v \omega_r \right) \qquad \text{for } \phi_v = \phi_{vMT} \qquad (7.6\text{-}35)$$

It can be seen from (7.6-35) that, for $\omega_r > 0$, the maximum torque will be positive when the term inside the parentheses is greater than zero. If $\tau_v > \tau_s$, then the maximum torque will be positive when

$$\omega_r < \sqrt{\frac{1}{\tau_v^2 - \tau_s^2}} \tag{7.6-36}$$

The maximum torque can be zero. In particular, when $\tau_v > \tau_s$, the rotor speed at which the maximum torque is zero may be determined by setting the terms inside the parentheses of (7.6-35) equal to zero, whereupon (7.6-36) becomes an equality. If, on the other hand, $\tau_s > \tau_v$, then the maximum torque is positive for all values of ω_r. We are beginning to see that this device has far more operating modes than a dc machine.

There is one more relationship that we should obtain. The rotor speeds at which maximum or minimum torque occurs for a given value of ϕ_v may be obtained by setting the derivative of (7.6-32), with respect to ω_r, to zero. In particular,

$$\omega_{rMT} = \frac{1}{\tau_s \sin \phi_v - \tau_v} \left(-\cos \phi_v \pm \frac{1}{\tau_s} \sqrt{\tau_s^2 + \tau_v^2 - 2\tau_s \tau_v \sin \phi_v} \right) \tag{7.6-37}$$

which reduces to (7.6-29) for $\phi_v = 0$.

The steady-state torque-speed characteristics of a brushless dc servomotor are shown in Fig. 7.6-3. The parameters, from Example 7A, are $r_s = 3.4\,\Omega$, $L_{ls} = 1.1\,\text{mH}$, and $L_{ms} = 11.0\,\text{mH}$; thus $L_{ss} = 12.1\,\text{mH}$. The motor is a four-

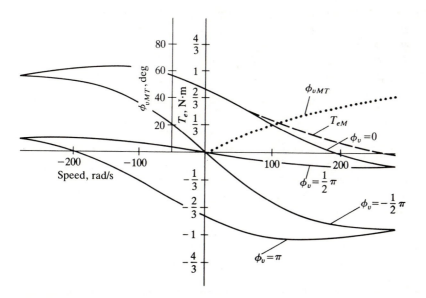

FIGURE 7.6-3
Torque-speed characteristics of a brushless dc motor.

pole device and λ_m'', determined from the open-circuit voltage, is 0.0826 V·s/rad. The voltage V_s is 11.25 V. For this machine, $\tau_s = 3.56$ ms and $\tau_v = 5.2$ ms. In Fig. 7.6-3 plots of the steady-state electromagnetic torque are shown for $\phi_v = 0$, $\pm\frac{1}{2}\pi$, π, and ϕ_{vMT}. The maximum torque T_{eM} for $\omega_r > 0$ is also plotted along with the angle ϕ_{vMT} which yields maximum torque. For $\omega_r > 0$, maximum torque occurs between $\phi_v = 0$ and $\phi_v = \frac{1}{2}\pi$. Note the fact that we can increase the torque considerably at higher rotor speeds compared to that with $\phi_v = 0$.

The influence of τ_s upon the steady-state torque-speed characteristics is illustrated in Figs. 7.6-4 and 7.6-5, wherein T_{eM} is shown by a dashed line for $\omega_r > 0$. τ_s is increased by a factor of 3 times by decreasing r_s. τ_s is increased by a factor of 3 by increasing L_{ss}. Note that in these cases $\tau_s > \tau_v$. Thus, maximum torque is always positive for $\omega_r > 0$, as shown in Figs. 7.6-4 and 7.6-5. A comparison of Figs. 7.6-3 through 7.6-5 reveals that a significantly larger torque can be achieved if the angle ϕ_v is maintained at the value to produce maximum torque, ϕ_{vMT}. The increase in torque is particularly significant if $\tau_s > \tau_v$. Considerable insight may be gained regarding the maximum-torque capability of the machine simply by comparing τ_s and τ_v. We may calculate τ_s from the machine parameters and τ_v from λ_m'' and V_s or from the inverse of the no-load speed. Thus, τ_s and τ_v may be readily determined, whereupon the torque characteristics caused by shifting the phase of the applied voltages may be anticipated without further calculation. However, in [1] it is shown that the

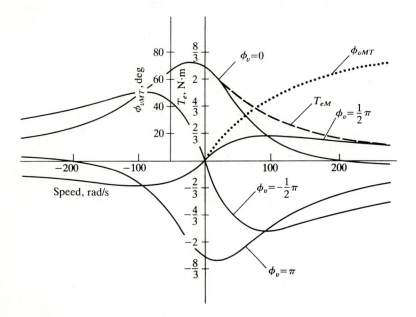

FIGURE 7.6-4. Torque-speed characteristics of a brushless dc motor with τ_s increased by a factor of 3 by decreasing r_s.

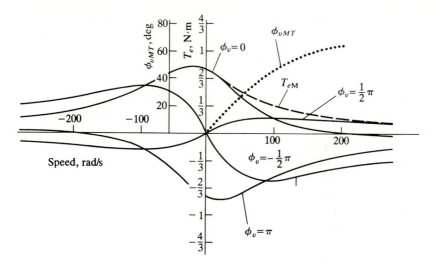

FIGURE 7.6-5
Same as Fig. 7.6-3 with τ_s increased by a factor of 3 by increasing L_{ss}.

increase in average torque due to shifting of the phase is achieved by the inverter at the expense of an increase in the level of the harmonic torque which comes about because of the harmonics introduced by the inverter. This may be prohibitive in some applications such as very precise speed control. Also, increased torque comes at the expense of increased current, which may become a limiting factor due to ohmic losses and the demagnetization effect upon the permanent magnet brought about by a large i_{ds}^r.

Before concluding our analysis of the steady-state operation of a brushless dc motor supplied from a voltage source with provisions for shifting the phase, it is interesting to consider the phase relation between $\tilde{V}_{as}$ and $\tilde{I}_{as}$ during maximum-torque operation. We found that, when ϕ_v was equal to the phase angle of the machine impedance, the machine would operate at maximum torque, (7.6-35). That is, with $\phi_v = \phi_{vMT}$, then

$$r_s + j\omega_r L_{ss} = |Z|e^{j\phi_{vMT}} \qquad \text{for } \phi_v = \phi_{vMT} \tag{7.6-38}$$

where
$$|Z| = \sqrt{r_s^2 + \omega_r^2 L_{ss}^2} \tag{7.6-39}$$

If (7.6-19) is solved for $\tilde{I}_{as}$, the phase of the current for the maximum-torque condition can be expressed as

$$\theta_{esi}(0) = \tan^{-1}\frac{\sin\phi_{vMT}}{(\sqrt{2}V_s)/(\omega_r\lambda_m^{\prime r}) - \cos\phi_{vMT}} \qquad \text{for } \phi_v = \phi_{vMT} \tag{7.6-40}$$

The phase angle of $\tilde{V}_{as}$ is ϕ_{vMT} for this condition. Hence, if $\theta_{esi}(0)$ calculated from (7.6-40) is larger than (less than) ϕ_{vMT}, the fundamental component of the current leads (lags) the fundamental component of the voltage. We realize that

$$\phi_{vMT} = \tan^{-1} \frac{\sin \phi_{vMT}}{\cos \phi_{vMT}} \tag{7.6-41}$$

If now we equate the denominators of (7.6-40) and (7.6-41), $\tilde{I}_{as}$ and $\tilde{V}_{as}$ will be in phase (unity power factor). That is, unity power factor and maximum torque will occur simultaneously when

$$\omega_r^* = \frac{V_s}{\sqrt{2} \, \lambda_m'' \cos \phi_{vMT}} \qquad \text{for } \theta_{esi}(0) = \phi_v = \phi_{vMT} \tag{7.6-42}$$

In the case of the example machine, this condition occurs at 102 rad/s. If the rotor speed is greater (less) than (7.6-42), the current $\tilde{I}_{as}$ leads (lags) the voltage $\tilde{V}_{as}$.

Example 7D. Consider the situation given in Example 7C. The load torque is removed. Calculate the value of ϕ_v so that the steady-state rotor speed is still 600 r/min.

We know from Example 7C that the electrical angular velocity of the rotor, ω_r, is 40π rad/s. Now, what shift in phase, ϕ_v, must occur to maintain this speed in steady-state operation with $T_e = T_L = 0$? From (7.6-32) and (7.6-33) we know that, for $T_e = 0$,

$$\cos \phi_v + \tau_s \omega_r \sin \phi_v = \tau_v \omega_r \tag{7D-1}$$

An analogous relationship may be obtained by setting the torque given by (7.6-9) equal to zero and substituting (7.6-4) and (7.6-5) for v_{qs}^r and v_{ds}^r. That is,

$$0 = V_s \cos \phi_v + \frac{\omega_r L_{ss}}{r_s} V_s \sin \phi_v - \frac{1}{\sqrt{2}} \omega_r \lambda_m'' \tag{7D-2}$$

In any event, we are faced with a trial-and-error type of a solution. From (7.6-27) and (7.6-28), respectively,

$$\tau_s = \frac{L_{ss}}{r_s} = \frac{12.1 \times 10^{-3}}{3.4} = 3.56 \text{ ms} \tag{7D-3}$$

$$\tau_v = \frac{\lambda_m''}{\sqrt{2} V_s} = \frac{0.0826}{\sqrt{2}(11.25)} = 5.2 \text{ ms} \tag{7D-4}$$

Substituting (7D-3) and (7D-4) into (7D-1) with $\omega_r = 40\pi$ rad/s, we have

$$\cos \phi_v + (3.56 \times 10^{-3})(40\pi) \sin \phi_v = (5.2 \times 10^{-3})(40\pi)$$

$$\cos \phi_v + 0.447 \sin \phi_v = 0.653 \tag{7D-5}$$

After a little work, we conclude that there are two values of ϕ_v which will satisfy this condition; $\phi_v = -29.3°$ or $77.5°$.

SP7.6-1. A four-pole two-phase brushless dc motor is operating with $\theta_r(0) = 0$, $V_{as} = 10 \cos \omega_r t$, and $I_{as} = \cos(\omega_r t - 20°)$. Calculate λ_m'' if $r_s = 4 \, \Omega$, $L_{ss} = 0.01$ H, and $\omega_r = 146$ rad/s. $[\lambda_m'' = 0.0393 \text{ V·s/rad}]$

SP7.6-2. Calculate the torque of the machine given in SP7.6-1 with $\omega_r = 0$ and $V_{as} = 10 \cos \omega_r t$. $[T_e = 0.191 \text{ N} \cdot \text{m}]$

SP7.6-3. Repeat SP7.6-1 neglecting $\omega_r^2 L_{ss}^2$. $[\lambda_m'' = 0.0427 \text{ V} \cdot \text{s/rad}]$

SP7.6-4. Calculate (a) the input power, (b) T_e, and (c) efficiency for the operating conditions given in SP7.6-1. $[(a)$ $P_{in} = 9.4 \text{ W};$ (b) $T_e = 0.074 \text{ N} \cdot \text{m};$ (c) eff. = 57.4 percent]

SP7.6-5. Calculate the current $\tilde{I}_{as}$ at T_{eM} for the brushless machine given in SP7.6-1. Assume $V_{as} = 10 \cos(\omega_r t + \phi_{vMT})$ and $\omega_r = 146 \text{ rad/s}$. $[\tilde{I}_{as} = 0.84\underline{/23.8°} \text{ A}]$

SP7.6-6. The parameters of the brushless dc machine are those given in SP7.6-1. Here, $\theta_r(0) = 0$, $V_{as} = \sqrt{2}V_s \cos(\omega_r t + \frac{1}{2}\pi)$. Determine a nonzero value of V_s so that $T_e = 0$ for $\omega_r > 0$. $[V_s = 11.1 \text{ V}]$

SP7.6-7. Calculate (a) the input power, (b) T_e, and (c) the efficiency for the operating conditions given in SP7.6-5. $[(a)$ $P_{in} = 11.8 \text{ W};$ (b) $T_e = 0.085 \text{ N} \cdot \text{m};$ (c) eff. = 52.9 percent]

7.7 STEADY-STATE OPERATION WITH CURRENT SOURCE INPUT

With the appropriate control, an inverter can be made to approximate a current source. In this case, the stator phase currents of the brushless dc motor are controlled relative to the rotor position so as to make $i_{ds}^r = 0$, thereby obtaining the maximum possible torque for a given stator current $[T_e = (P/2)\lambda_m''^r i_{qs}^r]$. The detailed operation of this type of inverter is well beyond the scope of this text. However, we can gain considerable insight into the operation of this type of a brushless dc motor drive by assuming that the stator currents are supplied by a current source of the form

$$i_{as} = \sqrt{2}i_s \cos\theta_{esi} \tag{7.7-1}$$

$$i_{bs} = \sqrt{2}i_s \sin\theta_{esi} \tag{7.7-2}$$

where
$$\theta_{esi} = \int_0^t \omega_e(\xi)\, d\xi + \theta_{esi}(0) \tag{7.7-3}$$

If we let $\omega_e = \omega_r$ in (7.7-3) and if we then substitute (7.7-1) and (7.7-2) into the equation of transformation, (7.5-1), we will obtain

$$i_{qs}^r = \sqrt{2}i_s \cos\phi_i \tag{7.7-4}$$

$$i_{ds}^r = -\sqrt{2}i_s \sin\phi_i \tag{7.7-5}$$

where
$$\boxed{\phi_i = \theta_{esi}(0) - \theta_r(0)} \tag{7.7-6}$$

Now let $\theta_r(0) = 0$ as before, and if we control the current source inverter so that $\theta_{esi}(0) = 0$, then $\phi_i = 0$ and, during steady-state operation,

$$\left. \begin{array}{l} I_{qs}^r = \sqrt{2}I_s \\ \\ I_{ds}^r = 0 \end{array} \right\} \quad \text{for } \phi_i = 0 \qquad \begin{array}{c} (7.7\text{-}7) \\ \\ (7.7\text{-}8) \end{array}$$

It is apparent that, since $T_e = (P/2)\lambda_m''^r i_{qs}^r$, the machine is supplying the maximum possible torque for a given current magnitude. In other words, $\sqrt{2}|\tilde{I}_{as}| = I_{qs}^r$.

Although we will not consider the brushless dc motor supplied from a current source inverter in detail, it is interesting to look at the steady-state voltage equations with the restriction that $I_{ds}^r = 0$. In particular, substituting $I_{ds}^r = 0$ into (7.6-7) and (7.6-8) yields

$$V_{qs}^r = r_s I_{qs}^r + \omega_r \lambda_m'' \tag{7.7-9}$$

$$V_{ds}^r = -\omega_r L_{ss} I_{qs}^r \tag{7.7-10}$$

From (7.6-4) and (7.6-5), we can write (7.7-9) and (7.7-10) as

$$\sqrt{2}V_s \cos\phi_v = r_s I_{qs}^r + \omega_r \lambda_m'' \tag{7.7-11}$$

$$\sqrt{2}V_s \sin\phi_v = \omega_r L_{ss} I_{qs}^r \tag{7.7-12}$$

from which

$$\phi_v = \tan^{-1}\frac{\omega_r \tau_s}{1 + (\omega_r \lambda_m'')/(I_{qs}^r r_s)} \tag{7.7-13}$$

Equation (7.7-13) tells us that the fundamental component of the resulting stator phase voltages are at a phase angle of ϕ_v with respect to the fundamental component of the stator current since $\phi_i = 0$. Also, (7.7-11) and (7.7-12) tell us that as the speed of the rotor increases, it requires a larger voltage V_s to maintain $\phi_i = 0$. This, of course, can be a limiting factor. It should be noted that we can achieve the given mode of steady-state operation with a conventional voltage source supply by shifting the phase of the applied voltages in accordance with (7.7-13). In this case, we can interpret the angle ϕ_v in (7.7-13) as the phase shift in voltage needed to make $I_{ds}^r = 0$, thereby maximizing the torque for a given current amplitude. This should not be confused with ϕ_{vMT}, defined previously, which maximizes torque for a given voltage amplitude. Although we can achieve the same steady-state operation with either a voltage or current source input, the dynamic characteristics may be markedly different. We will not consider current source, brushless dc motors further in this text since a background in switching circuits and control systems would be necessary to appreciate the details of operation.

SP7.7-1. A four-pole two-phase brushless dc motor with $r_s = 4\,\Omega$, $L_{ss} = 0.01$ H, and $\lambda_m'' = 0.0272$ V·s/rad is operating at $\omega_r = 146$ rad/s with $\theta_r(0) = 0$ and $I_{as} = 0.94 \cos\omega_r t$. Determine $\tilde{V}_{as}$. [$V_{as} = 5.58\underline{/10°}$]

SP7.7-2. Calculate (a) the input power, (b) T_e, and (c) the efficiency for the operating conditions given in SP7.7-1. [(a) $P_{in} = 7.26$ W; (b) $T_e = 0.051$ N·m; (c) eff. = 51 percent]

7.8 DYNAMIC AND STEADY-STATE PERFORMANCE

It is instructive to observe the variables of the brushless dc motor during free acceleration from stall and step changes in load torque with sinusoidal applied stator voltages (voltage source). The machine parameters used in this computer study are those given in Example 7A with $J = 1 \times 10^{-4}$ kg·m^2 and $B_m = 0$. The free-acceleration characteristics are shown in Fig. 7.8-1. The applied stator

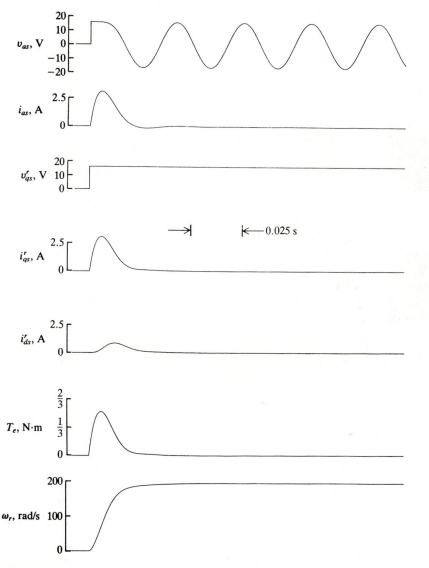

FIGURE 7.8-1
Free-acceleration characteristics of a brushless dc motor.

phase voltages are of the form given by (7.6-1) and (7.6-2) with $\theta_{esv}(0)$ and $\theta_r(0)$ equal to zero. Therefore, ϕ_v is zero, whereupon $v'_{ds} = 0$ and $v_s = 11.25$ V. The phase voltage v_{as}, phase current i_{as}, q-axis voltage v^r_{qs}, q-axis current i^r_{qs}, d-axis current i^r_{ds}, electromagnetic torque T_e, and rotor speed ω_r (in electrical rad/s) are plotted in Fig. 7.8-1. Since the device is a four-pole machine, 200 electrical rad/s is equivalent to 955 r/min. A plot of T_e versus ω_r is shown in Fig. 7.8-2 for the free acceleration depicted in Fig. 7.8-1. The steady-state torque-speed characteristic is superimposed for purposes of comparison.

The rapid acceleration of the brushless dc motor is apparent. In fact, the rotor reaches full speed in less than 0.05 s. The acceleration is so rapid that it is difficult to observe the change in frequency of the applied voltage v_{as} as the motor accelerates from stall. To illustrate the frequency change during the acceleration period, the free-acceleration characteristics given in Figs. 7.8-3 and 7.8-4 are for an inertia of five times the inertia of the rotor $(5 \times 10^{-4}$ kg$\cdot$m^2). It is important to note that the dynamic torque-speed characteristics shown in Figs. 7.8-2 and 7.8-4 differ from the steady-state torque-speed characteristics. One must be aware of this discrepancy if one chooses to use the expression for the steady-state torque in a transfer function formulation describing the dynamic characteristics of a brushless dc motor. Also, recall that, with $\phi_v = 0$, v'_{ds} is zero. However, from (7.6-22), steady-state i'_{ds} is zero only if either ω_r or i'_{qs} is zero. From Figs. 7.8-1 and 7.8-3, we see that the magnitude of i^r_{qs} is less than the amplitude of the phase currents except at $\omega_r = 0$ and $i'_{qs} = 0$. Although v^r_{qs} is equal to the peak value of a phase voltage, i^r_{qs} is not, in general, equal to the peak value of phase current.

The performance during step changes in load torque is illustrated in Fig. 7.8-5. Initially, the machine is operating with $T_L = 0.067$ N$\cdot$m. The load

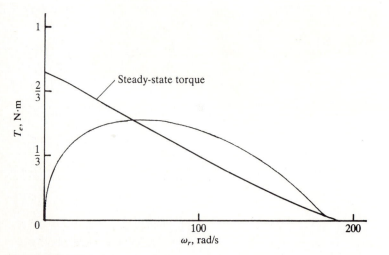

Steady-state torque

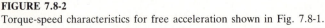

FIGURE 7.8-2
Torque-speed characteristics for free acceleration shown in Fig. 7.8-1.

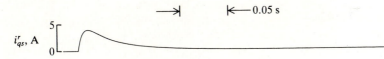

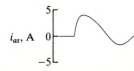

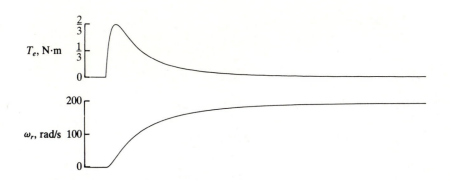

FIGURE 7.8-3
Free-acceleration characteristics of a brushless dc motor with inertia equal to five times the rotor inertia.

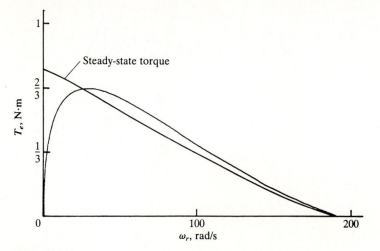

FIGURE 7.8-4
Torque-speed characteristics for free acceleration shown in Fig. 7.8-3.

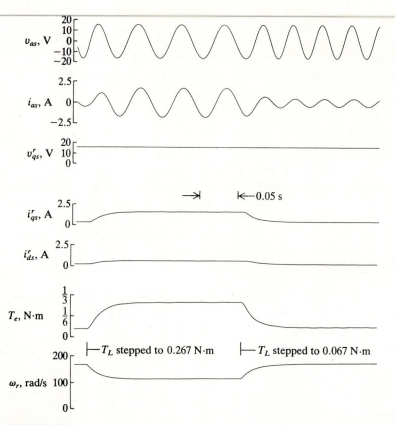

FIGURE 7.8-5
Dynamic performance of a brushless dc motor during step changes in load torque with total inertia twice the rotor inertia.

torque is suddenly stepped to $0.267 \, \text{N} \cdot \text{m}$. The machine slows down, and once steady-state operation is established, the load torque is stepped back to $0.067 \, \text{N} \cdot \text{m}$. In these studies, the inertia is $2 \times 10^{-4} \, \text{kg} \cdot \text{m}^2$, which is twice the inertia of the rotor.

SP7.8-1. Consider Fig. 7.8-3. Express ω_r by assuming that it may be approximated by an exponential increase. $[\omega_r \cong 193 \, (1 - e^{-25t})]$

SP7.8-2. Approximate T_e in Fig. 7.8-4 by a straight line between $T_e = 0.75 \, \text{N} \cdot \text{m}$ at $\omega_r = 0$ and $T_e = 0$ at $\omega_r = 193 \, \text{rad/s}$. Express ω_r for free acceleration with $J = (5)(1 \times 10^{-4}) \, \text{kg} \cdot \text{m}^2$. Compare with SP7.8-1. $[\omega_r \cong 193(1 - e^{-15.6t})]$

7.9 TIME-DOMAIN BLOCK DIAGRAMS AND STATE EQUATIONS

Although the analysis of control systems is not our intent, it is worthwhile to set the stage for this type of analysis by means of a "first look" at time-domain block diagrams and state equations. In this section, we will consider the brushless dc motor first described by the general nonlinear differential equations and then with $\phi_v = 0$ and i_{ds}^r neglected, whereupon the differential equations become linear. Some of the descriptive information regarding block diagrams and state variables given here is a repeat of that given in Sec. 3.6.

Nonlinear System Equations

Block diagrams, which portray the interconnection of the system equations, are used extensively in control system analysis and design. Although block diagrams are generally depicted with the use of the Laplace operator, we will not do this here since this would require a background in Laplace transformations. Instead, we will work with the time-domain equations by using the p operator to denote differentiation with respect to time and the operator $1/p$ to denote integration. Those familiar with Laplace transformations will have no trouble converting the time-domain block diagrams to transfer functions by using the Laplace operator.

Arranging the equations into block diagram representation is straightforward. From the voltage equations given by (7.5-17) and the relationship between torque and rotor speed, (7.4-3), we can write

$$v_{qs}^r = r_s(1 + \tau_s p)i_{qs}^r + r_s \tau_s \omega_r i_{ds}^r + \lambda_m'' \omega_r \qquad (7.9\text{-}1)$$

$$v_{ds}^r = r_s(1 + \tau_s p)i_{ds}^r - r_s \tau_s \omega_r i_{qs}^r \qquad (7.9\text{-}2)$$

$$T_e - T_L = \frac{2}{P}(B_m + Jp)\omega_r \qquad (7.9\text{-}3)$$

where $\tau_s = L_{ss}/r_s$. Solving (7.9-1) for i_{qs}^r, (7.9-2) for i_{ds}^r, and (7.9-3) for ω_r yields

$$i_{qs}^r = \frac{1/r_s}{\tau_s p + 1} (v_{qs}^r - r_s \tau_s \omega_r i_{ds}^r - \lambda_m'' \omega_r) \qquad (7.9\text{-}4)$$

$$i_{ds}^r = \frac{1/r_s}{\tau_s p + 1} (v_{ds}^r + r_s \tau_s \omega_r i_{qs}^r) \qquad (7.9\text{-}5)$$

$$\omega_r = \frac{P/2}{Jp + B_m} (T_e - T_L) \qquad (7.9\text{-}6)$$

A few comments are in order regarding these expressions. In (7.9-4), we see that the three voltage terms are multiplied by the operator $(1/r_s)/(\tau_s p + 1)$ to obtain i_{qs}^r. The fact that we are multiplying the voltages by an operator to obtain current is in no way indicative of the procedure that we might actually use to calculate i_{qs}^r. We are simply expressing the dynamic relationship between the voltage terms and the current i_{qs}^r in a form convenient for drawing block diagrams. The operator $(1/r_s)/(\tau_s p + 1)$ in (3.6-4) may also be interpreted as a transfer function relating the voltage terms and current. Those of you familiar with Laplace transform methods are likely accustomed to seeing transfer functions expressed in terms of the Laplace operator s instead of the differentiation operator p. In fact, the same transfer functions are obtained by using Laplace transform theory with p replaced by s.

The time-domain block diagram portraying (7.9-4) through (7.9-6) with $T_e = (P/2)\lambda_m'' i_{qs}^r$ is shown in Fig. 7.9-1. This diagram consists of a set of linear blocks, wherein the relationship between the input and corresponding output variable is depicted in transfer function form, and two multipliers which represent nonlinear blocks. Since the system is nonlinear, it is not possible to apply previously used techniques (or, for that matter, Laplace transform methods) for solving the differential equations implied by this block diagram. For this, we would use a computer. However, if we neglect i_{ds}^r as we will later, we will find that the multipliers in Fig. 7.9-1 are no longer needed and conventional methods of analyzing linear systems may be applied with relative ease.

The so-called state equations of a system are the formulation of the state variables into a matrix form convenient for computer implementation, particularly for linear systems. The state variables of a system are defined as a minimal set of variables such that knowledge of these variables at any initial time t_0 and information on the input excitation subsequently applied are sufficient to determine the state of the system at any time $t > t_0$ [3]. In the case of brushless dc machines, the stator currents i_{qs}^r, i_{ds}^r, the rotor speed ω_r, and the rotor position θ_r are the state variables. However, since θ_r can be established from ω_r using

$$\theta_r = \int_0^t \omega_r(\xi) \, d\xi + \theta_r(0) \qquad (7.9\text{-}7)$$

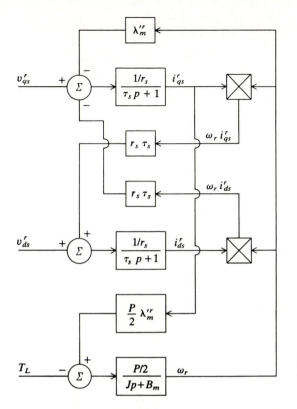

FIGURE 7.9-1
Time-domain block diagram of a brushless dc machine.

and since θ_r is considered a state variable only when shaft position is a controlled variable, we will omit θ_r from consideration in this development.

The formulation of the state equations for the brushless dc motor can be readily achieved by straightforward manipulation of the voltage equations given by (7.5-17) and the equation relating torque and rotor speed given by (7.4-3). In particular, solving the v_{qs}^r voltage equation (7.5-15) for di_{qs}^r/dt yields

$$\frac{di_{qs}^r}{dt} = -\frac{r_s}{L_{ss}} i_{qs}^r - \omega_r i_{ds}^r - \frac{\lambda_m''}{L_{ss}} \omega_r + \frac{1}{L_{ss}} v_{qs}^r \qquad (7.9\text{-}8)$$

Solving the v_{ds}^r voltage equation, (7.5-16), for di_{ds}^r/dt yields

$$\frac{di_{ds}^r}{dt} = -\frac{r_s}{L_{ss}} i_{ds}^r + \omega_r i_{qs}^r + \frac{1}{L_{ss}} v_{ds}^r \qquad (7.9\text{-}9)$$

Solving (7.4-3) for $d\omega_r/dt$ with $T_e = (P/2)\lambda_m'' i_{qs}^r$ yields

$$\frac{d\omega_r}{dt} = -\frac{B_m}{J} \omega_r + \left(\frac{P}{2}\right)^2 \frac{\lambda_m''}{J} i_{qs}^r - \frac{P}{2} \frac{1}{J} T_L \qquad (7.9\text{-}10)$$

All we have done is to solve the equations for the highest derivative of the state variables while substituting $T_e = (P/2)\lambda_m'' i_{qs}^r$ into (7.4-3). The state equations in matrix or vector matrix form are

$$
p\begin{bmatrix} i_{qs}^r \\ i_{ds}^r \\ \omega_r \end{bmatrix} = \begin{bmatrix} -\dfrac{r_s}{L_{ss}} & 0 & -\dfrac{\lambda_m''^r}{L_{ss}} \\ 0 & -\dfrac{r_s}{L_{ss}} & 0 \\ \left(\dfrac{P}{2}\right)^2 \dfrac{\lambda_m''^r}{J} & 0 & -\dfrac{B_m}{J} \end{bmatrix} \begin{bmatrix} i_{qs}^r \\ i_{ds}^r \\ \omega_r \end{bmatrix} + \begin{bmatrix} -\omega_r i_{ds}^r \\ \omega_r i_{qs}^r \\ 0 \end{bmatrix} + \begin{bmatrix} \dfrac{1}{L_{ss}} & 0 & 0 \\ 0 & \dfrac{1}{L_{ss}} & 0 \\ 0 & 0 & -\dfrac{P}{2}\dfrac{1}{J} \end{bmatrix} \begin{bmatrix} v_{qs}^r \\ v_{ds}^r \\ T_L \end{bmatrix}
$$

$$(7.9\text{-}11)$$

where p is the operator d/dt. Equation (7.9-11) is the state equation(s). Note however, that the second term (vector) on the right-hand side contains the products of state variables causing the system to be nonlinear.

Linear System Equations

As we know, if ϕ_v is set equal to zero and if i_{ds}^r is neglected, the brushless dc motor is described by equations identical in form to those of a permanent-magnet dc motor. In particular, the time-domain block diagram for the brushless dc motor with $\phi_v = 0$ and i_{ds}^r neglected is obtained from (7.9-4) and (7.9-6) with i_{ds}^r neglected in (7.9-4). The time-domain block diagram is shown in Fig. 7.9-2.

Usually, machine constants are given by the manufacturer. The voltage constant, generally denoted k_e or k_v, is our λ_m''. It is given either as a peak phase (winding) voltage or as an rms voltage. A torque constant k_t is also given. It is in the units of N·m/A and, hence, one would be led to believe from the torque equation (7.5-18) that

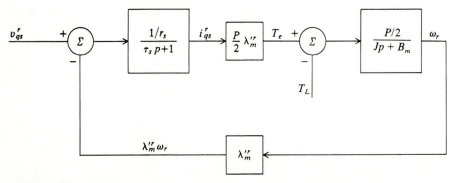

FIGURE 7.9-2
Time-domain block diagram of a brushless dc motor with $\phi_v = 0$ and i_{ds}^r neglected.

$$k_t = \frac{P}{2}\lambda_m'' = \frac{P}{2}k_e \qquad (7.9\text{-}12)$$

Unfortunately, k_t may be given with or without the number of poles taken into account.

A mechanical or inertia time constant is often used as in the case of dc machines. In the case of the brushless dc motor, we will define it as

$$\tau_m = \frac{Jr_s}{(P/2)^2(\lambda_m'')^2} \qquad (7.9\text{-}13)$$

This constant is often given by the manufacturers. However, one must realize that the manufacturer's value of τ_m will include the inertia of only the rotor whereas the J in (7.9-13) is, by our definition, the inertia of the rotor and connected load.

Let us return to the block diagram given in Fig. 7.9-2. As in Example 3C for a permanent-magnet dc machine, we can write the following transfer function:

$$\omega_r = \frac{\dfrac{1}{k_e\tau_s\tau_m}v_{qs}^r - \dfrac{P}{2}\dfrac{1}{J}\left(p + \dfrac{1}{\tau_s}\right)T_L}{p^2 + \left(\dfrac{1}{\tau_s} + \dfrac{B_m}{J}\right)p + \dfrac{1}{\tau_s}\left(\dfrac{1}{\tau_m} + \dfrac{B_m}{J}\right)} \qquad (7.9\text{-}14)$$

where $k_e = \lambda_m''$. The current i_{qs}^r is expressed as

$$i_{qs}^r = \frac{\dfrac{1}{\tau_s r_s}\left(p + \dfrac{B_m}{J}\right)v_{qs}^r + \dfrac{1}{k_e\tau_s\tau_m}T_L}{p^2 + \left(\dfrac{1}{\tau_s} + \dfrac{B_m}{J}\right)p + \dfrac{1}{\tau_s}\left(\dfrac{1}{\tau_m} + \dfrac{B_m}{J}\right)} \qquad (7.9\text{-}15)$$

With $\phi_v = 0$ and i_{ds}^r neglected, i_{qs}^r and ω_r are now the state variables. Thus, from (7.9-8),

$$\frac{di_{qs}^r}{dt} = -\frac{r_s}{L_{ss}}i_{qs}^r - \frac{\lambda_m''}{L_{ss}}\omega_r + \frac{1}{L_{ss}}v_{qs}^r \qquad (7.9\text{-}16)$$

Equation (7.9-10) still applies for ω_r. The system is described by a set of linear differential equations. In matrix form, the state equations become

$$p\begin{bmatrix} i^r_{qs} \\ \omega_r \end{bmatrix} = \begin{bmatrix} -\dfrac{r_s}{L_{ss}} & -\dfrac{\lambda''^r_m}{L_{ss}} \\ \left(\dfrac{P}{2}\right)^2 \dfrac{\lambda''_m}{J} & -\dfrac{B_m}{J} \end{bmatrix} \begin{bmatrix} i^r_{qs} \\ \omega_r \end{bmatrix} + \begin{bmatrix} \dfrac{1}{L_{ss}} & 0 \\ 0 & -\dfrac{P}{2}\dfrac{1}{J} \end{bmatrix} \begin{bmatrix} v^r_{qs} \\ T_L \end{bmatrix} \tag{7.9-17}$$

The state equations expressed in the form of (7.9-17) are called the *fundamental form*. In particular, the previous matrix equations may be expressed symbolically as

$$p\mathbf{x} = \mathbf{A}\mathbf{x} + \mathbf{B}\mathbf{u} \tag{7.9-18}$$

which is called the fundamental form, where p is the operator d/dt, $\mathbf{x}$ is the state vector (column matrix of state variables), and $\mathbf{u}$ is the input vector (column matrix of inputs to the system). We see that (7.9-17) and (7.9-18) are identical in form. Methods of solving equations of the fundamental form given by (7.9-18) are well known. Consequently, they are used extensively in control system analysis [3]. Before leaving this work, let us repeat the restrictions upon the linear differential equations given in this section. First, v^r_{ds} must be zero, i.e., $\phi_v = 0$; second i^r_{ds} is neglected, which means that v^r_{qs}, ω_r, and T_e should not be negative. This latter restriction is rather severe since it can easily be violated if the system variables are subjected to large excursions. One last comment. The stator (electrical) time constant τ_s is often neglected and the steady-state torque-speed characteristic is used in the transfer function. It was illustrated in Fig. 7.8-2 that this can lead to error in calculating the response of the rotor speed if the inertia of the mechanical system is small.

SP7.9-1. If $\phi_v = 0$ and i^r_{ds} is not neglected, are the system differential equations linear? Is i^r_{ds} a state variable? [No; yes, if state equations are written as (7.9-11)]

SP7.9-2. What is the difference between neglecting i^r_{ds} and neglecting the term $\omega_r^2 L^2_{ss}$ as we did in the analysis of steady-state operation with $\phi_v = 0$. [None]

7.10 THREE-PHASE PERMANENT-MAGNET SYNCHRONOUS MACHINE

A two-pole three-phase permanent-magnet synchronous machine is shown in Fig. 7.10-1. The stator windings are identical windings, displaced 120°. The windings are sinusoidally distributed, each with N_s equivalent turns and resistance r_s. Electromagnetic torque is produced by the interaction of the poles of the permanent-magnet rotor and the poles resulting from the rotating air gap mmf established by currents flowing in the stator windings. The rotating air gap mmf (mmf$_s$) established by symmetrical three-phase stator windings carrying balanced three-phase currents is given by (4.4-18).

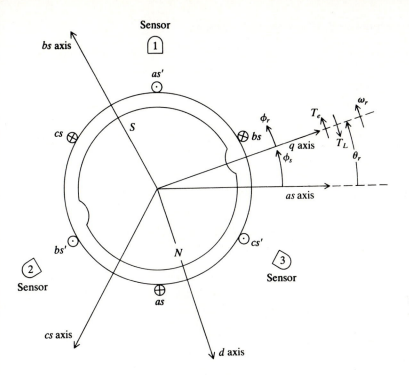

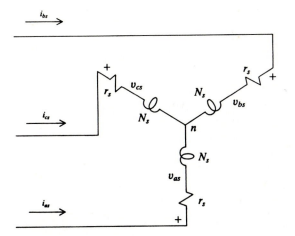

FIGURE 7.10-1
Two-pole three-phase permanent-magnet synchronous machine.

Voltage Equations and Winding Inductances

The voltage equations for the two-pole three-phase permanent-magnet synchronous machine shown in Fig. 7.10-1 may be expressed as

$$v_{as} = r_s i_{as} + \frac{d\lambda_{as}}{dt} \tag{7.10-1}$$

$$v_{bs} = r_s i_{bs} + \frac{d\lambda_{bs}}{dt} \tag{7.10-2}$$

$$v_{cs} = r_s i_{cs} + \frac{d\lambda_{cs}}{dt} \tag{7.10-3}$$

In matrix form,

$$\boxed{\mathbf{v}_{abcs} = \mathbf{r}_s \mathbf{i}_{abcs} + p\boldsymbol{\lambda}_{abcs}} \tag{7.10-4}$$

For voltages, currents, and flux linkages,

$$(\mathbf{f}_{abcs})^T = [f_{as} \quad f_{bs} \quad f_{cs}] \tag{7.10-5}$$

and

$$\mathbf{r}_s = \begin{bmatrix} r_s & 0 & 0 \\ 0 & r_s & 0 \\ 0 & 0 & r_s \end{bmatrix} \tag{7.10-6}$$

The flux linkage equations may be expressed as

$$\lambda_{as} = L_{asas} i_{as} + L_{asbs} i_{bs} + L_{ascs} i_{cs} + \lambda_{asm} \tag{7.10-7}$$

$$\lambda_{bs} = L_{bsas} i_{as} + L_{bsbs} i_{bs} + L_{bscs} i_{cs} + \lambda_{bsm} \tag{7.10-8}$$

$$\lambda_{cs} = L_{csas} i_{as} + L_{csbs} i_{bs} + L_{cscs} i_{cs} + \lambda_{csm} \tag{7.10-9}$$

In matrix form,

$$\boxed{\boldsymbol{\lambda}_{abcs} = \mathbf{L}_s \mathbf{i}_{abcs} + \boldsymbol{\lambda}'_m} \tag{7.10-10}$$

where $\boldsymbol{\lambda}'_m$ is the column vector

$$\boldsymbol{\lambda}'_m = \begin{bmatrix} \lambda_{asm} \\ \lambda_{bsm} \\ \lambda_{csm} \end{bmatrix} = \lambda'_m \begin{bmatrix} \sin\theta_r \\ \sin(\theta_r - \frac{2}{3}\pi) \\ \sin(\theta_r + \frac{2}{3}\pi) \end{bmatrix} \tag{7.10-11}$$

In (7.10-11), λ'_m is the amplitude of the flux linkages established by the permanent magnet as viewed from the stator phase windings. In other words, the magnitude of $p\lambda'_m$ would be the magnitude of the open-circuit voltage induced in each stator phase winding. The rotor displacement θ_r is defined by (7.3-10).

For our purposes, we will assume that the rotor is round (uniform air gap), whereupon $\mathbf{L}_s$ may be written as

$$\mathbf{L}_s = \begin{bmatrix} L_{ls} + L_{ms} & -\frac{1}{2}L_{ms} & -\frac{1}{2}L_{ms} \\ -\frac{1}{2}L_{ms} & L_{ls} + L_{ms} & -\frac{1}{2}L_{ms} \\ -\frac{1}{2}L_{ms} & -\frac{1}{2}L_{ms} & L_{ls} + L_{ms} \end{bmatrix} \tag{7.10-12}$$

where L_{ls} represents the leakage inductance and L_{ms} the magnetizing inductance. The off-diagonal terms are $-\frac{1}{2}L_{ms}$ since the mutual inductance between two stator windings displaced 120° is cos 120° times the mutual inductance if they were placed one on top of the other (L_{ms}).

Torque

An expression for the electromagnetic torque may be obtained by using the second entry in Table 2.5-1. Since the magnetic system is assumed to be linear, the field and coenergy are equal.

$$W_c = \tfrac{1}{2}L_{ss}(i_{as}^2 + i_{bs}^2 + i_{cs}^2) - \tfrac{1}{2}L_{ms}(i_{as}i_{bs} + i_{as}i_{cs} + i_{bs}i_{cs}) + \lambda'_m i_{as} \sin \theta_r$$

$$+ \lambda'_m i_{bs} \sin (\theta_r - \tfrac{2}{3}\pi) + \lambda'_m i_{cs} \sin (\theta_r + \tfrac{2}{3}\pi) + W_{pm} \tag{7.10-13}$$

where W_{pm} relates to the energy associated with the permanent magnet, which is constant for the machine shown in Fig. 7.10-1. Taking the partial derivative with respect to θ_r yields

$$T_e = \frac{P}{2} \lambda'_m \left[(i_{as} - \tfrac{1}{2}i_{bs} - \tfrac{1}{2}i_{cs}) \cos \theta_r + \frac{\sqrt{3}}{2} (i_{bs} - i_{cs}) \sin \theta_r \right] \tag{7.10-14}$$

The above expression is positive for motor action. The torque and speed may be related as given by (7.4-3).

Machine Equations in the Rotor Reference Frame

Since there are three stator variables (f_{as}, f_{bs}, and f_{cs}), we must use three substitute variables in the transformation of the stator variables to the rotor reference frame. In particular,

$$\boxed{\mathbf{f}^r_{qd0s} = \mathbf{K}^r_s \mathbf{f}_{abcs}} \tag{7.10-15}$$

where

$$(\mathbf{f}^r_{qd0s})^T = [f^r_{qs} \quad f^r_{ds} \quad f_{0s}] \tag{7.10-16}$$

$$\mathbf{K}^r_s = \frac{2}{3} \begin{bmatrix} \cos \theta_r & \cos (\theta_r - \tfrac{2}{3}\pi) & \cos (\theta_r + \tfrac{2}{3}\pi) \\ \sin \theta_r & \sin (\theta_r - \tfrac{2}{3}\pi) & \sin (\theta_r + \tfrac{2}{3}\pi) \\ \tfrac{1}{2} & \tfrac{1}{2} & \tfrac{1}{2} \end{bmatrix} \tag{7.10-17}$$

where the rotor displacement θ_r is defined by (7.3-10). The inverse of $\mathbf{K}^r_s$ is

$$(\mathbf{K}_s^r)^{-1} = \begin{bmatrix} \cos \theta_r & \sin \theta_r & 1 \\ \cos (\theta_r - \frac{2}{3}\pi) & \sin (\theta_r - \frac{2}{3}\pi) & 1 \\ \cos (\theta_r + \frac{2}{3}\pi) & \sin (\theta_r + \frac{2}{3}\pi) & 1 \end{bmatrix} \qquad (7.10\text{-}18)$$

A trigonometric interpretation of the above change of variables is shown in Fig. 6.9-2. It is important to note that the same notation $(\mathbf{K}_s^r)$ is used here for the three-phase transformation and for the two-phase transformation. This change of variables, (7.5-1), is Park's original transformation for synchronous machines [2].

The zero variable f_{0s} is the third substitute variable, and it is zero for balanced conditions. Also, f_{0s} is not a function of θ_r; therefore, the $0s$ variables are associated with stationary circuits. For this reason, a raised index is not incorporated with the zero variables.

Substituting the change of variables into the stator voltage equations given by (7.10-4) yields

$$(\mathbf{K}_s^r)^{-1}\mathbf{v}_{qd0s}^r = \mathbf{r}_s(\mathbf{K}_s^r)^{-1}\mathbf{i}_{qd0s}^r + p[(\mathbf{K}_s^r)^{-1}\boldsymbol{\lambda}_{qd0s}^r] \qquad (7.10\text{-}19)$$

Premultiplying each side by $\mathbf{K}_s^r$ yields

$$\boxed{\mathbf{v}_{qd0s}^r = \mathbf{r}_s\mathbf{i}_{qd0s}^r + \omega_r\boldsymbol{\lambda}_{dqs}^r + p\boldsymbol{\lambda}_{qd0s}^r} \qquad (7.10\text{-}20)$$

where

$$(\boldsymbol{\lambda}_{dqs}^r)^T = [\lambda_{ds}^r \quad -\lambda_{qs}^r \quad 0] \qquad (7.10\text{-}21)$$

If one wishes to work through the steps necessary to go from (7.10-19) to (7.10-20), the trigonometric relations given in Appendix A will be very helpful.

For a magnetically linear system, the stator flux linkages are expressed by (7.10-10). Substituting the change of variables into (7.10-10) yields

$$(\mathbf{K}_s^r)^{-1}\boldsymbol{\lambda}_{qd0s}^r = \mathbf{L}_{ss}(\mathbf{K}_s^r)^{-1}\mathbf{i}_{qd0s}^r + \boldsymbol{\lambda}_m' \qquad (7.10\text{-}22)$$

Premultiplying by $\mathbf{K}_s^r$ yields

$$\boldsymbol{\lambda}_{qd0s}^r = \begin{bmatrix} L_{ls} + \frac{3}{2}L_{ms} & 0 & 0 \\ 0 & L_{ls} + \frac{3}{2}L_{ms} & 0 \\ 0 & 0 & L_{ls} \end{bmatrix} \begin{bmatrix} i_{qs}^r \\ i_{ds}^r \\ i_{0s} \end{bmatrix} + \lambda_m'' \begin{bmatrix} 0 \\ 1 \\ 0 \end{bmatrix} \qquad (7.10\text{-}23)$$

To be consistent with our previous notation we have added the superscript r to λ_m'. In expanded form,

$$\boxed{\begin{aligned} v_{qs}^r &= r_s i_{qs}^r + \omega_r\lambda_{ds}^r + p\lambda_{qs}^r \\ v_{ds}^r &= r_s i_{ds}^r - \omega_r\lambda_{qs}^r + p\lambda_{ds}^r \\ v_{0s} &= r_s i_{0s} + p\lambda_{0s} \end{aligned}}$$

$$\qquad (7.10\text{-}24)$$
$$\qquad (7.10\text{-}25)$$
$$\qquad (7.10\text{-}26)$$

where

$$\lambda_{qs}^r = L_{ss} i_{qs}^r \qquad (7.10\text{-}27)$$

$$\lambda_{ds}^r = L_{ss} i_{ds}^r + \lambda_m'' \qquad (7.10\text{-}28)$$

$$\lambda_{0s} = L_{ls}i_{0s} \tag{7.10-29}$$

where

$$\boxed{L_{ss} = L_{ls} + \tfrac{3}{2}L_{ms}} \tag{7.10-30}$$

We must be careful here. When dealing with the two-phase machine, we also used L_{ss}; however, there it was $L_{ls} + L_{ms}$ rather than $L_{ls} + \tfrac{3}{2}L_{ms}$. Although we should probably use a different notation for the two- and three-phase machines, we will not. Instead we will call attention to this difference as we go along.

Substituting (7.10-27) through (7.10-29) into (7.10-24) through (7.10-26) and since λ_m'' is constant, $p\lambda_m'' = 0$, and we can write

$$v_{qs}^r = (r_s + pL_{ss})i_{qs}^r + \omega_r L_{ss}i_{ds}^r + \omega_r \lambda_m'' \tag{7.10-31}$$

$$v_{ds}^r = (r_s + pL_{ss})i_{ds}^r - \omega_r L_{ss}i_{qs}^r \tag{7.10-32}$$

$$v_{0s} = (r_s + pL_{ls})i_{0s} \tag{7.10-33}$$

Note that v_{qs}^r and v_{ds}^r, (7.10-31) and (7.10-32), are identical to v_{qs}^r and v_{ds}^r for the two-phase machine given by (7.5-10) and (7.5-11), respectively. The above equations may be rewritten in matrix form as

$$\begin{bmatrix} v_{qs}^r \\ v_{ds}^r \\ v_{0s} \end{bmatrix} = \begin{bmatrix} r_s + pL_{ss} & \omega_r L_{ss} & 0 \\ -\omega_r L_{ss} & r_s + pL_{ss} & 0 \\ 0 & 0 & r_s + pL_{ls} \end{bmatrix} \begin{bmatrix} i_{qs}^r \\ i_{ds}^r \\ i_{0s} \end{bmatrix} + \begin{bmatrix} \omega_r \lambda_m'' \\ 0 \\ 0 \end{bmatrix} \tag{7.10-34}$$

The expression for electromagnetic torque is obtained by expressing i_{as}, i_{bs}, and i_{cs}, in (7.10-14) in terms of i_{qs}^r and i_{ds}^r:

$$\boxed{T_e = \frac{3}{2}\frac{P}{2}\lambda_m'' i_{qs}^r} \tag{7.10-35}$$

which is positive for motor action.

Example 7E. The parameters of a four-pole permanent-magnet machine used as a brushless dc motor are $r_s = 3.4\,\Omega$, $L_{ls} = 1.1\,\text{mH}$, and $L_{ms} = 7.33\,\text{mH}$. When the device is driven at $1000\,\text{r/min}$, the open-circuit winding-to-winding voltage is sinusoidal with a peak-to-peak value of $60\,\text{V}$. Let us determine L_{ss} and λ_m''.

The first of these is obtained by straightforward substitution into (7.10-30).

$$L_{ss} = L_{ls} + \tfrac{3}{2}L_{ms}$$

$$= 1.1 + \tfrac{3}{2}(7.33) = 12.1\,\text{mH} \tag{7E-1}$$

Note that here we have selected L_{ms} so that L_{ss} is the same as in Example 7A. This will allow us to make a direct comparison between a two- and three-phase machine with essentially the same parameters.

Calculation of λ_m'' is a bit involved. The actual rotor speed in rad/s at which the measurement was taken is

$$\omega_{rm} = \frac{(r/\min)(\text{rad}/r)}{s/\min}$$

$$= \frac{(1000)(2\pi)}{60} = \frac{100}{3}\pi \text{ rad/s} \tag{7E-2}$$

The electrical angular velocity is

$$\omega_r = \frac{P}{2}\omega_{rm}$$

$$= \frac{4}{2}\frac{100\pi}{3} = \frac{200}{3}\pi \text{ rad/s} \tag{7E-3}$$

Let us assume that the open-circuit voltage is measured between a and b terminals; thus from (7.10-1) and (7.10-2) with i_{as} and $i_{bs} = 0$,

$$v_{ab} = v_{as} - v_{bs} = \frac{d\lambda_{as}}{dt} - \frac{d\lambda_{bs}}{dt} \tag{7E-4}$$

From (7.10-7), (7.10-8), and (7.10-11) and recalling that λ_m' and λ_m'' are the same quantity, we can write (7E-4) as

$$v_{ab} = \frac{d}{dt}\{\lambda_m''[\sin\theta_r - \sin(\theta_r - \tfrac{2}{3}\pi)]\}$$

$$= \lambda_m''\omega_r[\cos\theta_r - \cos(\theta_r - \tfrac{2}{3}\pi)]$$

$$= \lambda_m''\omega_r(\cos\theta_r - \cos\theta_r\cos\tfrac{2}{3}\pi - \sin\theta_r\sin\tfrac{2}{3}\pi)$$

$$= \lambda_m''\omega_r\left(\frac{3}{2}\cos\theta_r - \frac{\sqrt{3}}{2}\sin\theta_r\right)$$

$$= \sqrt{3}\lambda_m''\omega_r\cos(\theta_r + \tfrac{1}{6}\pi) \tag{7E-5}$$

We could have used phasor concepts (Appendix B) after taking the derivative. That is, from the second line of (7E-5) we can write

$$\tilde{V}_{ab} = \frac{\lambda_m''\omega_r}{\sqrt{2}}(1\underline{/0°} - 1\underline{/-120°})$$

$$= \frac{\lambda_m''\omega_r}{\sqrt{2}}\left(\frac{3}{2} + j\frac{\sqrt{3}}{2}\right) = \frac{\sqrt{3}\lambda_m''\omega_r}{\sqrt{2}}\underline{/30°} \tag{7E-6}$$

Now the peak-to-peak voltage is 60 V; hence, from either (7E-5) or (7E-6),

$$\frac{60}{2} = \sqrt{3}\lambda_m''\frac{200}{3}\pi \tag{7E-7}$$

The voltage must be divided by 2 since it is peak to peak. Solving for λ_m'' yields

$$\lambda_m'' = \frac{\frac{60}{2}}{\sqrt{3}(\frac{200}{3}\pi)} = 0.0826 \text{ V}\cdot\text{s/rad} \tag{7E-8}$$

Stator Voltages for Balanced Operation

When the three-phase permanent-magnet synchronous machine is supplied from a balanced three-phase voltage source, the phase voltages may be expressed as

$$v_{as} = \sqrt{2}v_s \cos \theta_{esv} \qquad (7.10\text{-}36)$$

$$v_{bs} = \sqrt{2}v_s \cos \left(\theta_{esv} - \tfrac{2}{3}\pi\right) \qquad (7.10\text{-}37)$$

$$v_{cs} = \sqrt{2}v_s \cos \left(\theta_{esv} + \tfrac{2}{3}\pi\right) \qquad (7.10\text{-}38)$$

The fundamental components of the stator applied voltages form a balanced three-phase set of *abc* sequence, where the amplitude v_s may be a function of time. Information regarding balanced sets is given in Appendix C. In (7.10-36) through (7.10-38),

$$\theta_{esv} = \int_0^t \omega_e(\xi)\, d\xi + \theta_{esv}(0) \qquad (7.10\text{-}39)$$

where, as always, $\theta_{esv}(0)$ is the time zero position of the applied voltages. If these voltages are substituted into the equation of transformation, (7.10-15), with $\omega_e = \omega_r$, we obtain

$$v_{qs}^r = \sqrt{2}v_s \cos \phi_v \qquad (7.10\text{-}40)$$

$$v_{ds}^r = -\sqrt{2}v_s \sin \phi_v \qquad (7.10\text{-}41)$$

and v_{0s} is zero for balanced conditions. For compactness,

$$\phi_v = \theta_{esv}(0) - \theta_r(0) \qquad (7.10\text{-}42)$$

where $\theta_r(0)$ is generally selected to be zero. Note that (7.10-40) through (7.10-42) are identical to (7.6-4) through (7.6-6), respectively, for the two-phase machine. It is interesting to note that the $\tfrac{2}{3}$ factor in $\mathbf{K}_s^r$, (7.10-17), makes the maximum amplitude of v_{qs}^r and v_{ds}^r the same as v_{as}, v_{bs}, and v_{cs}.

It is important to be aware that, as in the case of the two-phase machine, v_{qs}^r and v_{ds}^r may be changed in two ways. In particular, consider v_{qs}^r and v_{ds}^r given by (7.10-40) and (7.10-41), respectively. The amplitude of both v_{qs}^r and v_{ds}^r may be changed by changing v_s and the relative magnitude of v_{qs}^r and v_{ds}^r may be changed by changing ϕ_v.

Comparison of Equations for Two- and Three-Phase Machines

It seems appropriate to set out the important differences and similarities in the voltage and torque equations for the two- and three-phase permanent-magnet synchronous machines. First we note that the torque equation for the three-phase machine has the coefficient of $\tfrac{3}{2}$, whereas this coefficient is unity for the two-phase machine. Actually, it can be shown that this multiplier is $n_p/2$, where n_p is the number of stator phases.

Next, we introduced the 0s variables when defining the change of variables for the three-phase machine. However, these variables are zero when the stator variables are balanced or when a three-wire stator is being considered wherein $i_{as} + i_{bs} + i_{cs} = 0$ [1].

Even though we introduced a $0s$ variable, this did not change the form of the voltage equations in the rotor reference frame. Note that (7.10-31) and (7.10-32) are identical to (7.5-10) and (7.5-11), respectively. Hence, with $0s$ variables equal to zero, the voltage equations which describe the two- and three-phase machines are identical.

Moreover, when we transform a two-phase balanced set by using the two-phase $\mathbf{K}_s^r$, we obtain the same expressions as when we transform a three-phase balanced set by using the three-phase $\mathbf{K}_s^r$. For example, compare (7.10-40) and (7.10-41) with (7.6-4) and (7.6-5), respectively.

Perhaps the difference which is most easily overlooked is the fact that, for a two-phase machine, $L_{ss} = L_{ls} + L_{ms}$; but for a three-phase machine $L_{ss} = L_{ls} + \frac{3}{2}L_{ms}$. We saw a similar thing in the case of the two- and three-phase induction and/or synchronous machine.

Well, by this time you have most likely realized that we need not continue with a parallel development of the three-phase machine. Clearly, the material presented on the analysis of steady-state performance and transfer functions in Secs. 7.7 and 7.8, respectively, can be applied to a three-phase machine with little modification. These modifications for the three-phase machine are: (1) When you see L_{ss}, use $L_{ls} + \frac{3}{2}L_{ms}$; (2) multiply all torque (T_e) equations by $\frac{3}{2}$; and (3), which actually comes from (2), multiply k_t, given by (7.8-6), and the denominator of the expression for τ_m given in (7.8-7) both by $\frac{3}{2}$.

SP7.10-1. The voltage equations given for v_{qs}^r and v_{ds}^r for the two-phase brushless dc motor, (7.5-17), appear to be identical in form to those for a three-phase machine, (7.10-34). Is there a difference other than the zero variables? [Yes, L_{ss}]

SP7.10-2. Make a two-phase and a three-phase machine identical as far as T_e is concerned for a given phase voltage. $[L_{ls2} = L_{ls3}, L_{ms2} = \frac{3}{2}L_{ms3}, \lambda_{m2}''^r = \frac{3}{2}\lambda_{m3}''^r]$

7.11 BRUSHLESS DC MOTOR DRIVES

As we have mentioned on at least a dozen different occasions, when the stator of the permanent-magnet machine is supplied from an inverter which converts a dc voltage to ac voltages the frequency of which correspond to the speed of the rotor, the combination is a brushless dc motor. Speed control is achieved by varying the amplitude of the phase voltages by pulse width modulation (PWM), which is accomplished by electronically switching the phase voltages to zero at a high frequency relative to the fundamental frequency of the phase voltages. Before considering pulse width modulation in its elementary form, we will consider the operation of two types of brushless dc motors—a three-phase permanent-magnet synchronous machine supplied from a six-step 180° continuous-current inverter, where each phase of the machine is always connected to either the positive or negative terminal of the dc source and a six-step 120° discontinuous-current inverter, where each phase is open-circuited for essentially 120° of the cycle.

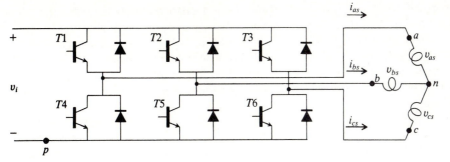

FIGURE 7.11-1
Brushless dc motor with six-step inverter configuration.

Six-Step 180° Continuous-Current Inverter

The circuit configuration of a 180° inverter is shown in Fig. 7.11-1 with the switching logic shown in Fig. 7.11-2. It is clear that in the case of the brushless dc motor inverter drive, the angular displacement of the applied stator voltages is θ_r defined by (7.3-10). Let us consider the action of one leg (phase) of the inverter shown in Fig. 7.11-1. The logic signals for the switching (commutation) of the transistors are shown in Fig. 7.11-2. $T1$ is turned on at $\theta_r = -90°$ and

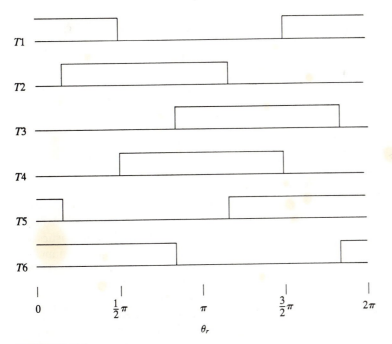

FIGURE 7.11-2
Transistor switching logic for the 180° continuous-current inverter.

turned off at $\theta_r = 90°$, at which time $T4$ is turned on. It is assumed that the turn-off time of the transistors is negligible whereupon the transistors become ideal switches. At the instant $T1$ is turned off, the current it was carrying is diverted to the diode in parallel with $T4$. The diode continues to conduct until the current decreases to zero. Once the current i_{as} reverses direction it is carried by $T4$. This mode of inverter operation is generally referred to as the continuous-voltage or continuous-current mode and, sometimes, the 180° conduction mode. We will consider the current flow in the transistor and diodes in more detail in a minute.

The following voltage equations may be written from Fig. 7.11-1:

$$v_{ap} = v_{as} + v_{np} \tag{7.11-1}$$

$$v_{bp} = v_{bs} + v_{np} \tag{7.11-2}$$

$$v_{cp} = v_{cs} + v_{np} \tag{7.11-3}$$

The stator is connected as a three-wire system, where $i_{as} + i_{bs} + i_{cs} = 0$ and, hence, the sum of v_{as}, v_{bs}, and v_{cs} is zero. Thus, by adding (7.11-1) through (7.11-3), we obtain

$$v_{np} = \tfrac{1}{3}(v_{ap} + v_{bp} + v_{cp}) \tag{7.11-4}$$

Hence,

$$v_{as} = \tfrac{2}{3}v_{ap} - \tfrac{1}{3}(v_{bp} + v_{cp}) \tag{7.11-5}$$

$$v_{bs} = \tfrac{2}{3}v_{bp} - \tfrac{1}{3}(v_{ap} + v_{cp}) \tag{7.11-6}$$

$$v_{cs} = \tfrac{2}{3}v_{cp} - \tfrac{1}{3}(v_{ap} + v_{bp}) \tag{7.11-7}$$

where v_{ap}, v_{bp}, and v_{cp} are either v_i or zero depending upon the state of $T1$ through $T6$. We will consider the waveform of the voltages in more detail later.

The free-acceleration characteristics of this type of brushless dc motor inverter drive is shown in Fig. 7.11-3. The machine parameters are those given previously with the exception that $L_{ms} = 7.3$ mH for the three-phase machine, which makes L_{ss} the same for the two- and three-phase example machines. For purposes of comparison, the inverter voltage is 25 V, where the constant component of v_{qs}^r is equal to the value used in the case of sinusoidal applied stator voltages. (We shall see this in a minute.) In addition to the variables plotted in Fig. 7.8-3, the winding-to-winding voltage v_{ab} and the d axis voltage v_{ds}^r are also shown in Fig. 7.11-3. The torque versus speed characteristics for this free acceleration are shown in Fig. 7.11-4. The dynamic performance of this system during load torque switching is shown in Fig. 7.11-5. As in Fig. 7.8-5, the inertia is twice the rotor inertia (2×10^{-4} kg·m^2), and the load torque is switched from 0.1 N·m to 0.4 N·m and then back to 0.1 N·m. Note that T_L for the three-phase machine is $\tfrac{3}{2}$ times that of the two-phase machine.

The steady-state and dynamic characteristics of the brushless dc motor driven from a continuous-current inverter are quite similar to those when

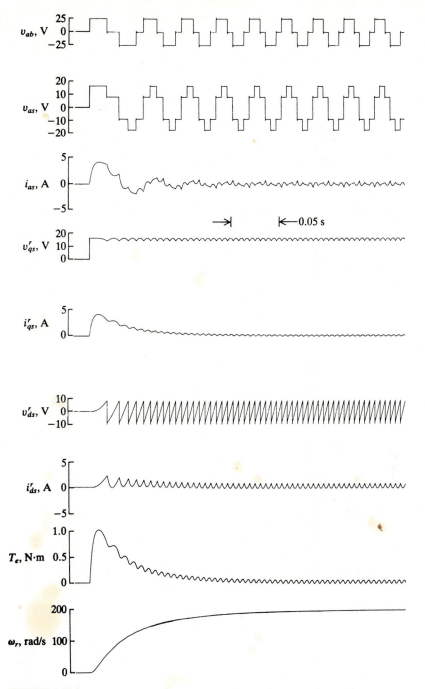

FIGURE 7.11-3
Free-acceleration characteristics of a brushless dc motor supplied from a six-step continuous-current inverter. Total inertia is five times the rotor inertia.

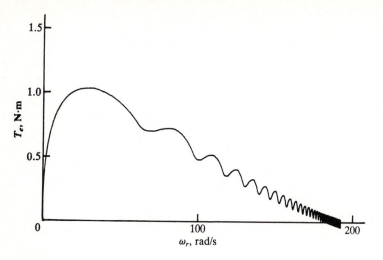

FIGURE 7.11-4
Torque-speed characteristics for free acceleration shown in Fig. 7.11-3.

two-phase sinusoidal voltages are applied to the two-phase machine. This follows from a comparision of the computer traces given in this section to those given in Sec. 7.9.

Before considering the discontinuous-current inverter, let us take a little closer look at the voltage and current waveforms of the continuous-current inverter. For this purpose, it is sufficient to consider only v_{as} and i_{as}. The voltage v_{as} is given by (7.11-5), wherein v_{ap} is equal to v_i when $T1$ is on and $T4$ is off, and zero when $T4$ is on and $T1$ is off. A similar situation exists for v_{bp} with regard to the states of $T2$ and $T5$ and for v_{cp} with regard to the states of $T3$ and $T6$. A plot of v_{as} as given by (7.11-5) for the switching logic given in Fig. 7.11-2 is shown in Fig. 7.11-6. Plotted below v_{as} in this figure is i_{as}. This waveform of i_{as} was taken from Fig. 7.11-5 with $T_L = 0.4 \, \text{N} \cdot \text{m}$. Plotted directly below i_{as} are the components of i_{as} which flow in $T1$, i_{aT1}, in $D1$, i_{aD1}, in $T4$, i_{aT4}, and in $D4$, i_{aD4}. $D1$ and $D4$ are the diodes in parallel with $T1$ and $T4$, respectively.

We have one more thing left to do. With $\phi_v = 0$, v'_{qs} can be approximated as $2v_i/\pi$ for a continuous-current inverter. To show this, let us assume that time zero is selected in Fig. 7.11-6 at the center of the maximum value of v_{as} and let us also assume that $\theta_r(0)$ is zero. Here, $\phi_v = 0$ since this selection of time zero will make $\theta_{esv}(0)$ of the fundamental component of v_{as} equal to zero. We can express v_{as} by Fourier series expansion as

$$v_{as} = \frac{2v_i}{\pi} (\cos \omega_r t + \tfrac{1}{5} \cos 5\omega_r t - \tfrac{1}{7} \cos 7\omega_r t + \cdots) \qquad (7.11\text{-}8)$$

(You are asked to prove this in a problem at the end of the chapter.) The voltages v_{bs} and v_{cs} may be expressed by substituting $\omega_r t - \tfrac{2}{3}\pi$ and $\omega_r t + \tfrac{2}{3}\pi$, respectively, for $\omega_r t$ in (7.11-8). If now v_{as}, v_{bs}, and v_{cs} are substituted into the

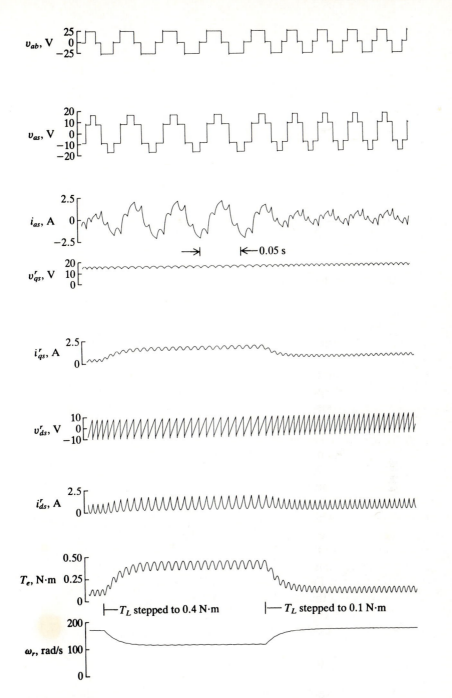

FIGURE 7.11-5
Dynamic performance during step changes in load torque of a brushless dc motor supplied from a continuous-current inverter. Total inertia is twice the rotor inertia.

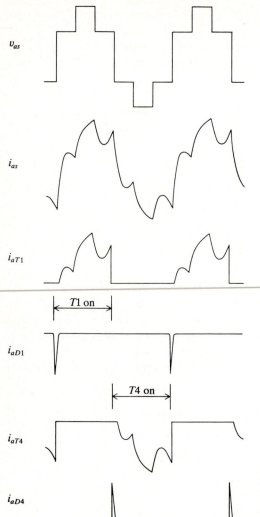

v_{as}

i_{as}

i_{aT1}

i_{aD1}

T1 on

i_{aT4}

T4 on

i_{aD4}

FIGURE 7.11-6
Plots of v_{as} and i_{as} and the components of i_{as} for steady-state operation of a continuous-current inverter drive.

equation of transformation, (7.10-15), we obtain

$$v^r_{qs} = \frac{2v_i}{\pi} \left(1 + \tfrac{2}{35} \cos 6\omega_r t - \tfrac{2}{143} \cos 12\omega_r t + \cdots;\right) \qquad (7.11\text{-}9)$$

$$v^r_{ds} = \frac{2v_i}{\pi} \left(\tfrac{12}{35} \sin 6\omega_r t - \tfrac{24}{143} \sin 12\omega_r t + \cdots\right) \qquad (7.11\text{-}10)$$

We see from (7.11-9) and (7.11-10) that, if we neglect the harmonics, v^r_{qs} becomes $2v_i/\pi$ and v^r_{ds} becomes zero. Take a minute to go back and look at the plots of v^r_{qs} and v^r_{ds} in Figs. 7.11-3 and 7.11-5. Is the average value of

$v_{qs}^r = 2v_i/\pi$ and is the average value of $v_{ds}^r = 0$? Is the dominant harmonic the sixth harmonic?

Although we are not going to go through the details, it can be shown that, with the provision for phase shifting incorporated, v_{as} may be expressed as [1]

$$v_{as} = \frac{2v_i}{\pi} \left[\cos\left(\omega_r t + \phi_v\right) + \tfrac{1}{5}\cos 5(\omega_r t + \phi_v) - \tfrac{1}{7}\cos 7(\omega_r t + \phi_v) + \cdots\right]$$

(7.11-11)

If the harmonics are neglected, v_{qs}^r and v_{ds}^r become

$$v_{qs}^r = \frac{2v_1}{\pi}\cos\phi_v$$

(7.11-12)

$$v_{ds}^r = -\frac{2v_i}{\pi}\sin\phi_v$$

(7.11-13)

Six-Step 120° Discontinuous-Current Inverter

The switching logic for the discontinuous-current inverter is shown in Fig. 7.11-7. To explain the operation of this inverter, let us again consider the leg containing $T1$ and $T4$ in Fig. 7.11-1. Transistor $T1$ is turned on at $\theta_r = -30°$ and turned off 120° later at $\theta_r = 90°$; however, $T4$ is not turned on until 60° later at $\theta_r = 150°$. Hence, the diode in parallel with $T4$ begins conducting when $T1$ is turned off and continues to conduct until the current i_{as} becomes zero. At

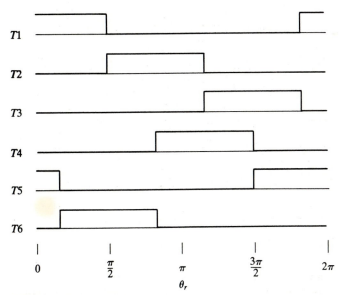

FIGURE 7.11-7
Transistor switching logic for the 120° discontinuous-current inverter.

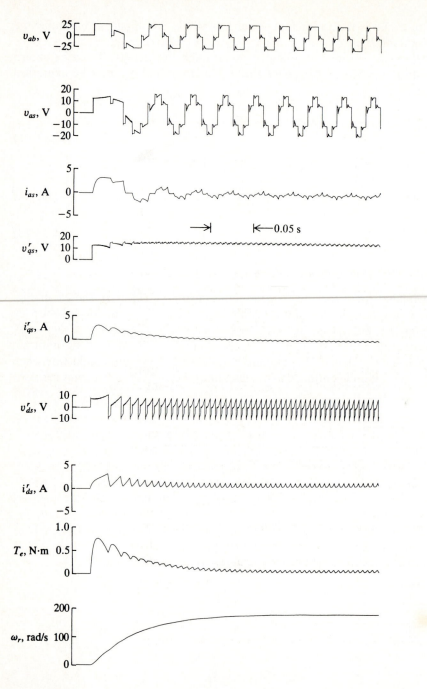

FIGURE 7.11-8

Free-acceleration characteristics of a brushless dc motor supplied from a six-step discontinuous-current inverter. Total inertia is five times the rotor inertia.

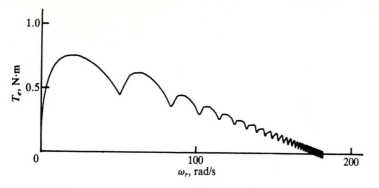

FIGURE 7.11-9
Torque-speed characteristics during free acceleration shown in Fig. 7.11-8.

this point, the *as* winding of the machine becomes open-circuited. The 60° delay in turning $T4$ on is done primarily for the purpose of lessening the possibility of a short circuit occurring through $T1$ and $T4$. The voltage v_{ap} in Fig. 7.11-1 is equal to the source voltage v_i if $T1$ is turned on or if the diode in parallel with $T1$ is conducting. When $T4$ is turned on or when the diode in parallel with $T4$ is conducting, v_{ap} is zero.

The computer traces shown in Figs. 7.11-8 through 7.11-10 depict the same modes of operation for the discontinuous-current inverter as those shown in Figs. 7.11-3 through 7.11-5, respectively, for the continuous-current inverter. We see from a comparison of the computer traces that the operation of the discontinuous-current inverter is considerably different from that of the continuous-current inverter. Since, in the case of the discontinuous-current inverter, the phase winding is open-circuited approximately one third of the time, we would expect a decrease in the average torque for a given speed when compared with a continuous-current inverter. An analysis of this type of operation is quite involved and beyond the scope of this text [1]. It is perhaps sufficient to warn that significant error may occur if the performance of a discontinuous-current inverter system is predicted from the equations that describe the operation of the continuous-current inverter system.

SP7.11-1. Express v_{ab} with v_{as} as given by (7.11-8). $\{v_{ab} = \sqrt{3}\, 2v_i/\pi[\cos{(\omega_r t + 30°)} + \frac{1}{5}\cos{(5\omega_r t + 30°)} - \frac{1}{7}\cos{(7\omega_r t + 30°)} + \cdots]\}$

SP7.11-2. Express v_{as} with $i_{as} = 0$ and i_{bs} and i_{cs} nonzero. $[v_{as} = \omega_r \lambda_m'' \cos{\theta_r}]$

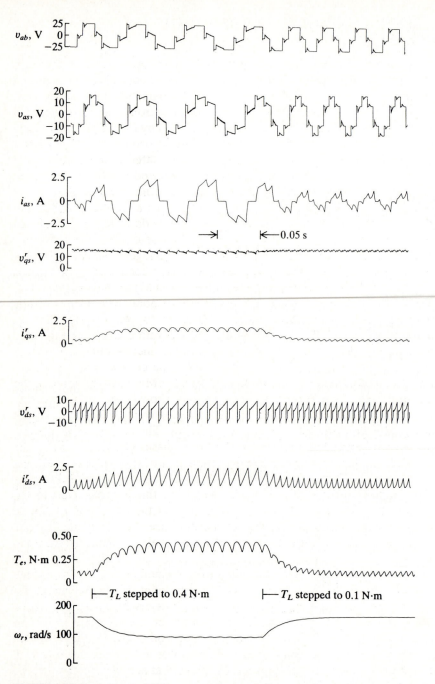

FIGURE 7.11-10
Dynamic performance during step changes in load torque of a brushless dc motor supplied from a discontinuous-current inverter. Total inertia is twice the rotor inertia.

7.12   SPEED CONTROL BY PULSE WIDTH MODULATION

By now we are well aware that the brushless dc motor operates much the same as a dc shunt motor with constant field current. Speed control can be achieved with both devices by varying the amplitude of the applied voltage. Consider, for a moment, the expression for steady-state electromagnetic torque of a brushless dc motor without phase shift ($\phi_v = 0$) given by (7.6-25). If we neglect the $\omega_r^2 L_{ss}^2$ term in the denominator, the torque is proportional to the difference between V_{qs}^r and $\omega_r \lambda_m''^r$. The voltage v_{qs}^r is directly related to the dc voltage of the inverter, (7.11-8). Hence, the control of rotor speed can be accomplished by changing v_{qs}^r so that load torque requirements, within the capability of the machine, may be satisfied ($T_e = T_L$) while maintaining a constant rotor speed. Clearly, v_{qs}^r may be changed by changing the inverter dc voltage v_i, which can be done if the inverter voltage is supplied from a phase-controlled rectifier, for example. In most cases, however, the inverter voltage is supplied from a constant source such as a battery. In this case, voltage control and, thus, speed control are achieved by pulse width modulation (PWM) of the inverter.

Although pulse width modulation may be incorporated into the inverters considered in the previous section, it is sufficient to consider only one to explain this technique. For this purpose, the operation of the continuous-current inverter (180°) with pulse width modulation incorporated is shown in Fig. 7.12-1. Therein, i_i, v_{as}, and i_{as} are plotted. The current i_i is the current flowing into the inverter from the dc source. In a PWM system, all three of the applied stator phase voltages (v_{bs} and v_{cs} not shown) are simultaneously switched to zero. The effective value of the phase voltages is controlled by controlling the time the voltages are held at zero. The zero voltage mode is achieved by instantaneously connecting all three phases of the machine to the same side of the dc source by appropriate switching of the transistors. During normal operation of the 180° inverter, two phases of the machine are connected to one terminal of the dc source while the third is connected to the other terminal. With all three phases connected to the same side of the dc source, the stator phases of the electric machine are short-circuited and the inverter current i_i is zero since the machine is effectively disconnected from the inverter.

The method of pulse width modulation depicted in Fig. 7.12-1 is one wherein the on and off times of the transistors are equal. Hence, to change the effective value of the stator phase voltages, it is necessary to change the rate or frequency of the pulse width modulation. Another common method is to maintain the frequency of the pulse width modulation constant, generally at a high frequency, and the effective value of the stator voltages is controlled by controlling the time that the phase voltages are held at zero during one cycle of the pulse width modulating frequency. There are many other aspects of pulse width modulation and speed control that are very important and interesting but well beyond the scope of this text.

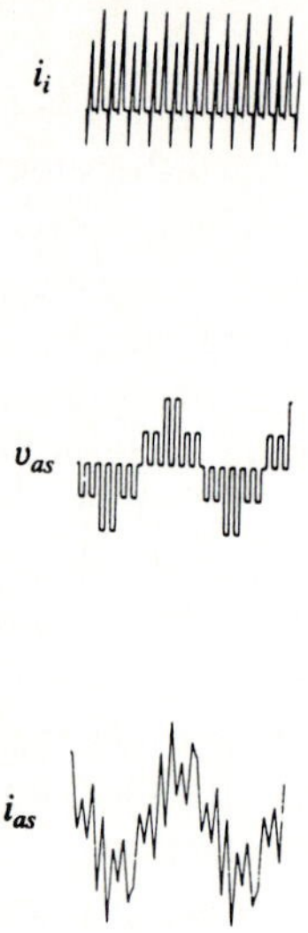

i_i

v_{as}

i_{as}

FIGURE 7.12-1
Pulse width modulation of a continuous-current inverter.

7.13 RECAPPING

We have learned that a brushless dc machine is a permanent-magnet synchronous machine supplied from a source the frequency of which is always equal to the speed of the rotor. Fortunately, we have been able to analyze this device from a very simplified point of view. Although this simplified analysis has limitations, it provides an excellent means of portraying the salient features of the operating characteristics of the brushless dc machine. In fact, with the exception of the discontinuous-current mode of operation, this simplified analysis provides an adequate means of analyzing this device. However, operating characteristics during the discontinuous-current mode are difficult to predict without the aid of a computer.

There is little doubt that the brushless dc motor will be used in an ever broadening range of applications. In fact, it will probably replace the permanent-magnet dc machine and the two-phase servomotor in many of their present-day applications. Although one reason for replacing the dc machine is that it does away with the brushes, we can see from the simplified analysis presented in this chapter that the brushless dc motor has a much broader range of operating characteristics than does a dc motor.

We will revisit the permanent-magnet synchronous machine when we study stepper motors in Chap. 8. As a stepper motor, the permanent-magnet synchronous machine is used widely as the positioning device in position-control systems.

7.14 REFERENCES

1. P. C. Krause, *Analysis of Electric Machinery*, McGraw-Hill Book Company, New York, 1986.
2. R. H. Park, "Two-Reaction Theory of Synchronous Machines—Generalized Method of Analysis—Part I," *AIEE Trans.*, vol. 48, July 1929, pp. 716–727.
3. B. C. Kuo, *Automatic Control Systems*, Prentice-Hall Inc., Englewood Cliffs, N.J., 1987.

7.15 PROBLEMS

1. It is found that $\lambda'_m = 0.1$ V·s for a permanent-magnet six-pole two-phase synchronous machine. Calculate the amplitude (peak value) of the open-circuit phase voltage measured when the rotor is turned at 60 revolutions per second (r/s).

2. Starting with (7.5-8) obtain (7.5-9).

3. Derive (7.5-18) from (7.4-2).

4. In the analysis of the brushless dc motor, we have selected $\theta_r(0) = 0$, where $\sqrt{2}\tilde{F}_{as} = F^r_{qs} - jF^r_{ds}$. Express the relationship between these same variables if we had selected $\theta_r(0) = \frac{1}{2}\pi$.

5. In Example 7C, an alternative method of calculating $\tilde{I}_{as}$ is suggested but not carried out. In particular, solution by appropriate combination of (7.5-18), (7.6-22), and (7.6-16) is suggested. Perform this suggested method of obtaining $\tilde{I}_{as}$ and draw the phasor diagram showing $\tilde{V}_{as}$, $\tilde{E}_a$, $r_s\tilde{I}_{as}$, $j\omega_r L_{ss}\tilde{I}_{as}$, and $\tilde{I}_{as}$.

6. A four-pole two-phase brushless dc motor is driven by a mechanical source at $\omega_{rm} = 3600$ r/min. The open-circuit voltage across one of the phases is 50 V (rms). (a) Calculate λ'_m. The mechanical source is removed and the following voltages are applied: $V_{as} = \sqrt{2}\, 25 \cos \theta_r$; $V_{bs} = \sqrt{2}\, 25 \sin \theta_r$, where $\theta_r = \omega_r t$. (b) Neglect friction ($B_m = 0$) and calculate the no-load rotor speed ω_r in rad/s.

7. Consider the brushless dc motor described in Example 7C. Calculate the (a) steady-state starting torque by assuming $v^r_{ds} = 0$ and (b) steady-state no-load speed in r/min.

8. The parameters of a two-pole two-phase brushless dc motor are as follows: $r_s = 2\ \Omega$, $\lambda''_m = 0.0707$ V·s/rad, $L_{ls} = 1$ mH, $L_{ms} = 9$ mH. The applied voltages are $V_{as} = \sqrt{2}\, 20 \cos \theta_r$ and $V_{bs} = \sqrt{2}\, 20 \sin \theta_r$, where $\theta_r = 200t$. (a) Calculate the steady-state electromagnetic torque T_e. (b) Determine I_{as}.

*9. For the values of ϕ_v determined in Example 7D, calculate (a) $\tilde{I}_{as}$ and (b) the steady-state starting torque.

10. Consider the brushless dc motor described in Prob. 8. Assume that the applied stator voltages are $V_{as} = \sqrt{2}\, 20 \cos (\omega_r t + \phi_{vMT})$ and $V_{bs} = \sqrt{2}\, 20 \sin (\omega_r t + \phi_{vMT})$. Calculate $\tilde{V}_{as}$, $\tilde{I}_{as}$, and T_e when $\omega_r = 200$ rad/s.

11. Speed control is to be achieved by voltage control with $\phi_v = 0$. Consider Example 7D where the load is removed and the speed of $\omega_r = 40\pi$ rad/s is to be maintained. Calculate V_s necessary to achieve this and the no-load current $\tilde{I}_{as}$.

*12. Calculate ω_r, $\tilde{E}_a$, and $\tilde{I}_{as}$ and construct the phasor diagram showing $\tilde{V}_{as}$, $\tilde{E}_a$, $r_s\tilde{I}_{as}$, $j\omega_r L_{ss}\tilde{I}_{as}$, and $\tilde{I}_{as}$ for the load conditions portrayed in Fig. 7.8-5. Neglect $\omega_r^2 L_{ss}^2$ in these calculations.

*13. For the load conditions given in Fig. 7.8-5, determine ω_r without neglecting $\omega_r^2 L_{ss}^2$. Compare with the values of ω_r calculated in Prob. 12.

14. Draw the time-domain block diagram of a brushless dc motor with provisions to shift the phase. Use ϕ_v as an input. Do not neglect i'_{ds}.

15. Derive (7.9-14) and (7.9-15).

16. Draw the time-domain block diagram with $\phi_v = 0$ and i'_{ds} not neglected.

17. Write the state equations for Prob. 16.

*18. Assume i'_{ds} may be neglected. Use Laplace transform methods to solve for ω_r for a change in load torque from $T_L = 0.267\,\mathrm{N \cdot m}$ to $T_L = 0.067\,\mathrm{N \cdot m}$ shown in Fig. 7.8-5.

19. Plot v_{as}, v_{bs}, v_{cs}, v_{ab}, and v_{np} for the 180° continuous-current inverter for $0 < \theta_r < 2\pi$ with the transistor switching shown in Fig. 7.11-2. Assume the inverter voltage v_i is constant.

*20. Derive (7.11-9) and (7.11-10).

*21. By using Fig. 7.11-10 with steady-state operation at $T_L = 0.4\,\mathrm{N \cdot m}$, plot v_{as}, i_{as}, and its components for the discontinuous-current inverter as was done in Fig. 7.11-6 for the continuous-current inverter. Identify, on the waveform of v_{as}, when an open-circuit condition occurs and which phase is open-circuited.

STEPPER MOTORS

8.1 INTRODUCTION

Stepper motors are electromechanical motion devices which are used primarily to convert information in digital form to mechanical motion. Although stepper motors were used as early as the 1920s, their use has skyrocketed with the advent of the digital computer. For example, when a logic output appears, the rotor is to rotate a predetermined angular displacement. These devices are found as the drivers for the paper in line printers and in other computer peripheral equipment such as in positioning of the magnetic-disk head. Whenever stepping from one position to another is required, whether the application is industrial, military, or medical, the stepper motor is generally used. Stepper motors come in various sizes and shapes but most fall into two types—the variable-reluctance stepper motor and the permanent-magnet stepper motor. Both types are considered in this chapter. We shall find that the operating principle of the variable-reluctance stepper motor is much the same as that of the reluctance machine which we have already discussed in earlier chapters, and the permanent-magnet stepper motor is similar in principle to the permanent-magnet synchronous machine considered at the beginning of Chap. 7.

8.2 BASIC CONFIGURATIONS
OF MULTISTACK VARIABLE-RELUCTANCE
STEPPER MOTORS

There are two general types of variable-reluctance stepper motors—single- and multistack. As a first approximation, the behavior of both types may be described from similar equations. Actually, the principle of operation of variable-reluctance stepper motors is the same as the reluctance machine which we considered in earlier chapters; only the mode of operation differs. There are, however, some new terms to define, and it is necessary for us to extend some of our previous definitions to fit the stepper motor. First, we will look at the multistack device in some detail, followed by a brief discussion of the single-stack variable-reluctance stepper motor.

In its most basic form, the multistack variable-reluctance stepper motor consists of three or more single-phase reluctance motors on a common shaft with their stator magnetic axes displaced from each other. The rotor of an elementary three-stack device is shown in Fig. 8.2-1. It has three cascaded two-pole rotors with a minimum-reluctance path of each aligned at the angular displacement θ_{rm}. In stepper motor language, each of the two-pole rotors is said to have two teeth. Now, visualize that each of these rotors has its own, separate, single-phase stator with the magnetic axes of the stators displaced from each other. In Fig. 8.2-1 we have labeled the individual rotors a, b, and c. The corresponding stators are shown in Fig. 8.2-2; the stator with the as winding is associated with the a rotor, the bs winding with the b rotor, etc. There are several things to note. First, we see that each of the single-phase stators has two poles, much the same in configuration as the dc machine, with the stator winding wound around both poles. In particular, positive current flows into as_1 and out as_1'; as_1' is connected to as_2 so that positive current flows into as_2 and out as_2'. Although we have shown only one circle for $as_1, \ldots, as_2'$, we realize that each would represent several turns, and that the number of turns from as_1 to as_1' (indicated by $N_s/2$ in Fig. 8.2-2) is the same as from as_2 to as_2'. Let us note one more thing; heretofore, we have referenced θ_{rm} (or θ_r) from the as axis to the maximum-reluctance path of a salient-pole rotor (Figs. 4.5-1, 4.5-4, and 6.2-1, for example). In Fig. 8.2-2, θ_{rm} is referenced to the minimum-reluctance path of the rotor. Since this is more or less standard in

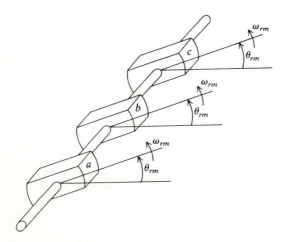

FIGURE 8.2-1
Rotor of an elementary two-pole three-stack variable-reluctance stepper motor.

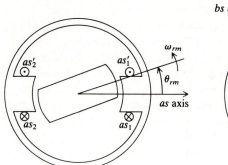

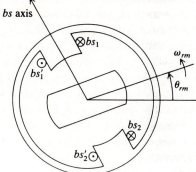

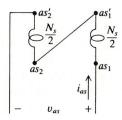

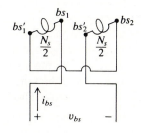

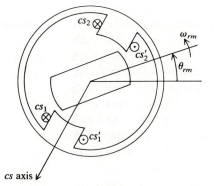

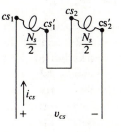

FIGURE 8.2-2
Stator configuration for an elementary two-pole three-stack variable-reluctance stepper motor.

stepper motor analysis, we will deviate from the convention we have established for reluctance and synchronous machines.

Each stack is often called a phase. In other words, a three-stack machine is a three-phase machine. This nomenclature can be misleading since we generally think of a three-phase ac device when we hear the words three-phase machine. We will find that a stepper motor is a discrete device, operated by switching a dc voltage from one stator winding to the other. Although more than three stacks (phases) may be used, perhaps as many as seven, three-stack variable-reluctance stepper motors are quite common. Our previous meaning of phase must be changed somewhat to accommodate the stepper motor.

Before writing any equations, let us see if we can gain some insight in regard to the operation of this device. To start, let the *bs* and *cs* windings be open-circuited, and let us apply a dc voltage to the *as* winding, whereupon we will assume that a constant i_{as} is immediately established. Now, since the magnetic systems of the three single-phase stators are separate, flux set up by one winding does not link the other windings, Hence, with only the *as* winding energized, flux exists only in the *as* axis. We know, from our work back in Chap. 2, that the minimum-reluctance path of the *a* part of the rotor (see Fig. 8.2-1) will align with the *as* axis. That is, at equilibrium with zero load torque, θ_{rm} in all parts of Fig. 8.2-2 is the same, either zero or 180°; let us say it is zero to make our discussion easier. (What would the rotor do if we could instantaneously reverse the direction of i_{as}?)

Stepper motors are used to convert digital or discrete information into a change in angular position. The spacing of a line printer or typewriter or the positioning of a robot arm are examples. Let us see how positioning (stepping) is achieved. For this, let us instantaneously deenergize the *as* winding and immediately establish a direct current in the *bs* winding. The minimum-reluctance path of the rotor will align itself with the *bs* axis. To do this, the rotor would rotate clockwise from $\theta_{rm} = 0$ to $\theta_{rm} = -60°$. Note that by advancing the mmf from the positive *as* axis to the positive *bs* axis, 120° counterclockwise, we have caused a 60° clockwise rotation of the rotor. There must be something wrong here. We recall from our work with rotating magnetic fields in Chap. 4 that, with the magnetic axes as shown in Fig. 8.2-2, an *abc* sequence of balanced sinusoidal currents will yield operation at synchronous speed with the rotor rotating counterclockwise. Therefore, it would seem that rotating the air gap mmf from the positive *as* axis to the positive *bs* axis would cause rotation in the counterclockwise direction. In the case of variable-reluctance stepper motors, we will find that the direction of stepping can be either in the same or opposite direction of the rotation of the air gap mmf depending upon the number of phases (stacks), the number of poles created by the stator windings, and the number of rotor teeth.

If, instead of energizing the *bs* winding, we energize the *cs* winding in Fig. 8.2-2, the rotor would have stepped counterclockwise from $\theta_{rm} = 0$ to $\theta_{rm} = 60°$. Thus, applying a dc voltage separately in the sequences *as*, *bs*, *cs*, *as*, . . . produces 60° steps in the clockwise direction, whereas the sequence *as*, *cs*, *bs*, *as*, . . . produces 60° steps in the counterclockwise direction. We need at least three stacks to achieve rotation (stepping) in both directions.

Before defining some stepper motor terms, let us think of one more thing. What if we energized both the *as* and *bs* windings? That is, assume that initially the *as* winding is energized with $\theta_{rm} = 0$ and the *bs* winding is energized without deenergizing the *as* winding. What happens? Well, the rotor rotates clockwise from $\theta_{rm} = 0$ to $\theta_{rm} = -30°$. We have reduced our step length by one half. This is referred to as *half-step operation*.

It is time to define terms. Let RT denote the number of rotor teeth per stack and ST the number of stator teeth per stack. The elementary device shown in Figs. 8.2-1 and 8.2-2 has two poles, two rotor teeth, and two stator teeth per stack; thus, $RT = ST = 2$. In fact, RT (rotor teeth per stack) always equals ST (stator teeth per stack) in a multistack variable-reluctance stepper motor. The number of stacks (phases) is denoted as N; here $N = 3$. Now, the tooth pitch, which we will denote as TP, is the angular displacement between rotor teeth. In this case, $TP = 180°$. We can write

$$TP = \frac{2\pi}{RT} \qquad (8.2\text{-}1)$$

We have one more term to define—the step length, denoted as SL. It is the angular rotation of the rotor as we change the excitation (dc voltage) from one phase to the other. In this case, the step length is 60°, $SL = 60°$. If we energize each stack separately, then going from *as* to *bs* to *cs* back to *as* causes the rotor to rotate one tooth pitch. In other words, the number of stacks (phases) times the step length is a tooth pitch. That is,

$$TP = N\,SL \qquad (8.2\text{-}2)$$

We can substitute (8.2-1) into (8.2-2) and obtain

$$SL = \frac{TP}{N} = \frac{2\pi}{RT\,N} \qquad (8.2\text{-}3)$$

We shall find use for all of these new terms as we go along.

Although the elementary device shown in Figs. 8.2-1 and 8.2-2 offers a good starting point in our analysis of stepper motors, it has limited application owing to its large step length. Let us consider the four-pole three-stack variable-reluctance stepper motor with four rotor teeth, as illustrated in Fig. 8.2-3. Here, $RT = 4$ and $N = 3$; therefore, from (8.2-1) the tooth pitch is $TP = 2\pi/RT = 90°$. From (8.2-2), the step length is $SL = TP/N = 30°$ and an *as*, *bs*, *cs*, *as*, . . . sequence produces 30° steps in the clockwise direction.

The device shown in Fig. 8.2-4 is a four-pole three-stack variable-reluctance stepper motor with eight rotor teeth. In this case, $RT = 8$ and $N = 3$, thus, $TP = 45°$ and $SL = 15°$. However, in this device an *as*, *bs*, *cs*, *as*, . . . sequence produces 15° steps in the counterclockwise direction. The pattern is clear; by increasing the number of rotor teeth we reduce the step length. The

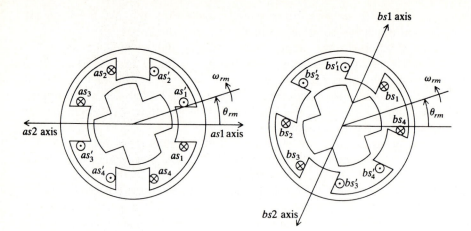

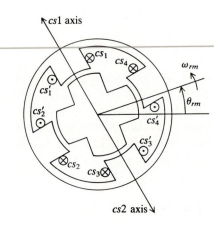

FIGURE 8.2-3
Four-pole three-stack variable-reluctance stepper motor with four rotor teeth.

step lengths of multistack variable-reluctance stepping motors typically range from 2 to 15°.

There appears to be an inconsistency in Fig. 8.2-4. In particular, θ_{rm} is referenced from the *as* axis to a position between rotor teeth. Earlier in this section, we established that, in the case of stepper motors, we would reference θ_{rm} from the *as* axis to the minimum-reluctance path of the rotor, whereupon the reluctance of the magnetic system associated with the *as* winding would be minimum when $\theta_{rm} = 0$. At first glance it appears that we have violated this stepper motor convention. However, when θ_{rm} is zero in Fig. 8.2-4, the reluctance of the magnetic system associated with the *as* winding is minimum.

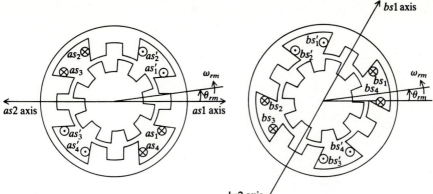

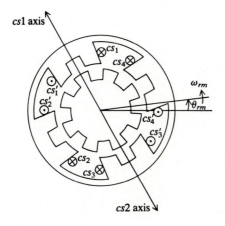

FIGURE 8.2-4
Four-pole three-stack variable-reluctance stepper motor with eight rotor teeth.

Hence, we must reference θ_{rm} from a position between rotor teeth to maintain the convention which we established earlier in this section. A cutaway view of a four-pole three-stack variable-reluctance stepper motor with eight rotor teeth is shown in Fig. 8.2-5.

SP8.2-1. Calculate the step length for an eight-pole three-stack variable-reluctance stepper motor with 16 rotor teeth. [$SL = 7.5°$]

SP8.2-2. Consider the two-pole two-phase reluctance motor shown in Fig. 4.6-1b. Calculate (a) TP, (b) SL, and (c) determine the direction of rotation when a dc voltage is switched from the as winding to the bs winding. [(a) $TP = 180°$; (b) $SL = 90°$; (c) either ccw or cw]

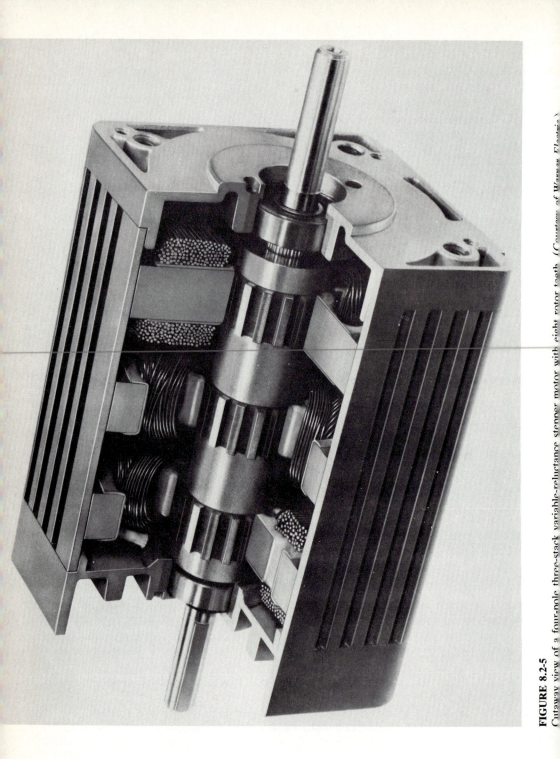

FIGURE 8.2-5
Cutaway view of a four-pole three-stack variable-reluctance stepper motor with eight rotor teeth. (*Courtesy of Warner Electric.*)

8.3   EQUATIONS FOR MULTISTACK VARIABLE-RELUCTANCE STEPPER MOTORS

The voltage equations for a three-stack variable-reluctance stepper motor may be written as

$$v_{as} = r_s i_{as} + \frac{d\lambda_{as}}{dt} \tag{8.3-1}$$

$$v_{bs} = r_s i_{bs} + \frac{d\lambda_{bs}}{dt} \tag{8.3-2}$$

$$v_{cs} = r_s i_{cs} + \frac{d\lambda_{cs}}{dt} \tag{8.3-3}$$

In matrix form,

$$\boxed{\mathbf{v}_{abcs} = \mathbf{r}_s \mathbf{i}_{abcs} + p\boldsymbol{\lambda}_{abcs}} \tag{8.3-4}$$

where p is the operator d/dt and, for voltages, currents, and flux linkages

$$(\mathbf{f}_{abcs})^T = [f_{as} \quad f_{bs} \quad f_{cs}] \tag{8.3-5}$$

with

$$\mathbf{r}_s = \begin{bmatrix} r_s & 0 & 0 \\ 0 & r_s & 0 \\ 0 & 0 & r_s \end{bmatrix} \tag{8.3-6}$$

Since magnetic coupling does not exist between phases, we can write the flux linkages as

$$\begin{bmatrix} \lambda_{as} \\ \lambda_{bs} \\ \lambda_{cs} \end{bmatrix} = \begin{bmatrix} L_{asas} & 0 & 0 \\ 0 & L_{bsbs} & 0 \\ 0 & 0 & L_{cscs} \end{bmatrix} \begin{bmatrix} i_{as} \\ i_{bs} \\ i_{cs} \end{bmatrix} \tag{8.3-7}$$

For the purpose of expressing the self-inductances L_{asas}, L_{bsbs}, and L_{cscs}, let us first consider the elementary two-pole device illustrated in Fig. 8.2-2. With the use of our work from Chap. 1, we can write as a first approximation,

$$L_{asas} = L_{ls} + L_A + L_B \cos 2\theta_{rm} \tag{8.3-8}$$

$$L_{bsbs} = L_{ls} + L_A + L_B \cos 2(\theta_{rm} - \tfrac{2}{3}\pi) \tag{8.3-9}$$

$$L_{cscs} = L_{ls} + L_A + L_B \cos 2(\theta_{rm} - \tfrac{4}{3}\pi) \tag{8.3-10}$$

From our previous work, we are aware that L_{ls} is the leakage inductance whereas L_A and L_B are constants with $L_A > L_B$. The rotor displacement is expressed as

$$\boxed{\theta_{rm} = \int_0^t \omega_{rm}(\xi)\, d\xi + \theta_{rm}(0)} \tag{8.3-11}$$

where ξ is a dummy variable of integration. We will use θ_{rm}, the actual rotor displacement, rather than θ_r, the electrical angular displacement. Although θ_{rm} and θ_r are related, $\theta_r = (P/2)\theta_{rm}$, where P is the number of poles, we will find it more convenient to use θ_{rm} in the analysis of stepper motors. We see that (8.3-8) is similar to (1.7-29) or (2.7-3), with θ_{rm} referenced to the minimum-reluctance path of the rotor. Equation (8.3-9) is easily developed once we realize that the self-inductance of the bs winding is the same as that of the as winding. However, since θ_{rm} is referenced from the as axis, the angular displacement to the bs axis from the as axis must be subtracted from θ_{rm} so that, when $\theta_{rm} = \frac{2}{3}\pi$, the argument of (8.3-9) is zero and (8.3-9) with $\theta_{rm} = \frac{2}{3}\pi$ becomes the same as (8.3-8) with $\theta_{rm} = 0$. Following this same line of reasoning, we would determine that the angular displacement of (8.3-10) is $-\frac{4}{3}\pi$. However, since $\cos 2(\theta_{rm} - \frac{4}{3}\pi) = \cos 2(\theta_{rm} + \frac{2}{3}\pi)$, we can use $\frac{2}{3}\pi$ as the angular displacement for L_{cscs}. It is obvious that we can express (8.3-8) through (8.3-10) in various forms. Later, we will find it advantageous to express the argument of (8.3-9) and (8.3-10) in terms of step length.

The self-inductances of the four-pole three-stack variable-reluctance device with four rotor teeth shown in Fig. 8.2-3 can be approximated as

$$L_{asas} = L_{ls} + L_A + L_B \cos 4\theta_{rm} \tag{8.3-12}$$

$$L_{bsbs} = L_{ls} + L_A + L_B \cos 4(\theta_{rm} - \tfrac{1}{3}\pi) \tag{8.3-13}$$

$$L_{cscs} = L_{ls} + L_A + L_B \cos 4(\theta_{rm} - \tfrac{2}{3}\pi) \tag{8.3-14}$$

Although we are using the same L_{ls}, L_A, and L_B to denote constants, we realize that these are not equal for the various machines.

For the four-pole three-stack variable-reluctance stepper motor with eight rotor teeth shown in Fig. 8.2-4, we can approximate the self-inductances as

$$L_{asas} = L_{ls} + L_A + L_B \cos 8\theta_{rm} \tag{8.3-15}$$

$$L_{bsbs} = L_{ls} + L_A + L_B \cos 8(\theta_{rm} - \tfrac{1}{3}\pi) \tag{8.3-16}$$

$$L_{cscs} = L_{ls} + L_A + L_B \cos 8(\theta_{rm} - \tfrac{2}{3}\pi) \tag{8.3-17}$$

By adding or subtracting multiples of 2π from the arguments, the above self-inductances may be expressed in terms of SL. In particular, for the devices shown in Figs. 8.2-2 and 8.2-3, where we have previously noted that a counterclockwise rotation of the stator mmf and stepping are in opposite directions, the inductances may be expressed as

$$L_{asas} = L_{ls} + L_A + L_B \cos (RT\,\theta_{rm}) \tag{8.3-18}$$

$$L_{bsbs} = L_{ls} + L_A + L_B \cos [RT\,(\theta_{rm} + SL)] \tag{8.3-19}$$

$$L_{cscs} = L_{ls} + L_A + L_B \cos [RT\,(\theta_{rm} - SL)] \tag{8.3-20}$$

For the device shown in Fig. 8.2-4, where rotation of the stator mmf and stepping are in the same direction, the self-inductances may be expressed as

$$L_{asas} = L_{ls} + L_A + L_B \cos{(RT\,\theta_{rm})} \qquad (8.3\text{-}21)$$

$$L_{bsbs} = L_{ls} + L_A + L_B \cos{[RT\,(\theta_{rm} - SL)]} \qquad (8.3\text{-}22)$$

$$L_{cscs} = L_{ls} + L_A + L_B \cos{[RT\,(\theta_{rm} + SL)]} \qquad (8.3\text{-}23)$$

An expression for the electromagnetic torque may be obtained from Table 2.5-1,

$$T_e = \frac{\partial W_c(\mathbf{i}, \theta_{rm})}{\partial \theta_{rm}} \qquad (8.3\text{-}24)$$

Since we are assuming a linear magnetic system, the field energy and coenergy are equal. Thus, since the mutual inductances are zero,

$$W_c = \tfrac{1}{2} L_{asas} i_{as}^2 + \tfrac{1}{2} L_{bsbs} i_{bs}^2 + \tfrac{1}{2} L_{cscs} i_{cs}^2 \qquad (8.3\text{-}25)$$

Substituting the self-inductances into (8.3-25) and taking the partial derivative with respect to θ_{rm} yields

$$
\begin{aligned}
T_e = &-\frac{RT}{2} L_B \{ i_{as}^2 \sin{(RT\,\theta_{rm})} + i_{bs}^2 \sin{[RT\,(\theta_{rm} \pm SL)]} \\
&+ i_{cs}^2 \sin{[RT\,(\theta_{rm} \mp SL)]} \}
\end{aligned}
\qquad (8.3\text{-}26)
$$

where the $+SL$ in the second argument and the $-SL$ in the third apply when the rotation of the stator mmf and the stepping are in opposite directions. The opposite signs of SL apply for the same direction of rotation. An alternate form of (8.3-26) using the tooth pitch TP is

$$
\begin{aligned}
T_e = &-\frac{RT}{2} L_B \left\{ i_{as}^2 \sin{\left(\frac{2\pi}{TP}\,\theta_{rm}\right)} + i_{bs}^2 \sin{\left[\frac{2\pi}{TP}\left(\theta_{rm} \pm \frac{TP}{3}\right)\right]} \right. \\
&\left. + i_{cs}^2 \sin{\left[\frac{2\pi}{TP}\left(\theta_{rm} \mp \frac{TP}{3}\right)\right]} \right\}
\end{aligned}
\qquad (8.3\text{-}27)
$$

It is important to note that the magnitude of the torque is proportional to the number of rotor teeth RT.

The torque and rotor angular position are related as

$$T_e = J\frac{d^2\theta_{rm}}{dt^2} + B_m\frac{d\theta_{rm}}{dt} + T_L \qquad (8.3\text{-}28)$$

where J is the total inertia in $\text{kg} \cdot \text{m}^2$ and B_m is a damping coefficient associated with the mechanical rotational system in $\text{N} \cdot \text{m} \cdot \text{s}$. The electromagnetic torque T_e is positive in the counterclockwise direction (positive direction of θ_{rm}) whereas the load torque T_L is positive in the clockwise direction.

SP8.3-1. The stator currents of a three-stack variable-reluctance machine are $i_{as} = I$, $i_{bs} = -I$, and $i_{cs} = 0$. Determine the no-load rotor position. [$\theta_{rm} = \pm TP/6$]

SP8.3-2. Repeat SP8.3-1 with $i_{as} = i_{bs} = i_{cs}$. [T_e is zero for all values of θ_{rm}.]

8.4 OPERATING CHARACTERISTICS OF MULTISTACK VARIABLE-RELUCTANCE STEPPER MOTORS

It is instructive to take a little closer look at the operating characteristics of a multistack variable-reluctance stepper motor from the standpoint of idealized, pseudosteady-state conditions. For this purpose, let us consider the expression for torque given by (8.3-27) for a three-stack motor with opposite directions of rotation of the stator mmf and stepping. In particular,

$$
T_e = -\frac{RT}{2} L_B \left\{ i_{as}^2 \sin\left(\frac{2\pi}{TP}\theta_{rm}\right) + i_{bs}^2 \sin\left[\frac{2\pi}{TP}\left(\theta_{rm} + \frac{TP}{3}\right)\right] \right.
$$

$$
\left. + i_{cs}^2 \sin\left[\frac{2\pi}{TP}\left(\theta_{rm} - \frac{TP}{3}\right)\right] \right\}
\tag{8.4-1}
$$

In Fig. 8.4-1, the three terms of (8.4-1) are plotted separately for equal, constant (steady-state) currents. Let us assume that there is no load torque, $T_L = 0$, and $i_{as} = I$ while i_{bs} and i_{cs} are zero. Only the first term of (8.4-1) is present; that is, only the steady-state torque due to i_{as} exists. The stable steady-state rotor position would be at $\theta_{rm} = 0$ denoted as point 1 on Fig. 8.4-1. Now, let us assume that i_{as} is instantaneously decreased from I to zero while i_{bs} is increased from zero to I. Hence, the steady-state torque plot due to i_{as} would instantaneously disappear from Fig. 8.4-1 and the torque due to i_{bs} would immediately appear. Now, we know that this cannot happen in practice since there would be electrical transients involved, but we are neglecting all transients in this discussion. Since, the torque at point 2 is negative, the rotor will rotate in the clockwise direction. We will then proceed along the i_{bs} torque plot until we have reached point 3. Note we have moved one step length in the clockwise direction. If, instead of energizing the *bs* winding after deenergizing the *as* winding, we energized the *cs* winding, then the torque at point 4 would appear. This is a positive T_e so the rotor will rotate in the counterclockwise direction and we will ride along the torque angle plot to point 5. Please realize that not only are we neglecting the electrical transients in this discussion but we are also neglecting the mechanical transients. Normally, there would be a damped oscillation about the new operating point, in this example, either point 3 or 5.

Half-step operation is depicted in Fig. 8.4-2. To explain this, let us again start at point 1 where $T_L = 0$ and only the *as* winding is energized ($i_{as} = I$). Instantaneously the *bs* winding is energized and $i_{bs} = I$. Now, both i_{as} and i_{bs} are I and only the *as* + *bs* torque plot, shown in Fig. 8.4-2, exists. Immediately, the torque at point 2 appears and the rotor starts to rotate in the clockwise direction coming to rest at point 3. The rotor has moved $SL/2$ clockwise.

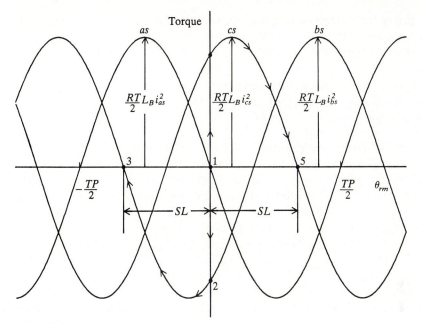

FIGURE 8.4-1
Stepping operation of a three-stack variable-reluctance stepper motor without load torque—steady-state torque-angle plots.

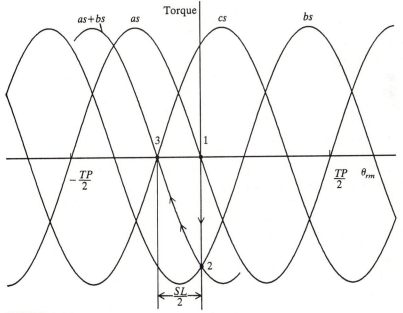

FIGURE 8.4-2
Half-step operation of a three-stack variable-reluctance stepper motor—steady-state torque-angle plots.

Stepping action with a load torque is shown in Fig. 8.4-3. Assume that initial operation is at point 1 with $i_{as} = I$ and $i_{bs} = i_{cs} = 0$. Recall that T_e is positive in the counterclockwise direction while T_L is positive in the clockwise direction, and stable operation occurs when $T_e = T_L$. Thus, at point 1, $T_e = T_L$. The as winding is deenergized while the bs winding is energized. Immediately, the negative T_e at point 2 appears and the rotor will move clockwise to point 3. If the cs winding is energized rather than the bs winding, the torque at point 4 would appear and the rotor would move to point 5. Note that the step length is still the same in both directions. However, the rotor will move more rapidly in the clockwise direction than in the counterclockwise direction since the load torque is in the clockwise direction. In other words, there is a larger torque to accelerate the rotor in the clockwise direction than in the counterclockwise direction.

The plots of i_{as}, i_{bs}, i_{cs}, and θ_{rm} shown in Fig. 8.4-4 allow us to view stepping operation from another standpoint. Initially, there is no load torque and $i_{as} = I$. The current i_{as} is stepped off and i_{bs} is stepped on. The rotor rotates clockwise to $\theta_{rm} = -SL$. Here we have indicated the presence of a damped mechanical oscillation which was not shown in the steady-state torque angle plots. Next, i_{bs} is switched off and i_{as} is switched back on. The rotor ends up back at $\theta_{rm} = 0$. Next we see half-step operation; i_{as} remains at I while i_{bs} is

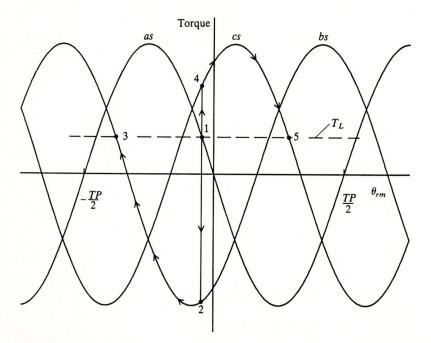

FIGURE 8.4-3
Stepping operation of a three-stack variable-reluctance stepper motor with load torque—steady-state torque-angle plots.

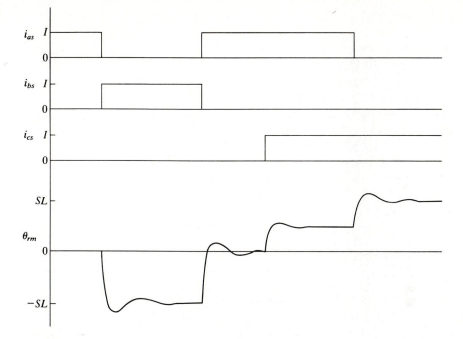

FIGURE 8.4-4
Stepping operation depicting θ_{rm} versus time—no load torque.

switched to I. The rotor advances to $\frac{1}{2}SL$. When i_{as} is switched to zero, the rotor again advances by $\frac{1}{2}SL$ to $\theta_{rm} = SL$.

SP8.4-1. In Fig. 8.4-3 the load torque is such that the initial operating point with $i_{as} = I$ and $i_{bs} = i_{cs} = 0$ is at $\theta_{rm} = -TP/8$. The current in the as winding is switched to zero and the current in the cs winding is switched to I. Determine the final value of θ_{rm}. Which direction will the rotor rotate? $[\theta_{rm} = -TP/8 - 2SL$; cw$]$

SP8.4-2. It is desirable to step from $\theta_{rm} = 0$ to $\theta_{rm} = -SL/3$ for the device shown in Fig. 8.2-3. Assume that we have the facility to control the winding currents. Let $i_{as} = I$; determine i_{bs}. $[i_{bs} = 0.81I]$

8.5 SINGLE-STACK VARIABLE-RELUCTANCE STEPPER MOTORS

As its name suggests, the single-stack variable-reluctance stepper motor has only one stack and all stator phases are arranged on this single stack. A three-phase single-stack variable-reluctance stepper motor is shown in Fig. 8.5-1. Here, it appears that we have taken the three two-pole single-phase stators shown in Fig. 8.2-2 and squeezed them into one stack. The magnetic axes of the stator windings are displaced 120° as in the case of the three-phase machines considered in earlier chapters; however, the stepper motor generally

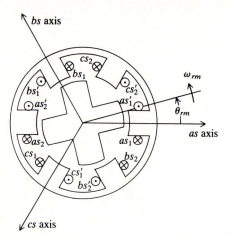

FIGURE 8.5-1
Two-pole three-phase single-stack variable-reluctance stepper motor with six stator teeth and four rotor teeth.

has stator teeth or poles which protrude rather than a circular inner stator surface.

Recall that in the case of the multistack variable-reluctance motor, the number of rotor and stator teeth per stack is the same. In the case of the single-stack stepper motor, the number of rotor teeth per stack, RT, is never equal to the number of stator teeth per stack, ST. If, for example, the rotor shown in Fig. 8.5-1 had the same number of teeth as the stator (6), then, when two diagonally opposite rotor teeth are aligned with two diagonally opposite stator teeth, all diagonally opposite rotor teeth would be aligned with diagonally opposite stator teeth and stepping action could not occur. The equations which we derived for the tooth pitch TP and step length SL for the multistack variable-reluctance stepper motor also apply for the single-stack stepper motor. For the two-pole three-phase stepper motor shown in Fig. 8.5-1, $RT = 4$ and,

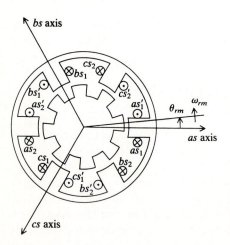

FIGURE 8.5-2
Two-pole three-phase single-stack variable-reluctance stepper motor with six stator teeth and eight rotor teeth.

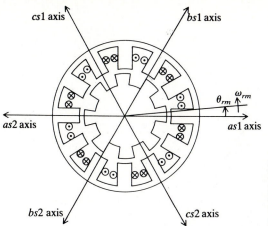

FIGURE 8.5-3
Four-pole three-phase single-stack variable-reluctance stepper motor with twelve stator teeth and eight rotor teeth.

thus, $TP = 2\pi/RT = 90°$ and $SL = TP/N = 30°$. Note that the sequence as, bs, cs, as, ... produces a counterclockwise stepping of the rotor.

Two other types of three-phase single-stack variable-reluctance stepper motors are shown in Figs. 8.5-2 and 8.5-3. The two-pole device shown in Fig. 8.5-2 has six stator teeth and eight rotor teeth. $TP = 45°$ and $SL = 15°$, and an as, bs, cs, as, ... sequence produces a clockwise stepping of the rotor. For the four-pole three-phase device shown in Fig. 8.5-3, $ST = 12$ and $RT = 8$. Thus, $TP = 45°$, $SL = 15°$. The step length is the same as for the stator with six teeth (Fig. 8.5-2); however, counterclockise stepping of the rotor occurs with the sequence as, bs, cs, as, ... In Fig. 8.5-3, the labeling of the coil sides of the windings is omitted because of lack of space.

The expressions given for the self-inductances of the three-stack (phase) variable-reluctance stepper motor, (8.3-18) through (8.3-23), also apply to the

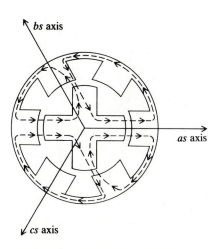

FIGURE 8.5-4
Two-pole three-phase single-stack variable-reluctance stepper motor given in Fig. 8.5-1 with $\theta_{rm} = 0$.

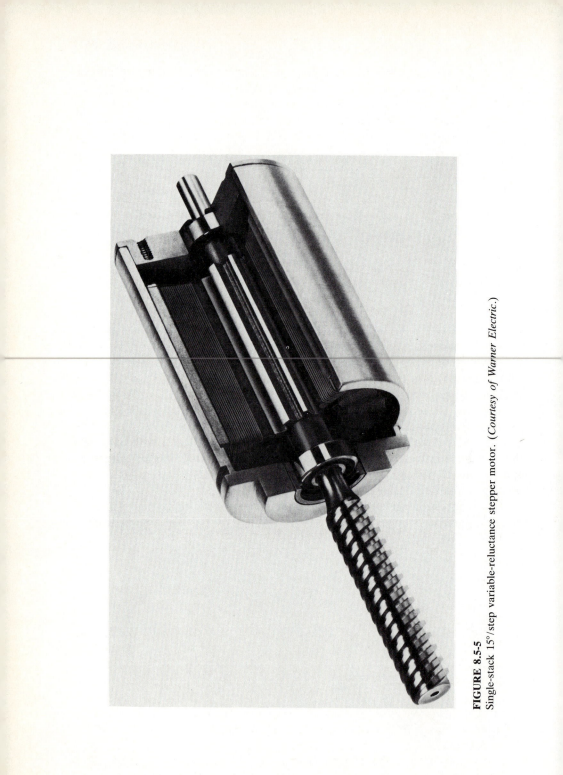

FIGURE 8.5-5
Single-stack 15°/step variable-reluctance stepper motor. (*Courtesy of Warner Electric.*)

three-phase single-stack variable-reluctance stepper motor. It would appear that the operation of the single-stack and multistack variable-reluctance stepper motors may be described by the same set of equations. Although this perception is essentially valid from an idealized point of view, it is not valid in the practical world. We see from Figs. 8.5-1 through 8.5-3, that the stator windings share the same magnetic system. Hence, there is a possibility of mutual coupling between stator phases. For the purposes of discussion, let us consider Fig. 8.5-4 (p. 347), which is Fig. 8.5-1 with $\theta_{rm} = 0$. The dashed lines shown therein depict the flux linking the bs winding due to positive current flowing in the as winding. If we assume that the reluctance of the iron is small so that it can be neglected, the flux linkages cancel, whereupon mutual coupling would not exist between stator phases. From an idealized standpoint, this is a valid line of reasoning; from a practical standpoint it is not.

Stepper motors are generally designed to operate at current levels which saturate the iron of the machine. Hence, owing to the increased reluctance of the saturated iron, less flux will be circulating around the longer paths through iron than through the short paths. Hence, a net mutual flux would exist between stator phases. For the case depicted in Fig. 8.5-4, there would be a net flux in the direction of the positive bs axis as a result of the saturation of the iron. Albeit relatively small in amplitude, a mutual inductance does exist in the practical application of single-stack variable-reluctance stepper motors. This complicates the analysis of these devices far beyond that which we care to deal with in this text. Instead, for our first look at stepper motors, we will consider it sufficient to neglect saturation and the mutual coupling it causes in single-stack variable-reluctance stepper motors. A single-stack variable-reluctance stepper motor is shown in Fig. 8.5-5. This device has a 15° step length and is equipped with an integral lead screw for translational motion.

SP8.5-1. Express the number of stator teeth possible for an N-phase single-stack variable-reluctance stepper motor. [$ST = n(2N)$, where $n = 1, 2, 3, \ldots$]

SP8.5-2. The rotor in Fig. 8.5-1 is replaced by the rotor from Fig. 8.5-3. Determine (a) TP, (b) SL, and (c) the direction of rotation with an $as, bs, cs, as, \ldots$ sequence. [(a) $TP = 45°$; (b) $SL = 15°$; (c) cw]

8.6 BASIC CONFIGURATION OF PERMANENT-MAGNET STEPPER MOTORS

The permanent-magnet stepper motor is quite common. Actually, it is a permanent-magnet synchronous machine and it may be operated either as a stepping motor or as a continuous-speed device. Here, we will concern ourselves only with its application as a stepping motor since continuous-speed operation is similar to the operation of a synchronous machine with a constant field excitation.

A two-pole two-phase permanent-magnet stepper motor with five rotor teeth is shown in Fig. 8.6-1. Most permanent-magnet stepper motors have

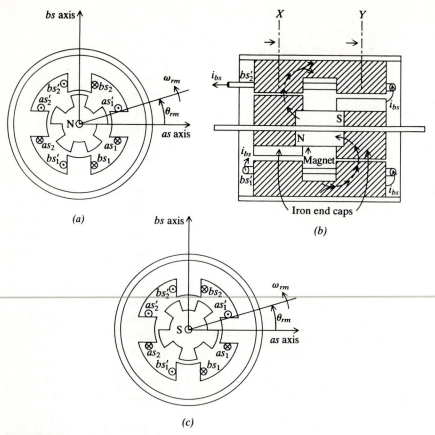

(a)

(b)

(c)

FIGURE 8.6-1

Two-pole two-phase permanent-magnet stepper motor. (a) Axial view at X; (b) side cross-sectional view; (c) axial view at Y.

more than two poles and more than five rotor teeth; some may have as many as eight poles and as many as fifty rotor teeth. Nevertheless, the elementary device shown in Fig. 8.6-1 is sufficient to illustrate the principle of operation of the permanent-magnet stepper motor. The axial cross-sectional view shown in Fig. 8.6-1b illustrates the permanent magnet which is mounted on the rotor. The permanent magnet magnetizes the iron end caps which are also mounted on the rotor and are slotted to form the rotor teeth. The view looking from left to right at X is shown in Fig. 8.6-1a. Figure 8.6-1c is the view from left to right at Y. The left end cap shown in Fig. 8.6-1a is magnetized as a north pole; the right end cap shown in Fig. 8.6-1c is magnetized as a south pole. Note that the rotor teeth of the left end cap are displaced one-half a tooth pitch from the teeth on the right end cap. Also, note that the stator windings are wound over the full axial length of the device; a part of the bs winding is shown in Fig. 8.6-1b.

Let us trace the main path of flux linking the *bs* winding for the rotor position shown in Fig. 8.6-1. This path is depicted by dashed lines in Fig. 8.6-1*b*. However, it is necessary to visualize the drawing in three dimensions. Flux leaves the left end cap through the rotor tooth at the top of the rotor that is starting to align with the stator tooth which has the bs_2 part of the *bs* winding. The flux travels up through the stator tooth in the stator iron. The flux then splits and travels around the circumference of the stator and returns to the south pole of the rotor through the stator tooth, positioned at the bottom in Fig. 8.6-1*c*, on which the bs_1 winding is wound. The main flux linking the *as* winding for the rotor position shown in Fig. 8.6-1 would enter the stator tooth on which the as_1 winding is wound from the rotor tooth on the right of Fig. 8.6-1*a*. The flux would travel around the circumference of the stator and return to the rotor through the stator pole upon which the as_2 winding is wound, Fig. 8.6-1*c*.

Stepping action can be explained by first assuming that the *bs* winding is open-circuited and a constant positive current is flowing in the *as* winding. As a result of this current, a south pole is established at the stator tooth on which the as_1 winding is wound, and a stator north pole is established at the stator tooth on which the as_2 winding is wound. The rotor would be positioned at $\theta_{rm} = 0$. Now let us simultaneously deenergize the *as* winding while energizing the *bs* winding with a positive current. The rotor will move one step length in the counterclockwise direction. To continue stepping in the counterclockwise direction, the *bs* winding is deenergized and the *as* winding is energized with a negative current. That is, counterclockwise stepping occurs with a current sequence of $i_{as}, i_{bs}, -i_{as}, -i_{bs}, i_{as}, \ldots$. Clockwise rotation is achieved by $i_{as}, -i_{bs}, -i_{as}, i_{bs}, \ldots$.

The tooth pitch *TP* can be calculated from (8.2-1); however, the *SL* for a permanent-magnet stepper motor cannot be calculated from (8.2-3). As we have mentioned, counterclockwise rotation of the device shown in Fig. 8.6-1 is achieved by a sequence of $i_{as}, i_{bs}, -i_{as}, -i_{bs}, i_{as}, \ldots$. We see that it takes four switchings (steps) to advance the rotor one tooth pitch. Thus,

$$TP = 2N\,SL \tag{8.6-1}$$

where *N* is the number of phases. Substituting (8.2-1) into (8.6-1) and solving for *SL* yields

$$SL = \frac{\pi}{RT\,N} \tag{8.6-2}$$

For the device shown in Fig. 8.6-1, $RT = 5$ and $N = 2$. From (8.6-2), $SL = 18°$.

Recall that in the case of variable-reluctance stepper motors, it is unnecessary to reverse the direction of the current in the stator windings to achieve rotation; therefore, the stator voltage source need only be unidirectional. However, in the case of a permanent-magnet stepper motor, it is necessary for the phase currents to flow in both directions to achieve rotation. Generally, stepper motors are supplied from a dc voltage source; hence, the electronic interface between the phase windings and the dc source must be bidirectional; that is, it must have the capability of applying a positive and negative voltage to each phase winding. This requirement markedly increases the cost of the electronic interface and its associated controls relative to a unidirectional source. As an alternative, permanent-magnet stepper motors are often equipped with what is referred to as *bifilar windings*. Rather than only one winding on each stator tooth, there are two identical windings with one wound opposite to the other, each having separate independent external terminals. With this type of winding configuration the direction of the magnetic field established by the stator windings is reversed, not by changing the direction of the current but by reversing the sense of the winding through which current is flowing. If, for example, the device shown in Fig. 8.6-1 is equipped with bifilar windings, there would be another *as* winding and another *bs* winding with separate, independent, external terminals wound opposite on the stator teeth to the windings shown. Although this increases the size and weight of the stepper motor, it eliminates the need for a bidirectional electronic interface. When this permanent-magnet stepper motor is equipped with bifilar windings as just described, it is (perhaps, inappropriately) called a four-phase device. Actually it has four windings, but it is still a two-phase device magnetically. Although we are not going to consider the bifilar-wound

FIGURE 8.6-2
Cutaway view of a permanent-magnet stepper motor. (*Courtesy of Sanyo Denki.*)

machine in detail, one should be aware of this somewhat ambiguous nomenclature. More specifically, care should be taken when using (8.6-2) to calculate the step length. The number of phases N in (8.6-2) is the number of phases magnetically rather than the number of windings. A cutaway view of a permanent-magnet stepper motor is shown in Fig. 8.6-2.

SP8.6-1. Consider the device shown in Fig. 8.6-1. The load torque is zero. Initially $i_{as} = I$ and $i_{bs} = 0$. From this condition, the following sequence occurs: $i_{as} = 0$ and $i_{bs} = I$, then $i_{as} = -I$ and $i_{bs} = I$. Determine the initial, intermediate, and final positions. $[\theta_{rm} = 0, 18°, 27°]$

SP8.6-2. A four-pole two-phase permanent-magnet stepper motor has 18 rotor teeth. Calculate TP and SL. $[TP = 20°; SL = 5°]$

8.7 EQUATIONS FOR PERMANENT-MAGNET STEPPER MOTORS

The voltage equations for a two-phase permanent-magnet stepper motor may be written as

$$v_{as} = r_s i_{as} + \frac{d\lambda_{as}}{dt} \tag{8.7-1}$$

$$v_{bs} = r_s i_{bs} + \frac{d\lambda_{bs}}{dt} \tag{8.7-2}$$

In matrix form,

$$\mathbf{v}_{abs} = \mathbf{r}_s \mathbf{i}_{abs} + p\boldsymbol{\lambda}_{abs} \tag{8.7-3}$$

where p is the operator d/dt, and for voltages, currents, and flux linkages

$$(\mathbf{f}_{abs})^T = [f_{as} \quad f_{bs}] \tag{8.7-4}$$

with

$$\mathbf{r}_s = \begin{bmatrix} r_s & 0 \\ 0 & r_s \end{bmatrix} \tag{8.7-5}$$

The flux linkages may be expressed as

$$\lambda_{as} = L_{asas} i_{as} + L_{asbs} i_{bs} + \lambda_{asm} \tag{8.7-6}$$

$$\lambda_{bs} = L_{bsas} i_{as} + L_{bsbs} i_{bs} + \lambda_{bsm} \tag{8.7-7}$$

In matrix form,

$$\boldsymbol{\lambda}_{abs} = \mathbf{L}_s \mathbf{i}_{abs} + \boldsymbol{\lambda}'_m \tag{8.7-8}$$

where

$$\mathbf{L}_s = \begin{bmatrix} L_{asas} & L_{asbs} \\ L_{bsas} & L_{bsbs} \end{bmatrix} \tag{8.7-9}$$

$$\boldsymbol{\lambda}'_m = \begin{bmatrix} \lambda_{asm} \\ \lambda_{bsm} \end{bmatrix} \tag{8.7-10}$$

From Fig. 8.6-1, we can write, as a first approximation,

$$\lambda'_m = \lambda'_m \begin{bmatrix} \cos\left(RT\,\theta_{rm}\right) \\ \sin\left(RT\,\theta_{rm}\right) \end{bmatrix} \tag{8.7-11}$$

where λ'_m, which may be thought of as a constant inductance times a constant current, is the amplitude of the flux linkages established by the permanent magnet as viewed from the stator phase windings. In other words, the magnitude of λ'_m is proportional to the magnitude of the open-circuit sinusoidal voltage induced in each stator phase winding. In (8.7-11),

$$\theta_{rm} = \int_0^t \omega_{rm}(\xi)\,d\xi + \theta_{rm}(0) \tag{8.7-12}$$

where ξ is a dummy variable of integration. Those who have read Chap. 7 on the brushless dc machines will recognize the similarity between (8.7-11) and (7.3-9). Also, the procedure for calculating λ'_m in the case of the stepper motor is identical to that illustrated in Example 7A.

From the idealized standpoint, the self-inductance of the stator phases of the device shown in Fig. 8.6-1 is constant, and the reluctance seen by the permanent magnet is also constant, independent of rotor position. However, in practice both the self-inductances and the reluctance vary with rotor position due to saturation of the stator iron and the differences from the idealized configuration which occur when shaping the poles. We shall disregard these departures from the idealized case and assume constant self-inductances and a constant reluctance seen by the permanent magnet independent of rotor position. When doing so, we are neglecting the reluctance torques caused by variation in self-inductances and the permanent magnet, both of which attempt to place the rotor in its minimum-reluctance position. The latter torque is often referred to as the *detent* or *retention torque*, since it exists whether or not the stator windings are excited, and, if the load torque is not too large, this detent torque will preserve the rotor position during a power failure. Nevertheless, the reluctance torques are small relative to the torque produced by the interaction of the permanent magnet and the stator currents and, although we are not looking at the complete picture when we neglect the reluctance torques, this approximation is certainly adequate for our first look at the permanent-magnet stepper motor.

With the assumption of constant self-inductances, we can write

$$L_{asas} = L_{ls} + L_{ms} = L_{ss} \tag{8.7-13}$$

$$L_{bsbs} = L_{ls} + L_{ms} = L_{ss} \tag{8.7-14}$$

Following a line of reasoning similar to that used in the case of the single-stack variable-reluctance stepper motor, it can be shown that stator mutual inductances do not exist if saturation is neglected. Thus,

$$\mathbf{L}_s = \begin{bmatrix} L_{ss} & 0 \\ 0 & L_{ss} \end{bmatrix} \tag{8.7-15}$$

An expression for the electromagnetic torque may be obtained by taking the partial derivative of the coenergy with respect to θ_{rm}, as given in Table 2.5-1 and (8.3-24). Since the stator mutual inductances are zero, the coenergy may be expressed as

$$W_c = \tfrac{1}{2}L_{asas}i_{as}^2 + \tfrac{1}{2}L_{bsbs}i_{bs}^2 + \lambda_{asm}i_{as} + \lambda_{bsm}i_{bs} + W_{pm} \qquad (8.7\text{-}16)$$

where L_{asas} and L_{bsbs} are given by (8.7-13) and (8.7-14), respectively, and λ_{asm} and λ_{bsm} are given by (8.7-11). The term W_{pm} is related to the energy associated with the permanent magnet. Since we are neglecting variations in the self-inductances and in W_{pm}, taking the partial derivative of W_c with respect to θ_{rm} yields

$$T_e = -RT\,\lambda_m'[i_{as}\sin(RT\,\theta_{rm}) - i_{bs}\cos(RT\,\theta_{rm})] \qquad (8.7\text{-}17)$$

The terms of (8.7-17) are plotted in Fig. 8.7-1, wherein it is assumed that constant currents are present in both phase windings. Each term of (8.7-17) is identified in Fig. 8.7-1. In particular, $\pm T_{eam}$ is the torque due to the interaction of the permanent magnet and $\pm i_{as}$, and $\pm T_{ebm}$ are due to the interaction of the permanent magnet and $\pm i_{bs}$.

The reluctance of the permanent magnet is large, approaching that of air. Since the flux established by the phase currents flows through the magnet, the reluctance of the flux path is relatively large. Hence, the variation in the

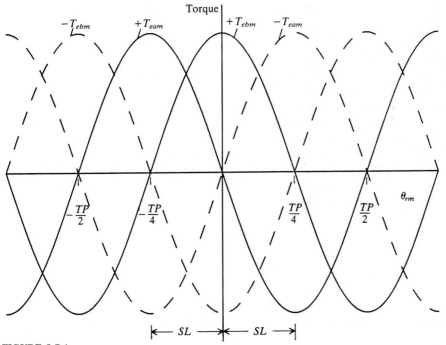

FIGURE 8.7-1
Plot of T_e for permanent-magnet stepper motor with constant phase currents.

reluctance due to rotation of the rotor is small and, consequently, the amplitudes of the reluctance torques are small relative to the torque produced by the interaction between the permanent magnet and the phase currents. For this reason, the reluctance torques are generally neglected, as we have done here, and the self-inductances are assumed to be constant. Therefore, the voltage equations for the permanent-magnet stepper motor become those of the permanent-magnet synchronous machine which we considered in Chap. 7, except for the use of θ_{rm} instead of θ_r and the difference in the referencing of θ_{rm} (to the minimum reluctance rather than the maximum reluctance). These equations are set forth in the following section.

Although a discussion of the stepping action of a permanent-magnet stepper motor using the steady-state torque-angle characteristics is appropriate, this explanation would be essentially a repeat of that given in Sec. 8.4 for the variable-reluctance devices. We will not do this, instead we will ask a few questions to help emphasize this similarity.

SP8.7-1. Express λ_{asm} in terms of SL rather than RT. Determine the number of step lengths in a period for the device shown in Fig. 8.6-1. $\{\lambda_{asm} = \lambda_m' \cos\left[\pi/(SL\ N)\right]\theta_{rm} ; 4\}$

SP8.7-2. Consider Fig. 8.7-1. The load torque is zero. Initially, $i_{as} = I$, $i_{bs} = 0$, then $i_{as} = 0$ and $i_{bs} = -3I$, and, finally, $i_{as} = -\frac{1}{2}I$ and $i_{bs} = 0$. Determine the three positions. $[\theta_{rm} = 0, -TP/4, -TP/2]$

8.8 EQUATIONS OF PERMANENT-MAGNET STEPPER MOTORS IN ROTOR REFERENCE FRAME—RELUCTANCE TORQUES NEGLECTED

Stepper motors often operate for a time in a continuous rotational mode especially when connected through a speed-reduction gear to the member being positioned. In some cases, the phase voltages may be stepped as fast as 500 to 1500 steps per second. When the permanent-magnet stepper motor is operated in this mode, it behaves very similar to either a permanent-magnet synchronous machine or a brushless dc machine depending upon the type of control employed. It is, therefore, advantageous to set forth the voltage equations of the permanent-magnet stepper motor in the rotor reference frame. This transformation is of advantage only if the self-inductances are assumed constant and the detent torque is neglected.

A change of variables which, in effect, transforms the stator variables to the rotor reference frame is

$$\begin{bmatrix} f_{qs}^r \\ f_{ds}^r \end{bmatrix} = \begin{bmatrix} -\sin\left(RT\ \theta_{rm}\right) & \cos\left(RT\ \theta_{rm}\right) \\ \cos\left(RT\ \theta_{rm}\right) & \sin\left(RT\ \theta_{rm}\right) \end{bmatrix} \begin{bmatrix} f_{as} \\ f_{bs} \end{bmatrix} \tag{8.8-1}$$

or

$$\boxed{\mathbf{f}_{qds}^r = \mathbf{K}_s^r \mathbf{f}_{abs}} \tag{8.8-2}$$

where f can represent either voltage, current, or flux linkage and θ_{rm} is defined by (8.7-12). It follows that

$$\mathbf{f}_{abs} = (\mathbf{K}_s^r)^{-1}\mathbf{f}_{qds}^r \tag{8.8-3}$$

where $(\mathbf{K}_s^r)^{-1} = \mathbf{K}_s^r$. The s subscript denotes stator variables and the r superscript indicates that the transformation is to a reference frame fixed in the rotor.

If you have studied Chaps. 6 and 7, you will recall that therein the stator variables of a synchronous machine (Chap. 6) and the stator variables of a permanent-magnet synchronous machine (Chap. 7) were transformed to the rotor reference frame by $\mathbf{K}_s^r$. However, the $\mathbf{K}_s^r$ used in Chaps. 6 and 7, (6.5-1) or (7.5-1), is different from that given by (8.8-1). Although both accomplish the same purpose of transforming the stator variables to the rotor reference frame, the expression for $\mathbf{K}_s^r$ used here for the permanent-magnet stepper is different from that used in Chaps. 6 and 7. There are two reasons for this. The actual rotor displacement θ_{rm} is used here rather than θ_r, the electrical angular displacement of the rotor, and this angular displacement is referenced from the positive as axis to the mimimum-reluctance path of the rotor rather than to the maximum-reluctance path as in Chaps. 6 and 7. Hence, if we maintain the convention of the q axis positioned at the maximum-reluctance path of the rotor, then the transformation is that given by (8.8-1). The trigonometric interpretation of (8.8-1) is given by Fig. 8.8-1.

Substituting (8.8-3) into (8.7-3) yields

$$(\mathbf{K}_s^r)^{-1}\mathbf{v}_{qds}^r = \mathbf{r}_s(\mathbf{K}_s^r)^{-1}\mathbf{i}_{qds}^r + p[(\mathbf{K}_s^r)^{-1}\boldsymbol{\lambda}_{qds}^r] \tag{8.8-4}$$

Multiplying (8.8-4) by $\mathbf{K}_s^r$ and simplifying yields

$$\boxed{\mathbf{v}_{qds}^r = \mathbf{r}_s\mathbf{i}_{qds}^r + RT\,\omega_{rm}\boldsymbol{\lambda}_{dqs}^r + p\boldsymbol{\lambda}_{qds}^r} \tag{8.8-5}$$

where

$$(\boldsymbol{\lambda}_{dqs}^r)^T = [\lambda_{ds}^r \quad -\lambda_{qs}^r] \tag{8.8-6}$$

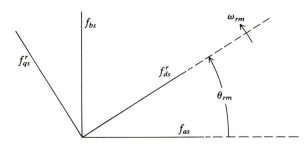

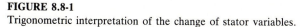

FIGURE 8.8-1
Trigonometric interpretation of the change of stator variables.

The last two terms of (8.8-5) are obtained from the last term of (8.8-4); in particular,

$$\mathbf{K}_s^r p[(\mathbf{K}_s^r)^{-1}\boldsymbol{\lambda}_{qds}^r] = \mathbf{K}_s^r[p(\mathbf{K}_s^r)^{-1}]\boldsymbol{\lambda}_{qds}^r + \mathbf{K}_s^r(\mathbf{K}_s^r)^{-1}p\boldsymbol{\lambda}_{qds}^r \qquad (8.8\text{-}7)$$

It is left to the reader to show that the right-hand side of (8.8-7) reduces to the last two terms of (8.8-5).

The stator flux linkages are expressed as

$$\boldsymbol{\lambda}_{abs} = \mathbf{L}_s \mathbf{i}_{abs} + \boldsymbol{\lambda}_m' \qquad (8.8\text{-}8)$$

where $\mathbf{L}_s$ is given by (8.7-15) and $\boldsymbol{\lambda}_m'$ is defined by (8.7-11). Substituting the change of variables into (8.8-8) yields

$$(\mathbf{K}_s^r)^{-1}\boldsymbol{\lambda}_{qds}^r = \mathbf{L}_s(\mathbf{K}_s^r)^{-1}\mathbf{i}_{qds}^r + \boldsymbol{\lambda}_m' \qquad (8.8\text{-}9)$$

Premultiplying by $\mathbf{K}_s^r$ and substituting (8.7-15) for $\mathbf{L}_s$ and (8.7-11) for $\boldsymbol{\lambda}_m'$ yield

$$\boldsymbol{\lambda}_{qds}^r = \begin{bmatrix} L_{ss} & 0 \\ 0 & L_{ss} \end{bmatrix}\begin{bmatrix} i_{qs}^r \\ i_{ds}^r \end{bmatrix} + \lambda_m'^{\,r}\begin{bmatrix} 0 \\ 1 \end{bmatrix} \qquad (8.8\text{-}10)$$

which is identical in form to (7.5-9). To be consistent with our previous notation, we have added the superscript r to λ_m'. We see from (8.8-10) that in our new system of variables, the flux linkage created by the permanent magnet appears constant. Hence, our fictitious circuits are fixed relative to the permanent magnet and, therefore, fixed in the rotor. We have accomplished the goal of eliminating flux linkages which vary with θ_{rm}.

In expanded form, the voltage equations are

$$\boxed{\begin{aligned} v_{qs}^r &= r_s i_{qs}^r + RT\,\omega_{rm}\lambda_{ds}^r + p\lambda_{qs}^r \\ v_{ds}^r &= r_s i_{ds}^r - RT\,\omega_{rm}\lambda_{qs}^r + p\lambda_{ds}^r \end{aligned}} \qquad \begin{aligned} (8.8\text{-}11) \\ (8.8\text{-}12) \end{aligned}$$

where

$$\lambda_{qs}^r = L_{ss}i_{qs}^r \qquad (8.8\text{-}13)$$

$$\lambda_{ds}^r = L_{ss}i_{ds}^r + \lambda_m'^{\,r} \qquad (8.8\text{-}14)$$

Substituting (8.8-13) and (8.8-14) into (8.8-11) and (8.8-12), and since $\lambda_m'^{\,r}$ is constant, $p\lambda_m'^{\,r} = 0$.

$$v_{qs}^r = (r_s + pL_{ss})i_{qs}^r + RT\,\omega_{rm}L_{ss}i_{ds}^r + RT\,\omega_{rm}\lambda_m'^{\,r} \qquad (8.8\text{-}15)$$

$$v_{ds}^r = (r_s + pL_{ss})i_{ds}^r - RT\,\omega_{rm}L_{ss}i_{qs}^r \qquad (8.8\text{-}16)$$

The above equations may be written in matrix form as

$$\boxed{\begin{bmatrix} v_{qs}^r \\ v_{ds}^r \end{bmatrix} = \begin{bmatrix} r_s + pL_{ss} & RT\,\omega_{rm}L_{ss} \\ -RT\,\omega_{rm}L_{ss} & r_s + pL_{ss} \end{bmatrix}\begin{bmatrix} i_{qs}^r \\ i_{ds}^r \end{bmatrix} + \begin{bmatrix} RT\,\omega_{rm}\lambda_m'^{\,r} \\ 0 \end{bmatrix}} \qquad (8.8\text{-}17)$$

The expression for the electromagnetic torque, with the reluctance torques neglected, is obtained by expressing i_{as} and i_{bs} in (8.7-17) in terms of i_{qs}^r and i_{ds}^r. In particular,

$$\boxed{T_e = RT \, \lambda_m'^r i_{qs}^r} \tag{8.8-18}$$

In Sec. 7.5, the qs and ds voltage equations are derived for a permanent-magnet synchronous machine. These equations are similar to those which we have derived in this section. In particular, (8.8-11), (8.8-12), and (8.8-17) are identical in form to (7.5-10), (7.5-11), and (7.5-17) with the exception that ω_r is replaced by $RT \, \omega_{rm}$. Also, the expression for the electromagnetic torque given by (8.8-18) is identical in form to (7.5-18) if $P/2$ in (7.5-18) is replaced by the rotor teeth RT .

When operated in the continuous rotational mode, the fundamental component of the voltages applied to the stator windings of the permanent-magnet stepper motor can be expressed:

$$v_{as} = -\sqrt{2}v_s \sin \theta_{esv} \tag{8.8-19}$$

$$v_{bs} = \sqrt{2}v_s \cos \theta_{esv} \tag{8.8-20}$$

These equations form a balanced two-phase set, where the amplitude v_s may be a function of time. One may question the form of these voltages since in the past we have selected v_{as} as the $\cos \theta_{esv}$. Recall that, in the case of the stepper motor, we referenced θ_{rm} as being different from that of the synchronous machine and the brushless dc machine. Consequently, the form of the change of variables, $\mathbf{K}_s^r$, is different. We have just pointed out that the equations given in Chap. 7 are identical to those for the permanent-magnet stepper motor with ω_r replaced with $RT \, \omega_{rm}$. To apply the material given in Secs. 7.6, 7.7, and 7.9 to the permanent-magnet stepper motor, it is necessary to select the instantaneous values of v_{as} and v_{bs} so that v_{qs}^r and v_{ds}^r are the same as in Secs. 7.6, 7.7, and 7.9. This is the case if v_{as} and v_{bs} are defined by (8.8-19) and (8.8-20) with

$$\theta_{esv} = \int_0^t \omega_e(\xi) \, d\xi + \theta_{esv}(0) \tag{8.8-21}$$

and $\theta_{esv}(0)$ is the time zero position of the applied voltages.

Now then, when the permanent-magnet stepper motor is operated in the continuous rotational mode between target positions, it is operated either in the open-loop control mode or the closed-loop control mode. In the so-called open-loop control mode, stator voltages of a constant stepping rate are applied until the desired position is reached. In the case of the closed-loop control scheme, the rotor position of the permanent-magnet stepper motor is detected, and this information is supplied to the driving inverter which generates the voltages applied to the stator phases. In this case, the frequency of the stator voltages is made to correspond to the rotor speed and the stepper motor is operated as a brushless dc motor between target positions. Therefore, when the permanent-magnet stepper motor is operated in the closed-loop control mode, it is a brushless dc motor and ω_e in (8.8-21) is made equal to $RT\,\omega_{rm}$. It would appear that we need only to replace ω_r with $RT\,\omega_{rm}$ in all voltage equations given in Secs. 7.6, 7.7, and 7.9 and they would apply to the permanent-magnet stepper motor with a closed-loop control. There is an exception, however. When converting between phasors and $\overset{r}{q}s$ and $\overset{r}{d}s$ quantities, it is necessary to replace $\tilde{F}_{as}$ in (7.6-16) through (7.6-21) with $\tilde{F}_{bs}$. In particular, we see from (8.8-19) and (8.8-20) that, during steady-state closed-loop control operation, F_{as} and F_{bs} are expressed as

$$F_{as} = -\sqrt{2}F_s \sin\left[RT\,\omega_{rm}t + \theta_{esf}(0)\right] \tag{8.8-22}$$

$$F_{bs} = \sqrt{2}F_s \cos\left[RT\,\omega_{rm}t + \theta_{esf}(0)\right] \tag{8.8-23}$$

wherein ω_e has been set equal to $RT\,\omega_{rm}$. Substituting F_{as} and F_{bs} into the transformation equation, (8.8-1), yields

$$F^r_{qs} = \sqrt{2}F_s \cos\left[\theta_{esf}(0) - RT\,\theta_{rm}(0)\right] \tag{8.8-24}$$

$$F^r_{ds} = -\sqrt{2}F_s \sin\left[\theta_{esf}(0) - RT\,\theta_{rm}(0)\right] \tag{8.8-25}$$

Now, from (8.8-23),

$$\tilde{F}_{bs} = F_s e^{j\theta_{esf}(0)}$$

$$= F_s \cos\theta_{esf}(0) + jF_s \sin\theta_{esf}(0) \tag{8.8-26}$$

If we let $\theta_{rm}(0) = 0$, then (8.8-26) may be expressed in terms of (8.8-24) and (8.8-25) as

$$\boxed{\sqrt{2}\tilde{F}_{bs} = F^r_{qs} - jF^r_{ds}} \tag{8.8-27}$$

This is (7.6-16) with $\tilde{F}_{as}$ replaced by $\tilde{F}_{bs}$. Thus, for the permanent-magnet stepper motor during closed-loop operation, (7.6-21) becomes

$$\boxed{\tilde{V}_{bs} = (r_s + j\,RT\,\omega_{rm}L_{ss})\tilde{I}_{bs} + \tilde{E}_b} \tag{8.8-28}$$

where $\tilde{E}_b$ is $\tilde{E}_a$ given by (7.6-20) with ω_r replaced by $RT\,\omega_{rm}$; in particular,

$$\boxed{\tilde{E}_b = \frac{1}{\sqrt{2}}\,RT\,\omega_{rm}\lambda_m''^r e^{j0}} \tag{8.8-29}$$

The stepper motor can be operated with or without phase shift, just as the brushless dc motor, and with $\theta_{rm}(0) = 0$ the phase shift for maximum torque is (7.6-34) with ω_r replaced with $RT\,\omega_{rm}$ [1]. In particular,

$$\phi_{vMT} = \tan^{-1}\left(\tau_s\,RT\,\omega_{rm}\right) \tag{8.8-30}$$

It follows that, with the appropriate substitutions of $RT\,\omega_{rm}$ for ω_r in the voltage equations and RT for $P/2$ in the torque equations, the time-domain block diagram given in Fig. 7.9-1 is valid for a permanent-magnet stepper motor operated with a closed-loop control scheme.

SP8.8-1. A permanent-magnet stepper motor with $RT = 7$ is operated at $\omega_{rm} = 100 \text{ rad/s}$. $\tilde{V}_{bs} = (10/\sqrt{2})\underline{/0°}$, $\tilde{I}_{bs} = (1/\sqrt{2})\underline{/-60.3°}$, $\theta_{rm}(0) = 0$, $r_s = 4\,\Omega$, and $L_{ss} = 0.01 \text{ H}$. Calculate $\lambda_m'^r$. Use equations from Sec. 7.6 with appropriate changes. $[\lambda_m'^r = 0.00277 \text{ V} \cdot \text{s/rad}]$

SP8.8-2. The stepper motor given in SP8.8-1 is operated at $\omega_{rm} = 100 \text{ rad/s}$ with $\tilde{I}_{bs} = (1/\sqrt{2})\underline{/-20°}$ and $\theta_{rm}(0) = 0$. Calculate ϕ_v. Use equations from Sec. 7.6 with appropriate changes. $[\phi_v = 32.8°]$

8.9 RECAPPING

We have considered the two types of electromechanical stepping devices used most often—the variable-reluctance and the permanent-magnet stepper motors. Although we have considered an important operating mode of the stepper motor, there are numerous aspects of stepper motor operation which we are unable to cover in an introductory text of this type. For example, the types and control of the voltage source and stability problems which can occur during continuous rotation of the stepper motor are beyond the scope of this textbook. Nevertheless, the information presented in this chapter sets the stage for study of these more advanced topics.

8.10 REFERENCE

1. P. P. Acarnley, *Stepping Motors*: *A Guide to Modern Theory and Practice*, Peter Pereginus Ltd. for the Institution of Electrical Engineers; Southgate House, Stevenage, Herts, SG1 1HQ, England, 1984.

8.11 PROBLEMS

1. Sketch the configuration of a two-pole four-stack variable-reluctance stepper motor with two rotor teeth. Use *as*, *bs*, *cs*, and *ds* to denote the phase windings. Calculate TP, SL, and give the excitation sequence for ccw rotation.

2. For Prob. 1, express the self-inductances and the torque using SL in the arguments.

*3. Those who have read Chap. 6 on synchronous and reluctance machines may wonder if Park's transformation can be used in the case of the variable-reluctance stepper motor to transform the stator circuits to fictitious circuits with constant inductances fixed in the rotor. The answer is negative, and the reason is that mutual inductances must exist between stator phases as given by (6.9-6). To show this, take a two-pole two-phase reluctance machine, assume $L_{asbs} = L_{bsas} = 0$, and evaluate $\mathbf{K}_s^r \mathbf{L}_s (\mathbf{K}_s^r)^{-1}$.

4. The four-pole three-stack variable-reluctance stepper motor shown in Fig. 8.2-3 is to be operated at a continuous speed of 30 rad/s. Neglect electrical transients and sketch the current i_{as} indicating the time it is zero and nonzero.

5. A four-pole five-stack variable-reluctance stepper motor has 8 rotor teeth as the rotor shown in Fig. 8.2-4. Its magnetic axes are arranged as, bs, cs, ds, and es, in the counterclockwise direction. Express the self-inductances with the constant angular displacement in terms of step length.

6. Express the self-inductances for the single-stack variable-reluctance stepper motor shown in Fig. 8.5-1 with the constant angular displacement in terms of step length.

7. A two-phase permanent-magnet stepper motor has 50 rotor teeth. When the rotor is driven by an external mechanical source at $\omega_{rm} = 100$ rad/s, the measured open-circuit phase voltage is 25 V, peak to peak. Calculate λ_m' and SL. If $i_{as} = 1$ A, $i_{bs} = 0$, express T_e.

8. Consider the two-phase permanent-magnet stepper motor of Fig. 8.6-1. Sketch i_{as} and i_{bs} versus time for the excitation sequence i_{as}, i_{bs}, $-i_{as}$, $-i_{bs}$, i_{as} .... Denote the time between steps as T_s and the stepping rate as $f_s = 1/T_s$. Establish a relationship between the fundamental frequency (ω_e) of i_{as} and i_{bs}, and the stepping rate f_s. Relate ω_{rm} to ω_e and to f_s.

9. A two-phase permanent-magnet stepper motor has 50 rotor teeth. The parameters are $\lambda_m' = 0.00226$ V·s/rad, $r_s = 10 \, \Omega$, and $L_{ss} = 1.1$ mH. The applied stator voltages form a balanced two-phase set with $V_s = 10$ V, $\omega_e = 314$ rad/s. Establish the steady-state rotor speed ω_{rm} and the maximum electromagnetic torque T_{eM} that can be developed at this speed.

*10. Consider a permanent-magnet stepper motor. Neglect the reluctance torques and assume the self-inductances are constant. Use the transformation to the rotor reference frame set forth in Chap. 7, (7.5-1), with θ_r replaced by $RT \, \theta_{rm}$ to express the voltage equations for v_{qs}' and v_{ds}' similar to (8.8-15) and (8.8-16).

*11. The permanent-magnet stepper motor shown in Fig. 8.6-1 is equipped with bifilar windings. Assume that the cs winding (ds winding) is wound on the same stator pole as the as winding (bs winding) but opposite to it, and also assume that the windings are tightly coupled (no leakage inductances). Express $v_{as} - v_{cs}$ and $v_{bs} - v_{ds}$.

CHAPTER
9

TWO-PHASE SERVOMOTORS AND SINGLE-PHASE INDUCTION MOTORS

9.1 INTRODUCTION

Although the voltage and torque equations derived in Chap. 5 for the induction machine are valid regardless of the mode of operation, we considered only balanced conditions. When the symmetrical two-phase induction motor is used as a control motor (servomotor) or as a single-phase motor, the stator phase voltages are normally unbalanced. As a servomotor, the two-phase induction motor is often used as a positioning device. In this application, a constant-amplitude voltage is generally applied to one of the stator phases whereas the amplitude of the other phase voltage is proportional to an error signal which is the difference between the desired position and the actual position of the driven member.

In single-phase applications, the induction motor is operated from a single-phase source. Household power is generally single-phase, and single-phase induction motors are used in washers, dryers, air conditioners, garbage disposals, etc. Symmetrical two-phase induction motors may be used in these single-phase applications. However, in order to develop a starting torque, it is necessary to make the symmetrical two-phase induction motor think it is being supplied from a two-phase source or, at least, something that resembles a two-phase source. We will find that this can be accomplished by placing a capacitor in series with one of the stator phase windings until the rotor has accelerated to 60 to 80 percent of normal operating speed, whereupon the capacitor and the phase winding are disconnected from the source. The motor then operates with only one of its phases connected to the single-phase source. Hence, we have two common modes of unbalanced operation of a symmetrical two-phase induction motor when used as a single-phase device: (1) During starting, the phase voltages are not a balanced two-phase set and the input impedance of one phase is different from the other owing to the series capacitor, and (2) during normal operation where one stator phase is open-circuited.

To analyze steady-state unbalanced operation of an induction machine, it is convenient to use the method of symmetrical components. This method is introduced in the following section and used to analyze unbalanced stator voltages, unequal stator impedances, and an open-circuited stator phase. All of these modes of operation occur either during the operation of the symmetrical two-phase induction motor as a position servomotor or as a single-phase induction motor.

In this chapter, the servomotor is considered as a part of an elementary positioning control system. Steady-state and dynamic characteristics are illustrated and the time-domain block diagram and state equations of a linear, simplified model of the servomotor are given. Single-phase applications of the symmetrical two-phase induction machine are analyzed with emphasis on the capacitor-start single-phase induction motor. The steady-state and free-acceleration characteristics of this device are illustrated by computer traces.

Only the symmetrical two-phase induction motor is considered in this chapter. Actually the unsymmetrical two-phase induction motor or the so-called *split-phase* induction motor is very often used rather than its symmetrical cousin. Although the last section of this chapter is devoted to a brief discussion of the split-phase machine, it is not analyzed. We have chosen to illustrate the salient features of single-phase induction machines by using the symmetrical device since it is far easier to analyze than the split-phase machine.

9.2 SYMMETRICAL COMPONENTS

We must deal with unbalanced conditions in the analysis of steady-state operation of the symmetrical two-phase induction motor when used either as a servomotor or in single-phase applications. This can be accomplished by using what is referred to as the method of symmetrical components. Although this is the first time those things have been mentioned, they have been around for some time. At about the end of the reign of the Red Baron in the skies over France and Germany, C. L. Fortescue [1] published the method of symmetrical components for the purpose of analyzing unbalanced multiphase systems. Since that time this method has been extended, modified, and used widely, some-

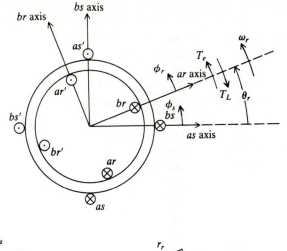

FIGURE 9.2-1
A two-pole two-phase symmetrical induction machine.

times inappropriately. Nevertheless, it is a powerful analytical tool. However, the derivation and the procedure for applying this method often seemed to be without theoretical basis. It has recently been shown that reference frame theory provides a rigorous derivation of the method of symmetrical components and sets clear guidelines for its application [2].

In this section, we will use reference frame theory to provide a theoretical basis for the method of symmetrical components. However, first we will discuss the concept of symmetrical components and, without derivation, set down the equations necessary to use this technique for those who do not wish to wade through the theoretical justification for using reference frame theory. For convenience, Fig. 5.2-1 of the two-pole two-phase symmetrical induction machine is repeated in Fig. 9.2-1.

Concept of Symmetrical Components

The method of symmetrical components allows us to represent an unbalanced two-phase set as two balanced sets or an unbalanced three-phase set as two balanced sets and a single phasor. The balanced sets are referred to as the *positive-sequence* and *negative-sequence* components whereas the single phasor is called the *zero-sequence* component. Here, we will deal only with a two-phase system. Also, the method of symmetrical components is valid only for steady-state conditions. In general, an unbalanced two-phase set may be expressed as

$$F_{as} = \sqrt{2}\, F_a \cos\left[\omega_e t + \theta_{efa}(0)\right] \tag{9.2-1}$$

$$F_{bs} = \sqrt{2}\, F_b \sin\left[\omega_e t + \theta_{efb}(0)\right] \tag{9.2-2}$$

We realize that (9.2-1) and (9.2-2) form a balanced set if $F_a = F_b$ and $\theta_{efa}(0) = \theta_{efb}(0)$. It is convenient for us to work with $\overset{s}{q}s$ and $\overset{s}{d}s$ variables rather than as and bs variables. Recall from Chap. 5, (5.5-1), that $f_{as} = f_{qs}^s$ and $f_{bs} = -f_{ds}^s$ or $\tilde{F}_{as} = \tilde{F}_{qs}^s$ and $\tilde{F}_{bs} = -\tilde{F}_{ds}^s$.

It can be shown that an unbalanced two-phase set may be broken up into two balanced sets as

$$\tilde{F}_{qs}^s = \tilde{F}_{qs+}^s + \tilde{F}_{qs-}^s \tag{9.2-3}$$

$$\tilde{F}_{ds}^s = \tilde{F}_{ds+}^s + \tilde{F}_{ds-}^s \tag{9.2-4}$$

Here, we are departing somewhat from tradition. Rather than using $\overset{s}{q}s$ and $\overset{s}{d}s$ variables, it has been customary to use as and bs variables and, hence, (9.2-3) and (9.2-4) are written, respectively, as $\tilde{F}_{as} = \tilde{F}_{as+} + \tilde{F}_{as-}$ and $\tilde{F}_{bs} = \tilde{F}_{bs+} + \tilde{F}_{bs-}$. Later, we will substitute $\tilde{F}_{as}$ for $\tilde{F}_{qs}^s$ and $\tilde{F}_{bs}$ for $-\tilde{F}_{ds}^s$; however, we will continue the break from tradition and use $\tilde{F}_{qs+}^s$, $\tilde{F}_{qs-}^s$, $\tilde{F}_{ds+}^s$, and $\tilde{F}_{ds-}^s$ rather than $\tilde{F}_{as+}$, $\tilde{F}_{as-}$, $-\tilde{F}_{bs+}$, and $-\tilde{F}_{bs-}$, respectively.

In (9.2-3) and (9.2-4), $\tilde{F}_{qs+}^s$ and $\tilde{F}_{ds+}^s$ form the positive-sequence set, where

$$\tilde{F}_{ds+}^s = j\tilde{F}_{qs+}^s \tag{9.2-5}$$

The negative-sequence set is $\tilde{F}_{qs-}^s$ and $\tilde{F}_{ds-}^s$, where

$$\tilde{F}_{ds-}^s = -j\tilde{F}_{qs-}^s \tag{9.2-6}$$

Both the positive- and negative-sequence sets are balanced, but in the case of the positive-sequence set $\tilde{F}_{ds+}^s$ leads $\tilde{F}_{qs+}^s$ ($\tilde{F}_{bs+}$ lags $\tilde{F}_{as+}$) by 90°, whereas in the case of the negative-sequence set $\tilde{F}_{ds-}^s$ lags $\tilde{F}_{qs-}^s$ ($\tilde{F}_{bs-}$ leads $\tilde{F}_{as-}$). Why is one set called the positive sequence and the other the negative sequence? Well, probably the best explanation is to point out that positive-sequence currents flowing in the stator windings (Fig. 9.2-1) will produce an air gap mmf which rotates counterclockwise whereas negative-sequence currents will produce an air gap mmf which rotates clockwise. It should be apparent that, if $\tilde{F}_{as}$ and $\tilde{F}_{bs}$ are balanced with $\tilde{F}_{bs} = -j\tilde{F}_{as}$, then the negative-sequence variables ($\tilde{F}_{qs-}^s$ and $\tilde{F}_{ds-}^s$) would not exist.

Substituting (9.2-5) and (9.2-6) into (9.2-3) and (9.2-4) yields

$$
\begin{bmatrix} \tilde{F}^s_{qs} \\ \tilde{F}^s_{ds} \end{bmatrix} = \begin{bmatrix} 1 & 1 \\ j & -j \end{bmatrix} \begin{bmatrix} \tilde{F}^s_{qs+} \\ \tilde{F}^s_{qs-} \end{bmatrix}
\tag{9.2-7}
$$

Now let us substitute $\tilde{F}_{as}$ for $\tilde{F}^s_{qs}$ and $\tilde{F}_{bs}$ for $-\tilde{F}^s_{ds}$, whereupon

$$
\begin{bmatrix} \tilde{F}_{as} \\ \tilde{F}_{bs} \end{bmatrix} = \begin{bmatrix} 1 & 1 \\ -j & j \end{bmatrix} \begin{bmatrix} \tilde{F}^s_{qs+} \\ \tilde{F}^s_{qs-} \end{bmatrix}
\tag{9.2-8}
$$

Solving for $\tilde{F}^s_{qs+}$ and $\tilde{F}^s_{qs-}$ yields

$$
\begin{bmatrix} \tilde{F}^s_{qs+} \\ \tilde{F}^s_{qs-} \end{bmatrix} = \frac{1}{2} \begin{bmatrix} 1 & j \\ 1 & -j \end{bmatrix} \begin{bmatrix} \tilde{F}_{as} \\ \tilde{F}_{bs} \end{bmatrix}
\tag{9.2-9}
$$

The symmetrical-component transformation matrix is defined from (9.2-9) as

$$
\mathbf{S} = \frac{1}{2} \begin{bmatrix} 1 & j \\ 1 & -j \end{bmatrix}
\tag{9.2-10}
$$

and

$$
(\mathbf{S})^{-1} = \begin{bmatrix} 1 & 1 \\ -j & j \end{bmatrix}
\tag{9.2-11}
$$

Since the equations needed for the analysis for the two-phase servomotor and single-phase induction motors have now been given, it is unnecessary for the reader to continue through the remainder of this section. However, for those who wish a rigorous development of the method of symmetrical components the following material will be very worthwhile and helpful.

Theory of Symmetrical Components

To establish a theoretical basis for the method of symmetrical components, it is convenient to employ reference frame theory. However, do not become apprehensive about using reference frame theory. We have already set the stage for this in our analysis of the induction, synchronous, and brushless dc motors. It is a painless extension of the change of variables used to analyze these machines. Also, in this derivation, we will repeat most of the equations already given. Hence, it will be unnecessary to refer back to this previous material.

Let us assume that the stator phase variables can be expressed as

$$
f_{as} = f_{as\alpha} \cos \omega_e t + f_{as\beta} \sin \omega_e t
\tag{9.2-12}
$$

$$
f_{bs} = f_{bs\alpha} \cos \omega_e t + f_{bs\beta} \sin \omega_e t
\tag{9.2-13}
$$

Although we need not place any restrictions upon the coefficients of the sinusoidal variations at this time, they ($f_{as\alpha}$, $f_{as\beta}$, $f_{bs\alpha}$, and $f_{bs\beta}$) are constant during steady-state operation. Equations (9.2-12) and (9.2-13) permit us to express any balanced or unbalanced situation one can imagine.

We will now define a change of variables as

$$\mathbf{f}_{qds} = \mathbf{K}_s \mathbf{f}_{abs} \qquad (9.2\text{-}14)$$

where

$$\mathbf{K}_s = \begin{bmatrix} \cos\theta & \sin\theta \\ \sin\theta & -\cos\theta \end{bmatrix} \qquad (9.2\text{-}15)$$

$$\theta = \int_0^t \omega(\xi)\, d\xi + \theta(0) \qquad (9.2\text{-}16)$$

where $(\mathbf{K}_s)^{-1} = \mathbf{K}_s$. Let us stop a second and recall the change of variables for the stator variables used in Chap. 5. There we set $f^s_{qs} = f_{as}$ and $f^s_{ds} = -f_{bs}$. Here, the change of variables for the stator variables is a function of an angular displacement which is not specified. Recall from Chap. 5 that ω_r was used in the transformation of rotor variables of the induction machine to the stationary reference frame. Here, we have not specified ω in the transformation of the stator variables. Moreover, we are not going to do that just yet; instead, we will let ω, the reference frame speed, be unspecified. In other words, the frame of reference to which we are transforming the stator variables may rotate at any constant or varying angular velocity or it may remain stationary. We shall call this reference frame the arbitrary reference frame since the angular velocity of the reference frame can be selected arbitrarily to expedite the solution of the system equations or to satisfy system constraints. Note that a raised index is not used for the arbitrary reference frame variables f_{qs} and f_{ds}.

Actually, this is really more reference frame theory than we need to know for the purposes at hand. We need only to understand that ω in (9.2-16) is unspecified. Later in this development, when we do give it a value, it will be zero. That sure does not make any sense. Why should we not set ω equal to zero now and forget it? The answer is about ready to appear. If we transform f_{as} and f_{bs} to the arbitrary reference frame by (9.2-14), we obtain

$$f_{qs} = f_{qsA} \cos(\omega_e t - \theta) + f_{qsB} \sin(\omega_e t - \theta)$$
$$+ f_{qsC} \cos(\omega_e t + \theta) + f_{qsD} \sin(\omega_e t + \theta) \qquad (9.2\text{-}17)$$

$$f_{ds} = f_{dsA} \cos(\omega_e t - \theta) + f_{dsB} \sin(\omega_e t - \theta)$$
$$+ f_{dsC} \cos(\omega_e t + \theta) + f_{dsD} \sin(\omega_e t + \theta) \qquad (9.2\text{-}18)$$

where

$$f_{qsA} = -f_{dsB} = \tfrac{1}{2}(f_{as\alpha} + f_{bs\beta}) \qquad (9.2\text{-}19)$$

$$f_{qsB} = f_{dsA} = \tfrac{1}{2}(f_{as\beta} - f_{bs\alpha}) \qquad (9.2\text{-}20)$$

$$f_{qsC} = f_{dsD} = \tfrac{1}{2}(f_{as\alpha} - f_{bs\beta}) \qquad (9.2\text{-}21)$$

$$f_{qsD} = -f_{dsC} = \tfrac{1}{2}(f_{as\beta} + f_{bs\alpha}) \qquad (9.2\text{-}22)$$

It is convenient to repeat (9.2-17) along with (9.2-18) expressed in terms of $f_{qsA}, \ldots, f_{qsD}$. Thus,

$$f_{qs} = f_{qsA} \cos(\omega_e t - \theta) + f_{qsB} \sin(\omega_e t - \theta)$$
$$+ f_{qsC} \cos(\omega_e t + \theta) + f_{qsD} \sin(\omega_e t + \theta) \qquad (9.2\text{-}23)$$

$$f_{ds} = f_{qsB} \cos(\omega_e t - \theta) - f_{qsA} \sin(\omega_e t - \theta)$$
$$- f_{qsD} \cos(\omega_e t + \theta) + f_{qsC} \sin(\omega_e t + \theta) \qquad (9.2\text{-}24)$$

Note that the first term on the right-hand side of (9.2-23) and the second term on the right-hand side of (9.2-24) form a balanced, two-phase set, so do the second term of (9.2-23) and the first term of (9.2-24), the third term of (9.2-23) and the fourth term of (9.2-24), and the fourth term of (9.2-23) and the third term of (9.2-24). Think back on what we have done. We started with an unbalanced set of stator variables given by (9.2-12) and (9.2-13). We defined a transformation to an unspecified reference frame and substituted the unbalanced set into this transformation and, much to our surprise, we obtained balanced sets which, by (9.2-19) through (9.2-22), are readily defined from the coefficients of the original unbalanced set. That is very interesting. Let us do one more thing before discussing this feature in more detail. Since we can assign any value to ω, let it and $\theta(0)$ of (9.2-16) both be zero, whereupon (9.2-23) and (9.2-24) become, respectively,

$$f_{qs}^s = f_{qsA} \cos\omega_e t + f_{qsB} \sin\omega_e t + f_{qsC} \cos\omega_e t + f_{qsD} \sin\omega_e t \qquad (9.2\text{-}25)$$

$$f_{ds}^s = -f_{qsA} \sin\omega_e t + f_{qsB} \cos\omega_e t + f_{qsC} \sin\omega_e t - f_{qsD} \cos\omega_e t \qquad (9.2\text{-}26)$$

Note that the order of the terms of (9.2-26) has been changed from that in (9.2-24) to allow a sequential, term-by-term arrangement of balanced sets. Note also that we have used the raised index s denoting the fact that the qs and ds variables are now in the stationary reference frame since $\omega = 0$. But so are f_{as} and f_{bs}; in fact, $f_{as} = f_{qs}^s$ and $f_{bs} = -f_{ds}^s$. We learned this back in Chap. 5 or, if we have forgotten, we can see it from the equation of transformation, (9.2-14), with θ set equal to zero in $\mathbf{K}_s$. What then has the transformation done for us? Well, it tells us how to divide up an unbalanced set into balanced sets. Once we have seen (9.2-25) and (9.2-26), we realize that we could have used several trigonometric manipulations and arrived at these same equations without using the transformation as a vehicle. Nevertheless, the transformation offers a direct, rigorous approach without any hand-waving.

Our present interest is the analysis of unbalanced steady-state operation of the induction motor. Hence, let us rewrite (9.2-25) and (9.2-26) by using capital letters to denote steady-state conditions (constant coefficients).

$$F^s_{qs} = F_{qsA} \cos \omega_e t + F_{qsB} \sin \omega_e t + F_{qsC} \cos \omega_e t + F_{qsD} \sin \omega_e t \qquad (9.2\text{-}27)$$

$$F^s_{ds} = -F_{qsA} \sin \omega_e t + F_{qsB} \cos \omega_e t + F_{qsC} \sin \omega_e t - F_{qsD} \cos \omega_e t \qquad (9.2\text{-}28)$$

Let us take a closer look at these four balanced sets. In particular, let us reconsider Fig. 5.5-1a which is given here as Fig. 9.2-2. Recall that the positive direction of f_{as} and f_{bs} is the same as the positive direction of the magnetic axes of the as and bs windings, respectively. Similarly, the positive direction of f^s_{qs} and f^s_{ds} is the same as the positive direction of the magnetic axes of the fictitious qs and ds windings in the stationary reference frame. Since $F_{as} = F^s_{qs}$ and $F_{bs} = -F^s_{ds}$, we can think either in terms of as and bs variables or in terms of qs and ds variables.

Our present objective is to determine the direction of rotation of the air gap mmf established by each of these four balanced sets. Consider the balanced set formed by the first terms on the right-hand side of (9.2-27) and (9.2-28) with the coefficient F_{qsA} and, for convenience, let us assume that we are dealing with currents (I_{qsA}). At $t = 0$, the positive maximum air gap mmf due to this set is aligned with the positive direction of f_{as} (as axis) or f^s_{qs} since $\cos \omega_e t = 1$ and $\sin \omega_e t = 0$. As time increases to where $\omega_e t = \frac{1}{2}\pi$, the positive maximum air gap mmf due to this balanced set is now aligned with the positive direction of f_{bs} (bs axis) or $-f^s_{ds}$. The air gap mmf due to this balanced set has rotated in the counterclockwise direction. We have assumed that I_{qsA} was positive. Would we get the same direction of rotation (counterclockwise) if I_{qsA} were negative? Yes. Now, we have always taken a counterclockwise direction of rotation of the rotor or counterclockwise displacement about the stator or rotor as positive. Hence, this air gap mmf rotating in the counterclockwise direction is a positively rotating mmf.

Now, the next balanced set is formed by the second terms on the right side of (9.2-27) and (9.2-28). If we consider F_{qsB} as I_{qsB}, then, by the same line

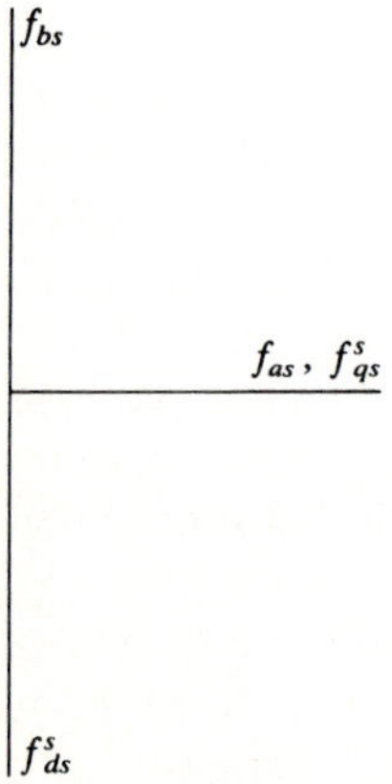

FIGURE 9.2-2
Trigonometric interpretation of the stator variables in stationary reference frame.

of reasoning as we used in the case of the F_{qsA} set, we find that this balanced set also produces an air gap mmf rotating in the counterclockwise direction. Since both the F_{qsA} set and the F_{qsB} set produce positively rotating air gap mmf's, we can combine the F_{qsA} and F_{qsB} variables as

$$F_{qsA} \cos \omega_e t + F_{qsB} \sin \omega_e t = F_{qsAB} \cos (\omega_e t + \phi) \tag{9.2-29}$$

$$-F_{qsA} \sin \omega_e t + F_{qsB} \cos \omega_e t = -F_{qsAB} \sin (\omega_e t + \phi) \tag{9.2-30}$$

where

$$F_{qsAB} = \sqrt{(F_{qsA})^2 + (F_{qsB})^2} \tag{9.2-31}$$

$$\phi = \tan^{-1}\left(-\frac{F_{qsB}}{F_{qsA}}\right) \tag{9.2-32}$$

In accordance with the convention established over the years, the variables (voltages, currents, or flux linkages) associated with the combined, positively rotating air gap mmf are referred to as the positive-sequence variables.

The balanced set formed by the third terms on the right-hand side of (9.2-27) and (9.2-28) with the F_{qsC} coefficient establishes an air gap mmf which rotates in the clockwise (negative) direction, opposite to the rotation of the mmf established by the first two sets. The F_{qsD} balanced set also produces a negatively rotating air gap mmf. The variables associated with the combination of these balanced sets are referred to as the negative-sequence variables.

Since we are considering steady-state conditions, we can express (9.2-27) and (9.2-28) in phasor form as

$$\tilde{F}_{qs}^s = \tilde{F}_{qs+}^s + \tilde{F}_{qs-}^s \tag{9.2-33}$$

$$\tilde{F}_{ds}^s = \tilde{F}_{ds+}^s + \tilde{F}_{ds-}^s \tag{9.2-34}$$

where the $+$ subscript denotes the phasors associated with the positive-sequence quantities and the $-$ subscript denotes negative-sequence quantities. We can express these positive- and negative-sequence phasors from (9.2-27) and (9.2-28) as

$$\sqrt{2}\,\tilde{F}_{qs+}^s = F_{qsA} - jF_{qsB} \tag{9.2-35}$$

$$\sqrt{2}\tilde{F}_{ds+}^s = F_{qsB} + jF_{qsA} \tag{9.2-36}$$

$$\sqrt{2}\tilde{F}_{qs-}^s = F_{qsC} - jF_{qsD} \tag{9.2-37}$$

$$\sqrt{2}\tilde{F}_{ds-}^s = -F_{qsD} - jF_{qsC} \tag{9.2-38}$$

Note that we can express the $\overset{s}{ds}$ variables in terms of the $\overset{s}{qs}$ variables. From (9.2-35) and (9.2-36) we see that

$$\tilde{F}_{ds+}^s = j\tilde{F}_{qs+}^s \tag{9.2-39}$$

From (9.2-37) and (9.2-38),

$$\boxed{\tilde{F}^s_{ds-} = -j\tilde{F}^s_{qs-}}$$

(9.2-40)

Substituting (9.2-39) and (9.2-40) into (9.2-34) and repeating (9.2-33), we have

$$\begin{bmatrix} \tilde{F}^s_{qs} \\ \tilde{F}^s_{ds} \end{bmatrix} = \begin{bmatrix} 1 & 1 \\ j & -j \end{bmatrix} \begin{bmatrix} \tilde{F}^s_{qs+} \\ \tilde{F}^s_{qs-} \end{bmatrix}$$

(9.2-41)

Since $\tilde{F}^s_{qs} = \tilde{F}_{as}$ and $\tilde{F}^s_{ds} = -\tilde{F}_{bs}$, we can write (9.2-41) as

$$\boxed{\begin{bmatrix} \tilde{F}_{as} \\ \tilde{F}_{bs} \end{bmatrix} = \begin{bmatrix} 1 & 1 \\ -j & j \end{bmatrix} \begin{bmatrix} \tilde{F}^s_{qs+} \\ \tilde{F}^s_{qs-} \end{bmatrix}}$$

(9.2-42)

Solving for $\tilde{F}^s_{qs+}$ and $\tilde{F}^s_{qs-}$ yields

$$\boxed{\begin{bmatrix} \tilde{F}^s_{qs+} \\ \tilde{F}^s_{qs-} \end{bmatrix} = \frac{1}{2}\begin{bmatrix} 1 & j \\ 1 & -j \end{bmatrix} \begin{bmatrix} \tilde{F}_{as} \\ \tilde{F}_{bs} \end{bmatrix}}$$

(9.2-43)

where

$$\boxed{\mathbf{S} = \frac{1}{2}\begin{bmatrix} 1 & j \\ 1 & -j \end{bmatrix}}$$

(9.2-44)

and

$$(\mathbf{S})^{-1} = \begin{bmatrix} 1 & 1 \\ -j & j \end{bmatrix}$$

(9.2-45)

The matrix $\mathbf{S}$ is the symmetrical-component transformation matrix for a two-phase system. Actually, (9.2-43) is not quite the form of the symmetrical-component transformation we commonly see. It is always written with $\tilde{F}_{as+}$ and $\tilde{F}_{as-}$ rather than $\tilde{F}^s_{qs+}$ and $\tilde{F}^s_{qs-}$, respectively. We know that this substitution can be made if we so choose but it tends to be somewhat confusing. It should be noted that in this development we have considered only one frequency, ω_e. Actually, we can show that this approach is valid regardless of the forms of F_{as} and F_{bs} as long as they are periodic functions the periods of which may be different; see [2] and [3].

Example 9A. The steady-state variables of an unbalanced two-phase system are

$$\tilde{F}_{as} = 1\underline{/45°}$$

(9A-1)

$$\tilde{F}_{bs} = \tfrac{1}{2}\underline{/-120°}$$

(9A-2)

Calculate $\tilde{F}^s_{qs+}$ and $\tilde{F}^s_{qs-}$.

There are actually two ways we can do this; one by using (9.2-9) and the other by using (9.2-35) and (9.2-37). Let us do both. First, by (9.2-9),

$$\begin{bmatrix} \tilde{F}^s_{qs+} \\ \tilde{F}^s_{qs-} \end{bmatrix} = \frac{1}{2} \begin{bmatrix} 1 & j \\ 1 & -j \end{bmatrix} \begin{bmatrix} 1\underline{/45^\circ} \\ \frac{1}{2}\underline{/-120^\circ} \end{bmatrix} \tag{9A-3}$$

From which

$$\begin{aligned} \tilde{F}^s_{qs+} &= \tfrac{1}{2}(1\underline{/45^\circ} + j\tfrac{1}{2}\underline{/-120^\circ}) \\ &= \tfrac{1}{2}\underline{/45^\circ} + \tfrac{1}{4}\underline{/-30^\circ} \\ &= 0.570 + j0.229 = 0.614\underline{/21.9^\circ} \end{aligned} \tag{9A-4}$$

$$\begin{aligned} \tilde{F}^s_{qs-} &= \tfrac{1}{2}(1\underline{/45^\circ} - j\tfrac{1}{2}\underline{/-120^\circ}) \\ &= \tfrac{1}{2}\underline{/45^\circ} - \tfrac{1}{4}\underline{/-30^\circ} \\ &= 0.137 + j0.479 = 0.498\underline{/74.0^\circ} \end{aligned} \tag{9A-5}$$

In order to use (9.2-35) and (9.2-37) to calculate $\tilde{F}^s_{qs+}$ and $\tilde{F}^s_{qs-}$ we must first write F_{as} and F_{bs} in expanded instantaneous form as given by (9.2-12) and (9.2-13). From (9A-1) and (9A-2),

$$F_{as} = \sqrt{2}\,\frac{1}{\sqrt{2}}\cos\omega_e t - \sqrt{2}\,\frac{1}{\sqrt{2}}\sin\omega_e t \tag{9A-6}$$

$$F_{bs} = -\sqrt{2}\,\frac{1}{4}\cos\omega_e t + \sqrt{2}\,\frac{\sqrt{3}}{4}\sin\omega_e t \tag{9A-7}$$

Comparing (9A-6) and (9A-7) with (9.2-12) and (9.2-13), we see that

$$F_{as\alpha} = \sqrt{2}\,\frac{1}{\sqrt{2}} \tag{9A-8}$$

$$F_{as\beta} = -\sqrt{2}\,\frac{1}{\sqrt{2}} \tag{9A-9}$$

$$F_{bs\alpha} = -\sqrt{2}\frac{1}{4} \tag{9A-10}$$

$$F_{bs\beta} = \sqrt{2}\,\frac{\sqrt{3}}{4} \tag{9A-11}$$

From (9.2-19) through (9.2-22),

$$\begin{aligned} F_{qsA} &= \tfrac{1}{2}(F_{as\alpha} + F_{bs\beta}) \\ &= \frac{\sqrt{2}}{2}\left(\frac{1}{\sqrt{2}} + \frac{\sqrt{3}}{4}\right) = \sqrt{2}\,0.570 \end{aligned} \tag{9A-12}$$

$$\begin{aligned} F_{qsB} &= \tfrac{1}{2}(F_{as\beta} - F_{bs\alpha}) \\ &= \frac{\sqrt{2}}{2}\left(-\frac{1}{\sqrt{2}} + \frac{1}{4}\right) = -\sqrt{2}\,0.229 \end{aligned} \tag{9A-13}$$

$$\begin{aligned} F_{qsC} &= \tfrac{1}{2}(F_{as\alpha} - F_{bs\beta}) \\ &= \frac{\sqrt{2}}{2}\left(\frac{1}{\sqrt{2}} - \frac{\sqrt{3}}{4}\right) = \sqrt{2}\,0.137 \end{aligned} \tag{9A-14}$$

$$\begin{aligned} F_{qsD} &= \tfrac{1}{2}(F_{as\beta} + F_{bs\alpha}) \\ &= \frac{\sqrt{2}}{2}\left(-\frac{1}{\sqrt{2}} - \frac{1}{4}\right) = -\sqrt{2}\,0.479 \end{aligned} \tag{9A-15}$$

Now we are ready to use (9.2-35) and (9.2-37), from which

$$\tilde{F}^s_{qs+} = \frac{1}{\sqrt{2}} (F_{qsA} - jF_{qsB})$$

$$= 0.570 + j0.229 = 0.614\underline{/21.9°} \tag{9A-16}$$

$$\tilde{F}^s_{qs-} = \frac{1}{\sqrt{2}} (F_{qsC} - jF_{qsD})$$

$$= 0.137 + j0.479 = 0.498\underline{/74.0°} \tag{9A-17}$$

SP9.2-1. $\tilde{F}_{as} = \tilde{F}_{bs} = F_s\underline{/0°}$, determine $\tilde{F}^s_{qs+}$ and $\tilde{F}^s_{qs-}$. $[\tilde{F}^s_{qs-} = \tilde{F}^{s*}_{qs+} = (\sqrt{2}/2) F_s\underline{/45°}]$
SP9.2-2. Express $\tilde{F}_{bs}$ in terms of $\tilde{F}_{as}$ so that the positive-sequence component is zero. $[\tilde{F}_{bs} = j\tilde{F}_{as}]$
SP9.2-3. $\tilde{F}^s_{ds+} = \tilde{F}^s_{ds-}$, determine $\tilde{F}_{as}$ and $\tilde{F}_{bs}$. $[\tilde{F}_{as} = 0, \tilde{F}_{bs} = -2j\tilde{F}^s_{qs+}]$

9.3 ANALYSIS OF UNBALANCED MODES OF OPERATION

The rotor windings of the two-phase servomotor are short-circuited (squirrel-cage windings). Hence, for steady-state unbalanced operation wherein the rotor speed is assumed constant, the equations which describe this constant-speed mode of operation are linear and the principle of superposition applies. Therefore, the substitute variables (F'^s_{qr} and F'^s_{dr}) for the rotor variables may be broken up into positive- and negative-sequence quantities in the same manner as F^s_{qs} and F^s_{ds}. In particular, we can write

$$\tilde{F}'^s_{qr} = \tilde{F}'^s_{qr+} + \tilde{F}'^s_{qr-} \tag{9.3-1}$$

$$\tilde{F}'^s_{dr} = \tilde{F}'^s_{dr+} + \tilde{F}'^s_{dr-} \tag{9.3-2}$$

and

$$\tilde{F}'^s_{dr+} = j\tilde{F}'^s_{qr+} \tag{9.3-3}$$

$$\tilde{F}'^s_{dr-} = -j\tilde{F}'^s_{qr-} \tag{9.3-4}$$

Armed with this information, let us see what we can derive in the way of voltage equations. For convenience, we will repeat (5.6-13), which are the steady-state voltage equations in q and d variables. Thus,

$$\begin{bmatrix} \tilde{V}^s_{qs} \\ \tilde{V}^s_{ds} \\ \tilde{V}'^s_{qr} \\ \tilde{V}'^s_{dr} \end{bmatrix} = \begin{bmatrix} r_s + j\omega_e L_{ss} & 0 & j\omega_e L_{ms} & 0 \\ 0 & r_s + j\omega_e L_{ss} & 0 & j\omega_e L_{ms} \\ j\omega_e L_{ms} & -\omega_r L_{ms} & r'_r + j\omega_e L'_{rr} & -\omega_r L'_{rr} \\ \omega_r L_{ms} & j\omega_e L_{ms} & \omega_r L'_{rr} & r'_r + j\omega_e L'_{rr} \end{bmatrix} \begin{bmatrix} \tilde{I}^s_{qs} \\ \tilde{I}^s_{ds} \\ \tilde{I}'^s_{qr} \\ \tilde{I}'^s_{dr} \end{bmatrix}$$

$$\tag{9.3-5}$$

Since at constant rotor speed ω_r the voltage equations are linear, superposition applies. That is, we can express the four equations of (9.3-5) twice; once for the positive-sequence variables and once for the negative-sequence variables. This gives two sets of four equations each. One set relates the positive-sequence voltages and currents, the other relates the negative-sequence voltages and currents. We have derived relationships between the q and d variables; that is, $\tilde{F}^s_{ds+} = j\tilde{F}^s_{qs+}$, $\tilde{F}'^s_{dr+} = j\tilde{F}'^s_{qr+}$, $\tilde{F}^s_{ds-} = -j\tilde{F}^s_{qs-}$, and $\tilde{F}'^s_{dr-} = -j\tilde{F}'^s_{qr-}$, which are (9.2-5), (9.3-3), (9.2-6), and (9.3-4), respectively. Therefore, the eight equations can be reduced back to four. If the d variables are expressed in terms of the q variables, the four equations are

$$
\begin{bmatrix} \tilde{V}^s_{qs+} \\ \dfrac{\tilde{V}'^s_{qr+}}{s} \\ \tilde{V}^s_{qs-} \\ \dfrac{\tilde{V}'^s_{qr-}}{2-s} \end{bmatrix} = \begin{bmatrix} r_s + jX_{ss} & jX_{ms} & 0 & 0 \\ jX_{ms} & \dfrac{r'_r}{s} + jX'_{rr} & 0 & 0 \\ 0 & 0 & r_s + jX_{ss} & jX_{ms} \\ 0 & 0 & jX_{ms} & \dfrac{r'_r}{2-s} + jX'_{rr} \end{bmatrix} \begin{bmatrix} \tilde{I}^s_{qs+} \\ \tilde{I}'^s_{qr+} \\ \tilde{I}^s_{qs-} \\ \tilde{I}'^s_{qr-} \end{bmatrix}
\qquad (9.3\text{-}6)
$$

where

$$ X_{ss} = \omega_e(L_{ls} + L_{ms}) \qquad (9.3\text{-}7) $$

$$ X'_{rr} = \omega_e(L'_{lr} + L_{ms}) \qquad (9.3\text{-}8) $$

$$ X_{ms} = \omega_e L_{ms} \qquad (9.3\text{-}9) $$

$$ s = \frac{\omega_e - \omega_r}{\omega_e} \qquad (9.3\text{-}10) $$

We realize that at any time we can change to the notation generally used by replacing $\tilde{F}^s_{qs+}$ with $\tilde{F}_{as+}$, $\tilde{F}'^s_{qr+}$ with $\tilde{F}'_{ar+}$, etc. However, we will not make this substitution in this text.

When we look at (9.3-6), we see that the positive- and negative-sequence quantities are decoupled. From this, we might be led to believe that the positive- and negative-sequence quantities may be considered separately regardless of the mode of operation of the induction motor. Although the voltage equations given by (9.3-6) provide a starting point, system constraints may cause the positive- and negative-sequence quantities to be coupled. In the modes of operation which we will consider, we shall find that the sequence quantities are decoupled when unbalanced voltages are applied to a symmetrical two-phase induction motor but coupled when an impedance is placed in series with one of the stator phase windings or when one of the stator phase windings is open-circuited.

The expression for the steady-state electromagnetic torque may be obtained by starting with (5.5-33), which is repeated here for convenience.

$$T_e = \frac{P}{2} L_{ms} (I_{qs}^s I_{dr}^{'s} - I_{ds}^s I_{qr}^{'s}) \qquad (9.3\text{-}11)$$

The instantaneous currents may each be expressed in terms of positive- and negative-sequence components. In particular, let

$$I_{qs}^s = \sqrt{2}\, I_{s+} \cos(\omega_e t + \phi_{s+}) + \sqrt{2}\, I_{s-} \cos(\omega_e t + \phi_{s-}) \qquad (9.3\text{-}12)$$

$$I_{ds}^s = -\sqrt{2}\, I_{s+} \sin(\omega_e t + \phi_{s+}) + \sqrt{2}\, I_{s-} \sin(\omega_e t + \phi_{s-}) \qquad (9.3\text{-}13)$$

$$I_{qr}^{'s} = \sqrt{2}\, I_{r+}' \cos(\omega_e t + \phi_{r+}) + \sqrt{2}\, I_{r-}' \cos(\omega_e t + \phi_{r-}) \qquad (9.3\text{-}14)$$

$$I_{dr}^{'s} = -\sqrt{2}\, I_{r+}' \sin(\omega_e t + \phi_{r+}) + \sqrt{2}\, I_{r-}' \sin(\omega_e t + \phi_{r-}) \qquad (9.3\text{-}15)$$

where the $+$ and $-$ subscripts denote positive- and negative-sequence quantities, respectively. If these expressions for the currents are substituted into (9.3-11) and with a few trigonometric identities, we can express the steady-state (constant-speed) unbalanced torque as

$$T_e = 2\frac{P}{2} L_{ms} [I_{s+} I_{r+}' \sin(\phi_{s+} - \phi_{r+}) - I_{s-} I_{r-}' \sin(\phi_{s-} - \phi_{r-})$$

$$+ I_{s+} I_{r-}' \sin(2\omega_e t + \phi_{s+} + \phi_{r-}) - I_{s-} I_{r+}' \sin(2\omega_e t + \phi_{s-} + \phi_{r+})]$$

$$(9.3\text{-}16)$$

It is interesting that with the assumption of symmetrical rotor circuits, the electromagnetic torque during unbalanced operation is made up of a constant and a sinusoidal component which pulsates at twice the frequency of the stator variables. Recall that we have assumed that the steady-state stator variables contain only one frequency, ω_e. Multiple frequencies are treated in [2] and [3].

The above equation for torque may be expressed in terms of positive- and negative-sequence current phasors. After considerable work,

$$T_e = 2\frac{P}{2} L_{ms} \{ \mathrm{Re}[\, j(\tilde{I}_{qs+}^{s*} \tilde{I}_{qr+}^{'s} - \tilde{I}_{qs-}^{s*} \tilde{I}_{qr-}^{'s})]$$

$$+ \mathrm{Re}[\, j(-\tilde{I}_{qs+}^{s} \tilde{I}_{qr-}^{'s} + \tilde{I}_{qs-}^{s} \tilde{I}_{qr+}^{'s})] \cos 2\omega_e t \qquad (9.3\text{-}17)$$

$$+ \mathrm{Re}[\tilde{I}_{qs+}^{s} \tilde{I}_{qr-}^{'s} - \tilde{I}_{qs-}^{s} \tilde{I}_{qr+}^{'s}] \sin 2\omega_e t \}$$

where the asterisk denotes the conjugate. The constant term [first term on right-hand side of (9.3-17)] is made up of the positive-sequence torque and the negative-sequence torque. The last two terms, which represent the pulsating torque component, could be combined; however, separate terms are somewhat more convenient. We note that the amplitude of the pulsating torque is related to the cross product of sequence currents.

Unbalanced Stator Voltages

As we mentioned previously, the stator voltages are unbalanced when the device is operated as a servomotor. We also mentioned that, during the starting period of a single-phase induction motor, a capacitor is placed in series with one of the windings. This type of unbalance, wherein the stator circuits appear unsymmetrical to the source because of the series capacitor, must be analyzed differently than when the stator circuits are symmetrical. We will consider the case of unbalanced source voltages applied to a symmetrical machine first and leave the series capacitor case for later.

Let us return to the voltage equations given by (9.3-6). We can apply these equations directly to solve for the sequence currents with unbalanced source voltages applied to the stator windings of a symmetrical machine. We need only determine $\tilde{V}^s_{qs+}$ and $\tilde{V}^s_{qs-}$ from $\tilde{V}_{as}$ and $\tilde{V}_{bs}$ by (9.2-9). We know that, since the rotor windings of the servomotor are short-circuited, $\tilde{V}'^s_{qr+}$ and $\tilde{V}'^s_{qr-}$ are zero.

Since this unbalanced mode of operation is described by positive- and negative-sequence quantities which are decoupled, it is instructive to portray the four voltage equations given by (9.3-6) in equivalent-circuit form as shown in Fig. 9.3-1. The positive-sequence equivalent circuit is identical in form to that given for balanced conditions in Fig. 5.6-2 with the rotor windings short-circuited. This was expected. The negative-sequence equivalent circuit differs only in that the slip s is replaced by $2 - s$. Recall that the negative-sequence voltages cause negative-sequence currents which, in turn, cause a negatively rotating air gap mmf. With respect to this negatively rotating air gap mmf, the slip is $(\omega_e + \omega_r)/\omega_e$, which is $2 - (\omega_e - \omega_r)/\omega_e$ or $2 - s$. That is the

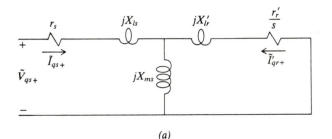

(a)

(b)

FIGURE 9.3-1
Equivalent-sequence circuits for unbalanced source voltages applied to a symmetrical two-phase induction motor. (*a*) Positive sequence; (*b*) negative sequence.

traditional way of looking at all this. In fact, this line of reasoning is often used to obtain the negative-sequence equivalent circuit in place of a derivation.

Equation (9.3-6) or the equivalent circuits which come from (9.3-6) can be used to solve for the sequence currents. The steady-state electromagnetic torque can then be calculated by appropriate substitution of the sequence currents into (9.3-17). Since the positive- and negative-sequence circuits are decoupled, the positive- and negative-sequence torques may be expressed from (5.6-28). In particular, the positive-sequence torque, which is due to the product of the positive-sequence currents in the first term on the right-hand side of (9.3-17), may be expressed as

$$T_{e+} = \frac{2(P/2)(X_{ms}^2/\omega_e)r_r's|\tilde{V}_{qs+}^s|^2}{[r_sr_r' + s(X_{ms}^2 - X_{ss}X_{rr}')]^2 + (r_r'X_{ss} + sr_sX_{rr}')^2} \qquad (9.3\text{-}18)$$

The negative-sequence torque, which is due to the product of the negative-sequence currents in the first term on the right-hand side of (9.3-17), may be expressed

$$T_{e-} = \frac{2(P/2)(X_{ms}^2/\omega_e)r_r'(2-s)|\tilde{V}_{qs-}^s|^2}{[r_sr_r' + (2-s)(X_{ms}^2 - X_{ss}X_{rr}')]^2 + [r_r'X_{ss} + (2-s)r_sX_{rr}']^2} \qquad (9.3\text{-}19)$$

Equation (9.3-18) was obtained from (5.6-28) with $\tilde{V}_{as}$ replaced by $\tilde{V}_{qs+}^s$, and (9.3-19) was obtained from (5.6-28) with $\tilde{V}_{as}$ replaced by $\tilde{V}_{qs-}^s$ and s replaced by $2-s$. The average torque, $T_{e,av}$, is the difference between the positive- and negative-sequence torques:

$$T_{e,av} = T_{e+} - T_{e-} \qquad (9.3\text{-}20)$$

Comparing the first two terms in (9.3-17) with (5.6-23), it is interesting to observe that, although torque, in general, is a nonlinear function of currents, we can use superposition to establish the total average torque by first calculating the positive- and negative-sequence currents from (9.3-6) or Fig. 9.3-1, then calculating individually the corresponding positive- and negative-sequence torques, and finally superimposing the results by using (9.3-20). However, it should be clear that, although superposition may be used to calculate the net average torque, the instantaneous torque (sum of average and pulsating components) cannot be calculated by using superposition since, from (9.3-17), the pulsating torque is related to the product of positive- and negative-sequence currents.

Although we could express the amplitude of the pulsating torque in terms of the sequence voltages, the algebraic manipulations necessary to do so are a bit prohibitive. It is sufficient, for our purposes, to take a little closer look at the phasor relationship $\tilde{I}_{qs+}^s\tilde{I}_{qr-}^{'s} - \tilde{I}_{qs-}^s\tilde{I}_{qr+}^{'s}$ which is common to the second and third terms of (9.3-17). With the rotor winding short-circuited, we can express

$$\tilde{I}'^s_{qr+} = -\frac{jX_{ms}}{r'_r/s + jX'_{rr}} \tilde{I}^s_{qs+} \tag{9.3-21}$$

$$\tilde{I}'^s_{qr-} = -\frac{jX_{ms}}{r'_r/(2-s) + jX'_{rr}} \tilde{I}^s_{qs-} \tag{9.3-22}$$

These equations are obtained from the equivalent circuits given in Fig. 9.3-1. Note the similarity between (9.3-21) and (5.6-24). Utilizing (9.3-21) and (9.3-22), we can write

$$\tilde{I}^s_{qs+}\tilde{I}'^s_{qr-} - \tilde{I}^s_{qs-}\tilde{I}'^s_{qr+} = -jX_{ms}\tilde{I}^s_{qs+} I^s_{qs-} \frac{2(1-s)/[s(2-s)]}{[r'_r/(2-s) + jX'_{rr}](r'_r/s + jX'_{rr})} \tag{9.3-23}$$

If we express the sequence currents in terms of the sequence voltages, we can write (9.3-23) as

$$\tilde{I}^s_{qs+}\tilde{I}'^s_{qr-} - \tilde{I}^s_{qs-}\tilde{I}'^s_{qr+} = -jX_{ms}\frac{\tilde{V}^s_{qs+}}{Z_+}\frac{\tilde{V}^s_{qs-}}{Z_-}\frac{2(1-s)/[s(2-s)]}{[r'_r/(2-s) + jX'_{rr}](r'_r/s + jX'_{rr})} \tag{9.3-24}$$

where Z_+ and Z_- are the input impedances of the positive- and negative-sequence equivalent circuits (Fig. 9.3-1), respectively.

The form of (9.3-24) allows a somewhat more direct means of calculating the amplitude of the pulsating torque. It is interesting, however, to evaluate (9.3-24) for the condition where the rotor speed is zero. With $\omega_r = 0$, $s = 1$ and (9.3-24) is zero. Hence, a pulsating torque does not exist at stall. Actually, the amplitude of the pulsating torque is zero at $\omega_r = 0$ regardless of the stator conditions. That is, an impedance may be in series with one of the stator windings or one winding may be opened-circuited. Regardless of the value of the sequence currents, (9.3-23) is zero when $s = 1$. The only requirement is that the rotor windings must be symmetrical. This is a very important observation. Since a pulsating torque does not occur at $\omega_r = 0$, the electromagnetic torque T_e will be constant when the rotor is at the desired position. Hence, position accuracy is not degraded by a pulsating torque giving rise to a pulsation in θ_r about a reference position.

Unbalanced Stator Impedances

When an impedance is placed in series with the as winding of the stator, we can write

$$e_{ga} = i_{as}z(p) + v_{as} \tag{9.3-25}$$

$$e_{gb} = v_{bs} \tag{9.3-26}$$

where v_{as} and v_{bs} are the voltages across the stator phase windings and e_{ga} and e_{gb} are the source voltages which may be unbalanced. In (9.3-25), $z(p)$ is the operational notation of the impedance; for example, a series rL would be expressed $z(p) = r + pL$. The phasor equivalents of (9.3-25) and (9.3-26) are

$$\tilde{V}_{as} = \tilde{E}_{ga} - \tilde{I}_{as}Z \tag{9.3-27}$$

$$\tilde{V}_{bs} = \tilde{E}_{gb} \tag{9.3-28}$$

We can apply (9.2-9) to determine $\tilde{V}^s_{qs+}$ and $\tilde{V}^s_{qs-}$ as

$$\begin{bmatrix} \tilde{V}^s_{qs+} \\ \tilde{V}^s_{qs-} \end{bmatrix} = \frac{1}{2} \begin{bmatrix} 1 & j \\ 1 & -j \end{bmatrix} \begin{bmatrix} \tilde{E}_{ga} - \tilde{I}_{as}Z \\ \tilde{E}_{gb} \end{bmatrix} \tag{9.3-29}$$

which yields

$$\tilde{V}^s_{qs+} = \tfrac{1}{2}(\tilde{E}_{ga} + j\tilde{E}_{gb} - \tilde{I}_{as}Z) \tag{9.3-30}$$

$$\tilde{V}^s_{qs-} = \tfrac{1}{2}(\tilde{E}_{ga} - j\tilde{E}_{gb} - \tilde{I}_{as}Z) \tag{9.3-31}$$

Now,

$$\tilde{I}_{as} = \tilde{I}^s_{qs} = \tilde{I}^s_{qs+} + \tilde{I}^s_{qs-} \tag{9.3-32}$$

Substituting (9.3-32) into (9.3-30) and (9.3-31) for $\tilde{I}_{as}$, then substituting the result into (9.3-6), and assuming the rotor windings are short-circuited, we obtain

$$\begin{bmatrix} \tilde{E}_1 \\ 0 \\ \tilde{E}_2 \\ 0 \end{bmatrix} = \begin{bmatrix} \tfrac{1}{2}Z + r_s + jX_{ss} & jX_{ms} & \tfrac{1}{2}Z & 0 \\ jX_{ms} & \dfrac{r'_r}{s} + jX'_{rr} & 0 & 0 \\ \tfrac{1}{2}Z & 0 & \tfrac{1}{2}Z + r_s + jX_{ss} & jX_{ms} \\ 0 & 0 & jX_{ms} & \dfrac{r'_r}{2-s} + jX'_{rr} \end{bmatrix} \begin{bmatrix} \tilde{I}^s_{qs+} \\ \tilde{I}'^s_{qr+} \\ \tilde{I}^s_{qs-} \\ \tilde{I}'^s_{qr-} \end{bmatrix}$$

$$\tag{9.3-33}$$

where

$$\tilde{E}_1 = \tfrac{1}{2}(\tilde{E}_{ga} + j\tilde{E}_{gb}) \tag{9.3-34}$$

$$\tilde{E}_2 = \tfrac{1}{2}(\tilde{E}_{ga} - j\tilde{E}_{gb}) \tag{9.3-35}$$

If the impedance is a series capacitor, then $Z = -j(1/\omega_e C)$. Also, note that the positive- and negative-sequence voltage equations are now coupled. Although we could derive an equivalent circuit to portray these voltage equations, it is not worth the work. We can use (9.3-33) directly; however, a computer would be helpful. We shall work more with this equation when we analyze the symmetrical two-phase induction motor used as a single-phase motor.

Open-Circuited Stator Phase

For the analysis of an open-circuited stator phase, let us assume that i_{as} (i^s_{qs}) is zero. Hence, from (5.5-24),

$$v_{qs}^s = p\lambda_{qs}^s \tag{9.3-36}$$

Now since $i_{qs}^s = 0$, λ_{qs}^s may be expressed from (5.5-28) as

$$\lambda_{qs}^s = L_{ms} i_{qr}^{'s} \tag{9.3-37}$$

Since $v_{as} = v_{qs}^s$, we can write

$$v_{as} = L_{ms} p i_{qr}^{'s} \tag{9.3-38}$$

$$v_{bs} = e_{gb} \tag{9.3-39}$$

where e_{gb} is the source voltage. Now, in phasor form,

$$\tilde{V}_{as} = jX_{ms} \tilde{I}_{qr}^{'s} \tag{9.3-40}$$

$$\tilde{V}_{bs} = \tilde{E}_{gb} \tag{9.3-41}$$

Substituting (9.3-40) and (9.3-41) into (9.2-9), we obtain

$$\tilde{V}_{qs+}^s = \tfrac{1}{2} jX_{ms} \tilde{I}_{qr}^{'s} + \tfrac{1}{2} j\tilde{E}_{gb} \tag{9.3-42}$$

$$\tilde{V}_{qs-}^s = \tfrac{1}{2} jX_{ms} \tilde{I}_{qr}^{'s} - \tfrac{1}{2} j\tilde{E}_{gb} \tag{9.3-43}$$

From (9.3-1),

$$\tilde{I}_{qr}^{'s} = \tilde{I}_{qr+}^{'s} + \tilde{I}_{qr-}^{'s} \tag{9.3-44}$$

and, since $\tilde{I}_{qs}^s = 0$, (9.2-3) becomes

$$\tilde{I}_{qs-}^s = -\tilde{I}_{qs+}^s \tag{9.3-45}$$

Substituting (9.3-44) into (9.3-42) and (9.3-43), then substituting the result into (9.3-6), with (9.3-45) incorporated, we can write (rotor windings short-circuited)

$$
\begin{bmatrix} \tfrac{1}{2} j\tilde{E}_{gb} \\ 0 \\ 0 \end{bmatrix}
=
\begin{bmatrix}
r_s + jX_{ss} & j\tfrac{1}{2} X_{ms} & -j\tfrac{1}{2} X_{ms} \\
jX_{ms} & \dfrac{r_r'}{s} + jX_{rr}' & 0 \\
-jX_{ms} & 0 & \dfrac{r_r'}{2-s} + jX_{rr}'
\end{bmatrix}
\begin{bmatrix} \tilde{I}_{qs+}^s \\ \tilde{I}_{qr+}^{'s} \\ \tilde{I}_{qr-}^{'s} \end{bmatrix}
\tag{9.3-46}
$$

From (9.3-40), the open-circuit voltage of the *as* winding is

$$\tilde{V}_{as} = jX_{ms}(\tilde{I}_{qr+}^{'s} + \tilde{I}_{qr-}^{'s}) \tag{9.3-47}$$

where $\tilde{I}_{qr+}^{'s}$ and $\tilde{I}_{qr-}^{'s}$ are calculated from (9.3-46). If the open-circuit voltage is not of interest, then $\tilde{I}_{qr+}^{'s}$ and $\tilde{I}_{qr-}^{'s}$ may be eliminated from (9.3-46).

SP9.3-1. Determine the rotor speed at which the negative-sequence rotor currents $\tilde{I}_{qr-}^{'s}$ and $\tilde{I}_{dr-}^{'s}$ are zero for unbalanced applied stator voltages. $[\omega_r = -\omega_e]$

SP9.3-2. Assume that the steady-state T_e versus ω_r plot shown in Fig. 5.6-3 is for $\tilde{V}_{as} = j\tilde{V}_{bs} = 1\underline{/0°}$. Plot the T_e versus ω_r for $\tilde{V}_{as} = -j\tilde{V}_{bs} = 1\underline{/0°}$. [Inverted mirror image]

SP9.3-3. Determine the rotor speed at which $Z_+ = Z_-$ for a symmetrical servomotor. $[\omega_r = 0]$

SP9.3-4. Express Z_+ and Z_-. $[Z_+ = (5.6\text{-}26),\ Z_- = (5.6\text{-}26)$ with s replaced by $2 - s]$

SP9.3-5. Express v_{bs} when $i_{bs} = 0$. $[v_{bs} = -L_{ms} p i_{dr}^{'s}]$

9.4 STEADY-STATE UNBALANCED OPERATION OF SERVOMOTOR

As we have mentioned previously, the two-phase induction motor is often used as a positioning servomotor. In this application, a constant-amplitude voltage is applied to one of the stator windings, which is generally referred to as the main winding, while the amplitude of the voltage applied to the other phase (control winding) is related to an error signal. Although the voltage applied to the control phase is in time quadrature with the voltage applied to the main winding, it is seldom equal in amplitude since it is related to the difference between the desired position and the actual position of the driven member. When the driven member is in the desired position, the voltage applied to the control phase is zero. This is the steady-state mode of operation since, during positioning, the error and, thus, the amplitude of the voltage applied to the control winding are changing continuously. Hence, one must be aware that the average torque versus speed characteristics for steady-state unbalanced operation will not give the complete picture of the positioning mode of operation. Nevertheless, the steady-state torque versus speed characteristics are instructive and provide us with a place from which to start.

Our objective is to obtain plots of the average electromagnetic torque versus speed for rated voltage applied to the main winding (the *as* or *bs* winding, whichever you choose) or a typical servomotor with the amplitude of the steady-state voltage applied to the control winding varied over a range from zero to ±120 percent of rated. The servomotor which we will use is similar to the one in Chap. 5. However, the rotor resistance is doubled to make its characteristics more closely portray those of a positioning servomotor. In particular, the example device is a four-pole two-phase $\frac{1}{10}$-hp 115-V (rms) 60-Hz servomotor with the following parameters: $r_s = 24.5\ \Omega$, $L_{ls} = 27.06$ mH, $L_{ms} = 273.7$ mH, $r_r' = 46\ \Omega$, $L_{lr}' = 27.06$ mH. For the purposes at hand, let the steady-state applied voltages be

$$V_{as} = k\sqrt{2}\,115\cos 377t \tag{9.4-1}$$

$$V_{bs} = \sqrt{2}\,115\sin 377t \tag{9.4-2}$$

Now, the positive-sequence torque T_{e+} may be calculated from (9.3-18), the negative-sequence torque T_{e-} from (9.3-19), and the magnitude of the double-frequency component $|T_{e,\text{pul}}|$ by appropriate use of (9.3-24) and (9.3-17). The positive-sequence torque T_{e+} versus rotor speed is shown in Fig.

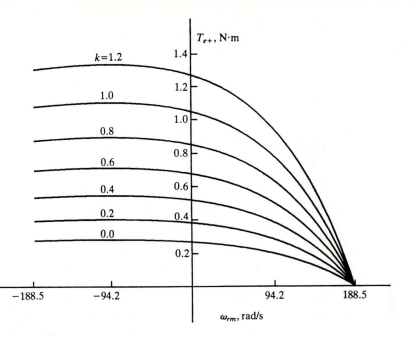

FIGURE 9.4-1
Positive-sequence electromagnetic torque T_{e+} for servomotor.

9.4-1. The plots of T_{e+} are shown only for positive values of k; negative values will produce an inverted mirror image. The negative-sequence torque, T_{e-} versus rotor speed is shown in Fig. 9.4-2 and the magnitude of the pulsating torque is shown in Fig. 9.4-3.

The average electromagnetic torque is the difference between T_{e+} and T_{e-}. The plot of the average electromagnetic torque $T_{e,av}$ versus rotor speed is shown in Fig. 9.4-4. Here, the average torque is plotted only for positive values of k. The torque versus speed characteristic for a negative value of k will be the negative mirror image about the zero-speed axis ($\omega_{rm} = 0$) of the torque-speed

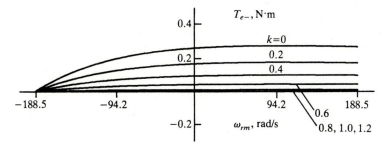

FIGURE 9.4-2
Negative-sequence electromagnetic torque T_{e-} for servomotor.

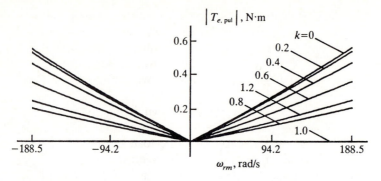

FIGURE 9.4-3
Magnitude of the double-frequency component of the electromagnetic torque for servomotor.

plot for the corresponding positive value of k. That is, for a given negative value of k and rotor speed ω_{rm}, $T_{e,av}$ will be equal to $-T_{e,av}$, evaluated at $-\omega_{rm}$ for the corresponding positive value of k.

It appears from Fig. 9.4-4 that $T_{e,av}$ at stall ($\omega_{rm} = 0$) is proportional to k. This is true in general and can be proved by noting that, at stall, $s = 1$ and $2 - s = 1$. Thus, the expressions for T_{e+} and T_{e-} given by (9.3-18) and (9.3-19) are identical, with the exception that T_{e+} is proportional to $|\tilde{V}^s_{qs+}|^2$ and T_{e-} is

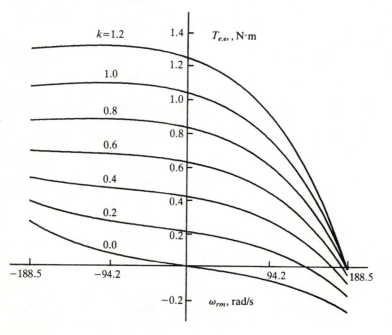

FIGURE 9.4-4
Average electromagnetic torque for servomotor.

proportional to $|\tilde{V}^s_{qs-}|^2$. If we evaluate $\tilde{V}^s_{qs+}$ and $\tilde{V}^s_{qs-}$ for the voltages given by (9.4-1) and (9.4-2), then substitute these into (9.3-18) and (9.3-19) to obtain T_{e+} and T_{e-}, we would find that $T_{e,\text{av}} = T_{e+} - T_{e-}$ is directly proportional to k at stall (see SP's at the end of this section). In other words, the stall torque is directly proportional to the amplitude of the voltage we apply to the control winding.

Althouth the torque-speed characteristics shown in Fig. 9.4-4 are for steady-state operation, one can start to get the picture of how a servomotor performs positioning. Let us assume that there is no load torque on the shaft and the driven member is correctly positioned. The average torque and rotor speed are both zero in this case. We are at the origin in Fig. 9.4-4. If it is desired to advance the position of the rotor in the counterclockwise direction, an error which is related to the difference between the desired position of the rotor and the actual position produces an applied voltage to the control phase in proportion to this error. An average torque is developed and the rotor advances in the counterclockwise direction, whereupon the error and thus, the magnitude of the voltage applied to the control winding decrease. Viewing this positioning from the steady-state torque-speed characteristics, we see that we will move on a trajectory from one torque-speed curve to the other along decreasing values of torque until we have reached the new desired position, whereupon we will again be at the origin of Fig. 9.4-4.

SP9.4-1. Express the instantaneous steady-state electromagnetic torque of the example servomotor with k in (9.4-1) equal to 0.8 and $\omega_{rm} = 6\pi$ rad/s. Assume the maximum instantaneous value of T_e occurs at $t = 0$. [$T_e = 0.8 + 0.02 \cos 754t$ N·m]

SP9.4-2. Determine the frequency of the rotor currents for the condition given in SP9.4-1. [108π and 132π rad/s]

SP9.4-3. Determine the frequency of the rotor currents at stall if $V_{as} = 0$ and $V_{bs} =$ (9.4-2). [120π]

SP9.4-4. Determine the rotor positions at which the instantaneous rotor currents will be equal in SP9.4-3. [$\theta_r(0) = \pm 45$ or $\pm 135°$]

SP9.4-5. Suppose $V_{as} = k\sqrt{2}\,V_s \cos \omega_e t$ and $V_{bs} = \sqrt{2}\,V_s \sin \omega_e t$. Express $\tilde{V}^s_{qs+}$ and $\tilde{V}^s_{qs-}$. [$\tilde{V}^s_{qs+} = \frac{1}{2}(k+1)V_s$ and $\tilde{V}^s_{qs-} = \frac{1}{2}(k-1)V_s$]

SP9.4-6. For the conditions of SP9.4-5, show that the average torque $T_{e,\text{av}}$ at stall ($\omega_{rm} = 0$) is proportional to k. [At $\omega_{rm} = 0$, $s = 1$, $2 - s = 1$, and $T_{e,\text{av}} =$ (9.3-18) through (9.3-19) with $\tilde{V}^s_{qs+}$, $\tilde{V}^s_{qs-}$ given by SP9.4-5]

9.5 DYNAMIC AND STEADY-STATE PERFORMANCE OF A POSITIONING SYSTEM

It is instructive to observe the dynamic characteristics of a two-phase servomotor when used in an elementary positioning system. The positioning system considered is depicted in Fig. 9.5-1. The servomotor used in this example is that described earlier with the total inertia of the rotor and connected mechanical load of 1×10^{-2} kg·m^2.

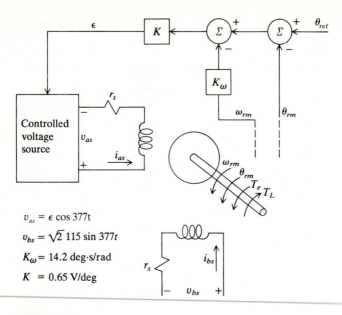

$$v_{as} = \epsilon \cos 377t$$

$$v_{bs} = \sqrt{2}\,115 \sin 377t$$

$$K_\omega = 14.2 \text{ deg·s/rad}$$

$$K = 0.65 \text{ V/deg}$$

FIGURE 9.5-1
Elementary positioning system.

It may be somewhat inappropriate to use a control system as an example since a course in automatic control systems is not a prerequisite for this text. Even though the salient features of the control system are quite straightforward, we need not become involved with the details of the control in order to observe and appreciate the dynamic characteristics of the servomotor when used in this application. In fact, it is difficult to visualize the dynamic performance of the positioning servomotor without incorporating it into actual control system setting.

The servomotor is a four-pole device, hence, from (4.5-9), $\theta_{rm} = \frac{2}{4}\theta_r$. The reference position θ_{ref} is compared to a signal proportional to θ_{rm} to make up part of the error signal ϵ. The error signal is the amplitude of the voltage applied to the control winding, in this case, the as winding. In particular,

$$v_{as} = \epsilon \cos 377t \tag{9.5-1}$$

$$v_{bs} = \sqrt{2}\,115 \sin 377t \tag{9.5-2}$$

A signal proportional to the rotor speed is also a part of the error signal. This signal is necessary to achieve a well-damped response of θ_{rm} following changes in the reference position θ_{ref}. As shown in Fig. 9.5-1,

$$\epsilon = K(\theta_{\text{ref}} - \theta_{rm} - K_\omega \omega_{rm}) \tag{9.5-3}$$

The values of K and K_ω are given to provide information to those who wish to consider the control system in more detail.

$$K = 37.24 \text{ V/rad} = 0.65 \text{ V/deg} \qquad (9.5\text{-}4)$$

$$K_\omega = 14.2 \text{ deg} \cdot \text{s/rad} \qquad (9.5\text{-}5)$$

The dynamic performance of the elementary positioning system is depicted in Figs. 9.5-2 through 9.5-5. In Fig. 9.5-2 the system is initially operating with zero load torque and $\theta_{ref} = 0$, thus $v_{as} = 0$ and $v_{bs} = \sqrt{2} \, 115 \sin t$. The following variables are plotted: θ_{ref}, θ_{rm}, ϵ, v_{as}, T_e, ω_{rm}, i_{as}, and i_{bs}. As indicated, θ_{ref} is stepped from zero to 200°. Instantaneously, the error ϵ increases and the amplitude of the voltage applied to the control winding increases accordingly. The rotor position θ_{rm} begins to increase. The error signal decreases and crosses through zero for the first time before θ_{rm} has

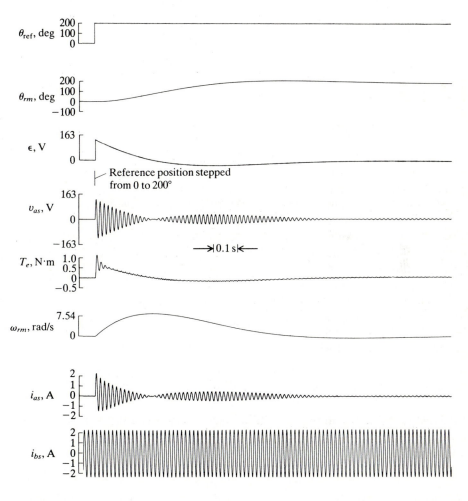

FIGURE 9.5-2
Dynamic performance of positioning system following a step increase in reference position; $T_L = 0$.

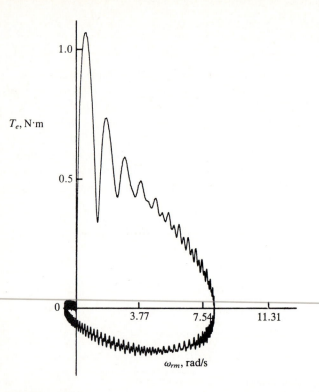

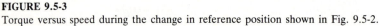

FIGURE 9.5-3
Torque versus speed during the change in reference position shown in Fig. 9.5-2.

reached the desired value of 200°. Recall that the error signal is a function of θ_{ref}, θ_{rm}, and ω_{rm}, (9.5-3). When ϵ becomes negative the average torque becomes negative at essentially the same time and the rotor speed begins to decrease back to zero. The control parameters are adjusted so that only a slight transient overshoot of the desired position, θ_{ref}, occurs. A plot of T_e versus ω_{rm} for the change in reference position portrayed in Fig. 9.5-2 is shown in Fig. 9.5-3.

Let us take a moment to consider the waveform of the transient T_e immediately following the step increase in θ_{ref}. Recall that in Chap. 5 the transient torque-speed characteristics of several induction machines were given; Figs. 5.7-2, 5.7-4, and 5.7-6. In each of these cases, the machines accelerated from stall with $v_{as} = \sqrt{2}\,V_s \cos \omega_e t$ and $v_{bs} = \sqrt{2}\,V_s \sin \omega_e t$ applied at $t = 0$. In the case of the positioning system shown in Fig. 9.5-1, the voltage is always applied to the bs winding. Therefore, the waveform of transient torque immediately following the step change in θ_{ref} depends upon the instantaneous value of i_{bs} at the time the step change is made. In Figs. 9.5-2 and 9.5-3 the step increase in θ_{ref} was made so that the maximum positive value of v_{as} occurred at the time of the step.

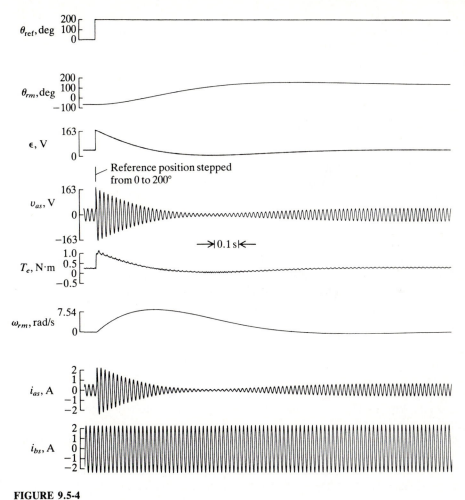

FIGURE 9.5-4
Same as Fig. 9.5-2 with $T_L = 0.25\,\text{N} \cdot \text{m}$.

The plots shown in Figs. 9.5-4 and 9.5-5 are for conditions identical to those in Figs. 9.5-2 and 9.5-3, respectively, with the exception that a load torque of $T_L = 0.25\,\text{N} \cdot \text{m}$ is applied to the rotor shaft and the change in θ_{ref} was made when $\cos 377t$ was passing through zero going positive. Initially, with $\theta_{\text{ref}} = 0$, a voltage must be applied to the control winding to maintain the rotor at stall. Hence, an error signal is established. Even though $\theta_{\text{ref}} = 0$, θ_{rm} is approximately $-62°$ in order to establish the necessary control winding applied voltage. Since the load torque is constant, it will always be necessary for θ_{rm} to be approximately $62°$ less than the reference position. It follows that a constant angular position of $62°$ may be added inside the parentheses of (9.5-3), whereupon θ_{ref} and the steady-state value of θ_{rm} would correspond. Note that

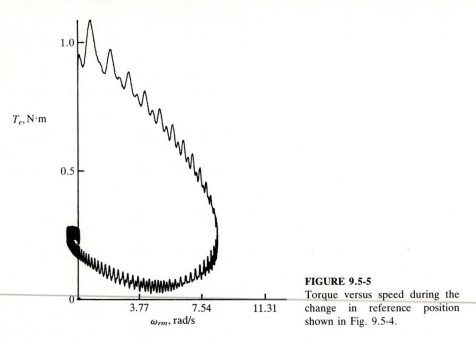

FIGURE 9.5-5
Torque versus speed during the change in reference position shown in Fig. 9.5-4.

the amplitude of the transient torque following the step change in θ_{ref} is considerably less in Figs. 9.5-4 and 9.5-5 than in Figs. 9.5-2 and 9.5-3.

SP9.5-1. Why in Fig. 9.5-2 does T_e become negative when ϵ becomes negative? $[|\tilde{V}^s_{qs-}| > |\tilde{V}^s_{qs+}|$ and $s \cong 2 - s]$

SP9.5-2. Consider Fig. 9.5-4. At maximum ω_{rm}, read ω_{rm} and θ_{rm}. Verify the value of ϵ. $\{\epsilon = 0.65[200 - 20 - (14.2)(8.5)]\}$

SP9.5-3. In Fig. 9.5-4, $\epsilon \cong 39$ V during steady-state operation. Determine $\tilde{V}^s_{qs+}$ and $\tilde{V}^s_{qs-}$. $[\tilde{V}^s_{qs+} = 71.3\underline{/0°}, \tilde{V}^s_{qs-} = -43.7\underline{/0°}]$

9.6   LINEAR APPROXIMATION OF SERVOMOTOR TORQUE-SPEED CHARACTERISTICS

We are aware that the voltage and torque equations for an induction machine are nonlinear. Therefore, we are unable to apply linear system theory in the design of the positioning control system with the voltage and torque equations as we know them. Fortunately, there is a simplified linear expression for torque in terms of the voltage applied to the control winding (or the error signal ϵ) and ω_{rm} which is quite accurate in portraying the dynamic and steady-state performance of the positioning system considered in the previous section. For this purpose, consider the plots of the average steady-state electromagnetic torque versus rotor speed given in Fig. 9.4-4. Let us approximate these plots as straight lines expressed by

$$\boxed{T_e = C_t \epsilon - C_\omega \omega_{rm}} \tag{9.6-1}$$

The constant C_t is determined from the torque at $\omega_{rm} = 0$. With $\epsilon = \sqrt{2}\,115$ V, the steady-state value of T_e is 1.07 N·m where $\omega_{rm} = 0$. Thus, for the example servomotor,

$$C_t = \frac{1.07}{\sqrt{2}\,115} = 6.58 \times 10^{-3} \text{ N} \cdot \text{m/V} \tag{9.6-2}$$

The value of C_ω is selected so that with $\epsilon = \sqrt{2}\,115$ V, T_e is zero when $\omega_{rm} = (2/P)377$ rad/s. Thus, with $P = 4$,

$$C_\omega = \frac{1.07}{(\frac{2}{4})377} = 5.68 \times 10^{-3} \text{ N} \cdot \text{m} \cdot \text{s/rad} \tag{9.6-3}$$

Therefore, (9.6-1) for the example servomotor becomes

$$T_e = 6.58 \times 10^{-3}\epsilon - 5.68 \times 10^{-3}\omega_{rm} \tag{9.6-4}$$

At first consideration one would not expect this linear approximation to yield a satisfactory portrayal of the torque versus speed characteristics of a servomotor. All electrical transients and the pulsating torque have been neglected with this approximation; however, this would seem to be acceptable unless the combined inertia of the rotor and mechanical system was unusually small. Perhaps the most questionable aspect of this approximation is the inaccuracy which occurs as the rotor speed increases from stall. For example, with ϵ equal to zero and ω_{rm} equal to synchronous speed, $(2/P)377$ rad/s, T_e calculated from the linear relationship, (9.6-4), is -1.07 N·m. The value from Fig. 9.4-4 is -0.27 N·m. Nevertheless, this linearized approximation of the T_e versus ω_{rm} characteristic is still quite accurate if the rotor speed does not become too large. Note that, in the positioning system considered in the previous section, the rotor speed was quite small; in particular, ω_{rm} did not exceed 10 rad/s during the positioning illustrated. Therefore, one might conclude that (9.6-4) would adequately portray the torque versus speed characteristics of a servomotor when used in a positioning system if the inertia of the rotor and mechanical system is not unusually small and if the excursions in rotor speed during positioning are less than, say, 10 to 20 percent of synchronous speed. The computer traces shown in Figs. 9.6-1 and 9.6-2 justify this conclusion. Here, (9.6-4) is used to describe the performance of the servomotor rather than the detailed voltage and torque equations used in the previous section. Figures 9.6-1 and 9.6-2 should be compared with Figs. 9.5-2 and 9.5-3, respectively. The responses are nearly identical and sufficiently accurate for use in control system design as long as the previously mentioned inertia and speed restrictions are not violated.

SP9.6-1. Compare the value of maximum ω_{rm} and the time to maximum ω_{rm} (first zero crossing of T_e) in Figs. 9.5-2 and 9.6-1. [$\omega_{rm} \cong 8.5$ rad/s in 0.24 s, $\omega_{rm} \cong 7.9$ rad/s in 0.23 s]

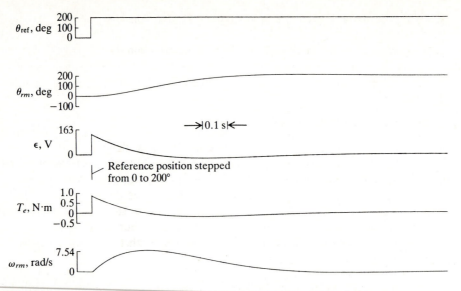

FIGURE 9.6-1
Dynamic performance of positioning system following a step increase in reference position with $T_L = 0$; linear approximation of T_e versus speed characteristic of servomotor.

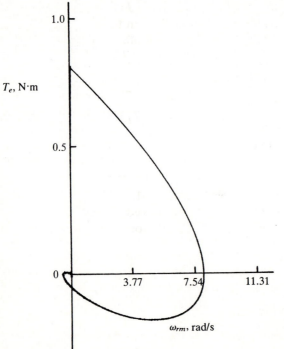

FIGURE 9.6-2
Torque versus speed during the change in reference position shown in Fig. 9.6-1.

SP9.6-2. Determine C_ω if ω_r is used in (9.6-4) rather than ω_{rm}. [$C_\omega = 2.84 \times 10^{-3}$ N·m·s/rad]

SP9.6-3. Relate ϵ and k in (9.4-1) for steady-state operation. [$\epsilon = k\sqrt{2}$ 115]

9.7 TIME-DOMAIN BLOCK DIAGRAM AND STATE EQUATIONS

It is instructive to develop the time-domain block diagram and the state equations for the linearized approximation of the T_e versus speed characteristics of the servomotor. We have talked about time-domain block diagrams and state equations before (Secs. 3.6 and 7.9). Therein we defined these terms in detail. We shall not repeat that information here since it will take the interested reader only a moment to refer back to either Sec. 3.6 or 7.9.

The equations that describe the servomotor, with the linearized approximation for T_e, may be expressed as

$$T_e = C_t\epsilon - C_\omega \omega_{rm} \tag{9.7-1}$$

$$T_e - T_L = (B_m + Jp)\omega_{rm} \tag{9.7-2}$$

$$\omega_{rm} = \tfrac{1}{180}\pi\, p\theta_{rm} \tag{9.7-3}$$

where p is the operator d/dt. Note that (9.7-1) is (9.6-1) and (9.7-2) is the familiar relationship between torque and speed, where J is the inertia and B_m is the damping coefficient. Note also that we have chosen to use ω_{rm} and θ_{rm}, in degrees, rather than ω_r and θ_r. First, we will substitute T_e, given by (9.7-1), into (9.7-2) and solve for ω_{rm}. Next, we will solve (9.7-3) for θ_{rm}. Thus, the equations for the time-domain block diagram are

$$\omega_{rm} = \frac{1}{Jp + B_m}\,(C_t\epsilon - C_\omega\omega_{rm} - T_L) \tag{9.7-4}$$

$$\theta_{rm} = \frac{1}{p}\,\frac{180}{\pi}\,\omega_{rm} \tag{9.7-5}$$

The time-domain block diagram is given in Fig. 9.7-1.

The formulation of the state equations is achieved by solving (9.7-2) and (9.7-3) for $p\omega_{rm}$ and $p\theta_{rm}$, respectively. Thus, from (9.7-2) with (9.7-1) substituted for T_e,

$$p\omega_{rm} = \left(-\frac{B_m}{J} - \frac{C_\omega}{J}\right)\omega_{rm} + \frac{C_t}{J}\,\epsilon - \frac{1}{J}\,T_L \tag{9.7-6}$$

$$p\theta_{rm} = \frac{180}{\pi}\,\omega_{rm} \tag{9.7-7}$$

The state equations in matrix or vector matrix form become

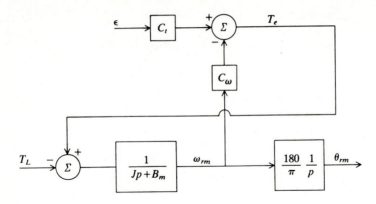

FIGURE 9.7-1
Time-domain block diagram of linearized approximation of a positioning servomotor.

$$p \begin{bmatrix} \omega_{rm} \\ \theta_{rm} \end{bmatrix} = \begin{bmatrix} -\dfrac{B_m}{J} & -\dfrac{C_\omega}{J} & 0 \\ \dfrac{180}{\pi} & & 0 \end{bmatrix} \begin{bmatrix} \omega_{rm} \\ \theta_{rm} \end{bmatrix} + \begin{bmatrix} \dfrac{C_t}{J} & -\dfrac{1}{J} \\ 0 & 0 \end{bmatrix} \begin{bmatrix} \epsilon \\ T_L \end{bmatrix} \qquad (9.7\text{-}8)$$

SP9.7-1. Explain how to incorporate the positioning control into (9.7-8). [Replace ϵ with (9.5-3)]
SP9.7-2. Determine $\mathbf{u}$ for SP9.7-1. [$\mathbf{u}^T = (\theta_{\text{ref}}, T_L)$]

9.8 SINGLE-PHASE INDUCTION MOTORS

In Chap. 4 we talked briefly about single-phase induction motors. It would probably be worthwhile to review this material in Sec. 4.6 before proceeding with this section. Although we will find that we must provide some means of starting the device, the single-phase induction motor has only one stator winding energized during normal operation. With this in mind, let us calculate the steady-state torque versus speed characteristics with voltage applied to only one stator winding of a symmetrical two-phase induction motor with the other winding open-circuited. Recall that we have already derived the voltage equations necessary to make these calculations. In particular, (9.3-46) can be used to determine the sequence currents with the *as* winding open-circuited and a voltage source connected to the *bs* winding. Once these calculations are made, the sequence currents may be substituted into (9.3-17) to determine the average and pulsating components of the steady-state electromagnetic torque.

The steady-state torque versus speed characteristics are shown in Fig. 9.8-1 for a symmetrical two-phase induction motor with rated voltage applied to one phase and the other phase open-circuited. The symmetrical two-phase

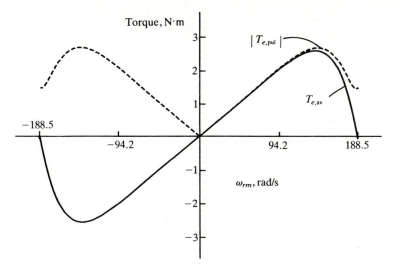

FIGURE 9.8-1
Steady-state torque versus speed characteristics for single-phase induction motor.

induction machine is a four-pole $\frac{1}{4}$-hp 110-V 60-Hz motor with the following parameters: $r_s = 2.02\ \Omega$, $X_{ls} = 2.79\ \Omega$, $X_{ms} = 66.8\ \Omega$, $r'_r = 4.12\ \Omega$, $X'_{lr} = 2.12\ \Omega$. The total inertia is $J = 1.46 \times 10^{-2}\ \text{kg} \cdot \text{m}^2$.

The average steady-state electromagnetic torque $T_{e,av} = T_{e+} - T_{e-}$ and the magnitude of the double-frequency component of the torque $|T_{e,\text{pul}}|$ are plotted in Fig. 9.8-1. There are at least two features worth mentioning. First, the average plot of the torque $T_{e,av}$ for $\omega_{rm} < 0$ is the negative mirror image of that for $\omega_{rm} > 0$. Secondly, the plot of the pulsating torque $|T_{e,\text{pul}}|$ is symmetrical about the zero speed axis. Finally, we see verification of our earlier claim that the starting torque is zero; $T_e = 0$ at $\omega_{rm} = 0$.

SP9.8-1. Determine the frequency of the rotor currents when ω_{rm} is equal to synchronous speed in Fig. 9.8-1. [120 Hz]

SP9.8-2. For the torque-speed characteristics shown in Fig. 9.8-1, determine the approximate rotor speed at which the steady-state instantaneous torque first pulsates to a negative value. [$\omega_{rm} \cong 1000\ \text{r/min}$]

9.9 CAPACITOR-START INDUCTION MOTOR

As we know, the single-phase induction motor will not develop a starting torque since two equal and oppositely rotating air gap mmf's are generated by a sinusoidal winding current. If, now, we take a two-phase symmetrical induction motor and apply the same single-phase voltage to both phases, the net torque at stall will still be zero since the winding currents will be instantaneously equal and the air gap mmf will pulsate along an axis midway between the *as* and *bs* axes. Consequently, two equal and oppositely rotating air gap mmf's again result. If, however, we cause the current in one of the phases to be different instantaneously from that in the other phase, a starting torque can be developed since this would cause one of the rotating air gap mmf's to be larger than the other. One way of doing this is to place a capacitor in series with one of the windings of a two-phase symmetrical induction motor. This will cause the current in the phase with the series capacitor to lead the current in the other winding when the same voltage is applied to both.

We have already derived the equations necessary to calculate the component currents with an impedance in series with the *as* winding. In particular, (9.3-33) can be used to calculate the component currents with a capacitor in series with the *as* winding. In (9.3-33), we set $Z = -j1/\omega_e C$ and let $\tilde{E}_{ga} = \tilde{E}_{gb}$, the single-phase source voltage. Counterclockwise rotation will occur since $\tilde{I}_{as}$ will lead $\tilde{I}_{bs}$. Note that, for the assumed positive direction of the magnetic axes and for a balanced two-phase set, we have had $\tilde{I}_{as}$ leading $\tilde{I}_{bs}$ by 90° for counterclockwise rotation of the air gap mmf.

Once the component currents are calculated, (9.3-17) can be used to determine the average steady-state electromagnetic torque $T_{e,av}$ and the magnitude of the double-frequency component $|T_{e,pul}|$. These steady-state torque versus speed characteristics are shown in Fig. 9.9-1 for $C = 530.5$ μF.

In capacitor-start single-phase induction motors, the winding with the series capacitor is disconnected from the source after the rotor has reached 60 to 80 percent of synchronous speed. This is generally accomplished by a centrifugal switching mechanism located inside the housing of the motor. Once

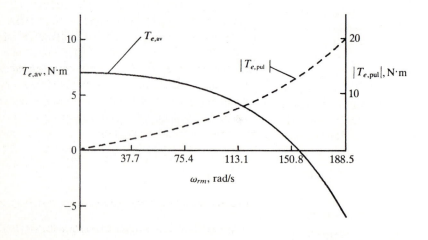

FIGURE 9.9-1
Steady-state torque versus speed characteristics with a capacitor in series with one winding of the two-phase induction machine.

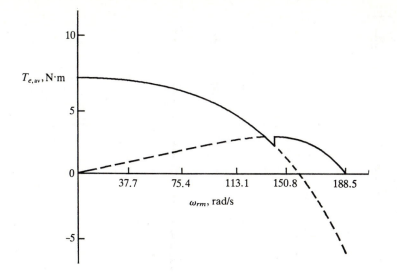

FIGURE 9.9-2
Average steady-state torque versus speed characteristics of a capacitor-start single-phase induction motor.

the winding with the series capacitor is disconnected, the device then operates as a single-phase induction motor. In Fig. 9.9-2, the plot of average torque versus speed with a series capacitor in one phase (Fig. 9.9-1) is superimposed upon the plot of average torque versus speed with a single-phase winding (Fig. 9.8-1). The transition from capacitor-start to single-phase operation at 75 percent of synchronous speed is illustrated.

Although the capacitor-start single-phase induction motor is by far the most common type of single-phase induction motor, a capacitor-start capacitor-run induction motor is sometimes used. In this case, both phase windings are energized during normal operation. The value of the series capacitance is changed from the start value to the run value once the rotor reaches 60 to 80 percent of synchronous speed. This is accomplished using two capacitors connected in parallel with provision to open-circuit one of the parallel paths. The purpose of the run capacitor is to establish a leading current during normal loads, thereby increasing the torque capability over that which is possible with only one stator winding energized. Since two capacitors are needed, this device is somewhat more expensive and often the application does not justify this added cost.

SP9.9-1. In Fig. 9.9-2 the device switches from capacitor-start to single-phase operation at a rotor speed of 75 percent of synchronous speed. Will the rotor accelerate faster or slower immediately following the switching? [Faster]

SP9.9-2. If the value of the capacitor was decreased, would you expect the starting torque to decrease or increase? Why? [Decrease, less leading component of current]

9.10 DYNAMIC AND STEADY-STATE PERFORMANCE OF A CAPACITOR-START SINGLE-PHASE INDUCTION MOTOR

The free-acceleration characteristics of the example capacitor-start single-phase induction motor are shown in Fig. 9.10-1. The variables v_{as}, i_{as}, v_{bs}, i_{bs}, v_c, T_e, and ω_{rm} are plotted. The voltage v_c is the instantaneous voltage across the capacitor which is connected in series with the bs winding. The machine variables are shown with an expanded scale in Fig. 9.10-2 to illustrate the switching out of the bs winding, which is disconnected from the source at a normal current zero once the rotor reaches 75 percent of synchronous speed.

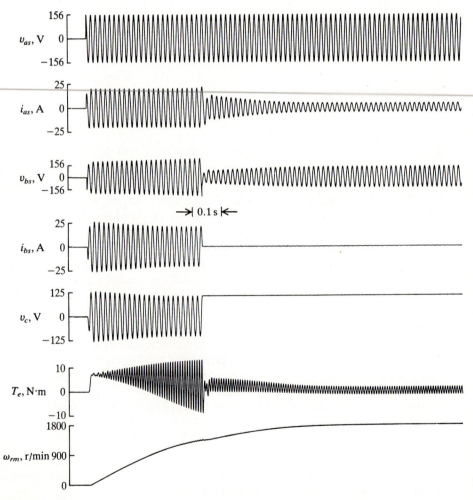

FIGURE 9.10-1
Free-acceleration characteristics of a capacitor-start single-phase induction motor.

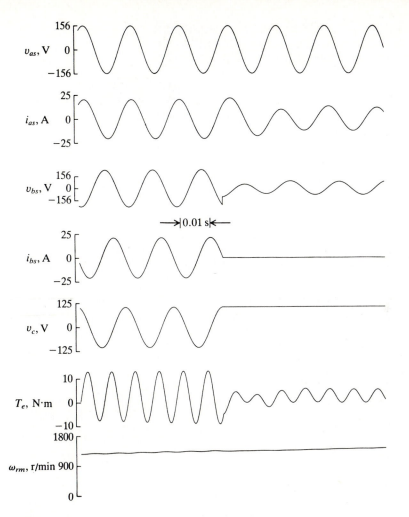

FIGURE 9.10-2
Expanded plot of Fig. 9.10-1 illustrating the disconnecting of the capacitor and *bs* winding.

The voltage across the capacitor is shown to remain constant at its value when the *bs* winding is disconnected from the source. In practice, this voltage would slowly decay owing to leakage currents within the capacitor which are not considered in this analysis. The torque versus speed characteristics given in Fig. 9.10-3 are for the free acceleration shown in Fig. 9.10-1. The dynamic and steady-state characteristics following changes in load torque are illustrated in Fig. 9.10-4. Therein v_{as}, i_{as}, v_{bs} (open-circuited), i'_{ar}, T_e, ω_{rm}, and T_L are plotted.

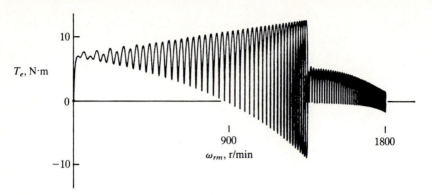

FIGURE 9.10-3
Torque versus speed characteristics for Fig. 9.10-1.

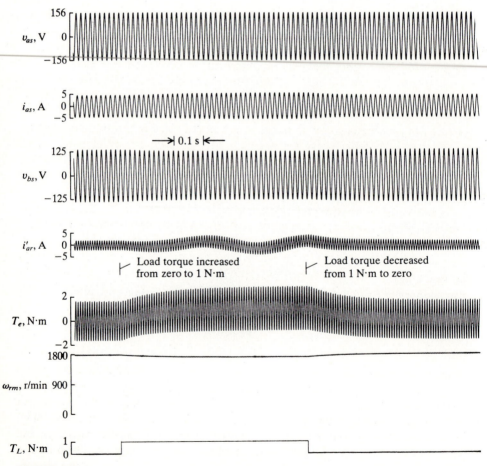

FIGURE 9.10-4
Step changes in load torque of single-phase induction motor.

SP9.10-1. In Fig. 9.10-1, the capacitor is in series with the *bs* winding and the voltage applied to the *as* winding is $V_{as} = \sqrt{2}\, 110 \cos \omega_e t$. Calculate the steady-state stator currents I_{as} and I_{bs} at stall ($\omega_{rm} = 0$). Compare with the traces of i_{as} and i_{bs} in Fig. 9.10-1. Neglect the magnetizing reactance X_{ms} in these calculations. [$I_{as} \cong 19.8 \cos (377t - 38.6°)$; $I_{bs} \cong -25.3 \cos (377t - 0.8°)$]

SP9.10-2. Determine the frequency of $I_{qr}^{'s}$ and $I_{dr}^{'s}$ for the loaded condition ($T_L = 1\,\text{N} \cdot \text{m}$) in Fig. 9.10-4. [60 Hz]

9.11   SPLIT-PHASE INDUCTION MOTOR

Although we have considered only the symmetrical two-phase induction motor as a single-phase induction motor, it is probably used less widely than the split-phase induction motor. The split-phase induction motor is an unsymmetrical two-phase induction machine. That is, the stator windings are different. The main or run winding remains energized during normal operation while the start or auxiliary winding is switched out after the rotor reaches 60 to 80 percent of synchronous speed. The *r* to *X* ratio of the run winding would be much the same as that of the stator windings of a two-phase symmetrical machine; however, the start winding has a higher *r* to *X* ratio. Hence, with the same voltage applied to the start and run windings, the current flowing in the start winding would lead the current flowing in the run winding. We see the logic behind all of this. Rather than using only a capacitor to shift the phase of one of the winding currents in order to develop a starting torque, the machine is designed with different stator windings so that one current leads the other due to the difference in the winding impedances. Depending on the application and the design of the machine, a capacitor may or may not be used in series with the start winding.

We will not analyze the split-phase induction machine. The analysis is rather involved since the mutual inductances between the rotor windings and the run winding are different from those between the rotor windings and the start winding. Actually we have established the main operating characteristics of single-phase induction motors with the least amount of effort by considering the symmetrical two-phase machine. If, however, one wishes to consider the split-phase device in more detail, this analysis is given in [3].

9.12   RECAPPING

Even though the analysis of unbalanced steady-state operation of a symmetrical two-phase induction machine is rather involved, it is the easiest of all electromechanical devices to analyze in the unbalanced mode of operation. We have taken advantage of this feature to provide a "first look" at the method of symmetrical components to analyze unbalanced modes of operation which not only occur in the operation of the servomotor and a single-phase machine but also in three-phase devices.

The dynamic characteristic of the positioning servomotor along with the time-domain block diagram and state equations of the linear, simplified representation of the servomotor when used to position provide a meaningful introduction to electromechanical motion control. This material is presented primarily for the student interested in control systems.

Although the split-phase induction motor was not analyzed, the application of the symmetrical two-phase induction motor as a single-phase device is sufficient coverage of the single-phase induction motor for our purposes. Even with this rather brief treatment of single-phase induction machines, we have been able to introduce the major operating features of most present-day single-phase induction motors.

9.13 REFERENCES

1. C. L. Fortescue, "Method of Symmetrical Co-ordinates Applied to the Solution of Polyphase Networks," *AIEE Trans.*, vol. 37, 1918, pp. 1027–1115.
2. P. C. Krause, "Method of Symmetrical Components Derived by Reference Frame Theory," *IEEE Trans.*, *Power Apparatus and Systems*, vol. 104, June 1985, pp. 1492–1499.
3. P. C. Krause, *Analysis of Electric Machinery*, McGraw-Hill Book Company, New York, 1986.

9.14 PROBLEMS

1. Calculate $\tilde{F}^s_{qs+}$ and $\tilde{F}^s_{qs-}$ for the following sets using (9.2-9).
   (a) $\tilde{F}_{as} = 10\underline{/30°}$, $\tilde{F}_{bs} = 30\underline{/-60°}$.
   (b) $\tilde{F}_{as} = 10\underline{/0°}$, $\tilde{F}_{bs} = 0$.
   (c) $F_{as} = \cos(\omega_e t + 45°)$, $F_{bs} = \cos(\omega_e t - 45°)$.
*2. Repeat Prob. 1 using (9.2-35) and (9.2-37).
3. Start with (9.3-5) and derive (9.3-6).
4. Derive (9.3-16).

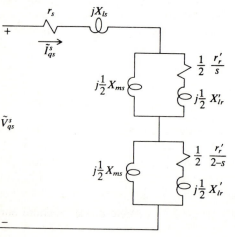

FIGURE 9.14-1
Equivalent circuit for single-phase stator winding.

*5. Show that (9.3-16) and (9.3-17) are equivalent.

6. Express (9.3-46) with $\tilde{I}^s_{qr+}$ and $\tilde{I}'^s_{qr-}$ eliminated.

*7. By using information from SP9.5-3 determine I_{as}, I_{bs}, I'_{ar} and I'_{br} for the steady-state mode of operation in Fig. 9.5-4. Do not neglect the current flowing in X_{ms}.

8. Use the plots of $T_{e,av}$ in Fig. 9.4-4 for $k = 1$ and $k = 0$ for $\omega_{rm} = 188.5\,\text{rad/s}$ to approximate $f(\epsilon)$ in $T_e = C_t\epsilon - f(\epsilon)\omega_{rm}$. Use C_t given by (9.6-2).

9. Develop the time-domain block diagram and the state equations when the positioning control described by (9.5-3) is incorporated with the linear approximation of the servomotor derived in Section 9.6.

*10. The equivalent circuit for steady-state operation of an induction motor with only one stator winding is shown in Fig. 9.19-1. Show that this equivalent circuit is the same as that given by (9.3-46).

APPENDIX
A

ABBREVIATIONS, CONSTANTS AND CONVERSION FACTORS, TRIGONOMETRIC IDENTITIES, AND LAPLACE TRANSFORM PAIRS

Abbreviations

alternating current	ac
ampere	A
ampere-turn	At
coulomb	C
direct current	dc
foot	ft
gauss	G
gram	g
henry	H
hertz	Hz
horsepower	hp
inch	in
joule	J

kilogram	kg
kilovar	kvar
kilovolt	kV
kilovoltampere	kVA
kilowatt	kW
magnetomotive force	mmf
maxwell	Mx
megawatt	MW
meter	m
microfarad	μF
millihenry	mH
newton	N
newton meter	N $\cdot$ m
oersted	Oe
pound	lb
poundal	pdl
power factor	pf
pulse width modulation	PWM
radian	rad
revolution per minute	r/min (rpm)
second	s
voltampere reactive	var
volt	V
voltampere	VA
watt	W
weber	Wb

Constants and Conversion Factors

permeability of free space	$\mu_0 = 4\pi \times 10^{-7}$ Wb/A $\cdot$ m
permittivity (capacitivity) of free space	$\epsilon_0 = 8.854 \times 10^{-12}$ C^2/N $\cdot$ m^2
acceleration of gravity	$g = 9.807$ m/s^2
length	1 m = 3.218 ft = 39.37 in
mass	1 kg = 0.0685 slug = 2.205 lb (mass)
force	1 N = 0.225 lb = 3.6 oz
torque	1 N $\cdot$ m = 0.738 lb $\cdot$ ft
energy	1 J (W $\cdot$ s) = 0.738 lb $\cdot$ ft
power	1 W = 1.341 $\times$ 10^{-3} hp
moment of inertia	1 kg $\cdot$ m^2 = 0.738 slug $\cdot$ ft^2
	= 23.7 lb $\cdot$ ft^2
magnetic flux	1 Wb = 10^8 Mx (lines)
magnetic flux density	1 Wb/m^2 = 10,000 G
	= 64.5 klines/in^2
magnetizing force	1 At/m = 0.0254 At/in =
	0.0126 Oe

Trigonometric Identities

$e^{j\alpha} = \cos \alpha + j \sin \alpha$

$a \cos x + b \sin x = \sqrt{a^2 + b^2} \cos (x + \phi)$ $\phi = \tan^{-1} (-b/a)$

$\cos^2 x + \sin^2 x = 1$

$\sin 2x = 2 \sin x \cos x$

$\cos 2x = \cos^2 x - \sin^2 x = 2 \cos^2 x - 1 = 1 - 2 \sin^2 x$

$\cos x \cos y = \frac{1}{2} \cos (x + y) + \frac{1}{2} \cos (x - y)$

$\sin x \sin y = \frac{1}{2} \cos (x - y) - \frac{1}{2} \cos (x + y)$

$\sin x \cos y = \frac{1}{2} \sin (x + y) + \frac{1}{2} \sin (x - y)$

$\cos (x \pm y) = \cos x \cos y \mp \sin x \sin y$

$\sin (x \pm y) = \sin x \cos y \pm \cos x \sin y$

$\cos^2 x + \cos^2 (x - \frac{2}{3}\pi) + \cos^2 (x + \frac{2}{3}\pi) = \frac{3}{2}$

$\sin^2 x + \sin^2 (x - \frac{2}{3}\pi) + \sin^2 (x + \frac{2}{3}\pi) = \frac{3}{2}$

$\sin x \cos x + \sin (x - \frac{2}{3}\pi) \cos (x - \frac{2}{3}\pi) + \sin (x + \frac{2}{3}\pi) \cos (x + \frac{2}{3}\pi) = 0$

$\cos x + \cos (x - \frac{2}{3}\pi) + \cos (x + \frac{2}{3}\pi) = 0$

$\sin x + \sin (x - \frac{2}{3}\pi) + \sin (x + \frac{2}{3}\pi) = 0$

$\sin x \cos y + \sin (x - \frac{2}{3}\pi) \cos (y - \frac{2}{3}\pi) + \sin (x + \frac{2}{3}\pi) \cos (y + \frac{2}{3}\pi)$
$\quad = \frac{3}{2} \sin (x - y)$

$\sin x \sin y + \sin (x - \frac{2}{3}\pi) \sin (y - \frac{2}{3}\pi) + \sin (x + \frac{2}{3}\pi) \sin (y + \frac{2}{3}\pi)$
$\quad = \frac{3}{2} \cos (x - y)$

$\cos x \sin y + \cos (x - \frac{2}{3}\pi) \sin (y - \frac{2}{3}\pi) + \cos (x + \frac{2}{3}\pi) \sin (y + \frac{2}{3}\pi)$
$\quad = -\frac{3}{2} \sin (x - y)$

$\cos x \cos y + \cos (x - \frac{2}{3}\pi) \cos (y - \frac{2}{3}\pi) + \cos (x + \frac{2}{3}\pi) \cos (y + \frac{2}{3}\pi)$
$\quad = \frac{3}{2} \cos (x - y)$

$\sin x \cos y + \sin (x + \frac{2}{3}\pi) \cos (y - \frac{2}{3}\pi) + \sin (x - \frac{2}{3}\pi) \cos (y + \frac{2}{3}\pi)$
$\quad = \frac{3}{2} \sin (x + y)$

$\sin x \sin y + \sin (x + \frac{2}{3}\pi) \sin (y - \frac{2}{3}\pi) + \sin (x - \frac{2}{3}\pi) \sin (y + \frac{2}{3}\pi)$
$\quad = -\frac{3}{2} \cos (x + y)$

$\cos x \sin y + \cos (x + \frac{2}{3}\pi) \sin (y - \frac{2}{3}\pi) + \cos (x - \frac{2}{3}\pi) \sin (y + \frac{2}{3}\pi)$
$\quad = \frac{3}{2} \sin (x + y)$

$$\cos x \cos y + \cos \left(x + \tfrac{2}{3}\pi\right) \cos \left(y - \tfrac{2}{3}\pi\right) + \cos \left(x - \tfrac{2}{3}\pi\right) \cos \left(y + \tfrac{2}{3}\pi\right)$$
$$= \tfrac{3}{2} \cos (x + y)$$

Laplace Transform Pairs

$f(t) = \mathcal{L}^{-1}\{F(s)\}$	$F(s) = \mathcal{L}\{f(t)\}$
$\delta(t)$	1
$u(t)$	$\dfrac{1}{s}$
$tu(t)$	$\dfrac{1}{s^2}$
$\dfrac{t^{n-1}}{(n-1)!}\, u(t), \; n = 1, 2, \ldots$	$\dfrac{1}{s^n}$
$e^{-\alpha t} u(t)$	$\dfrac{1}{s + \alpha}$
$te^{-\alpha t} u(t)$	$\dfrac{1}{(s + \alpha)^2}$
$\dfrac{t^{n-1}}{(n-1)!}\, e^{-\alpha t} u(t), \; n = 1, 2, \ldots$	$\dfrac{1}{(s + \alpha)^n}$
$\dfrac{1}{\beta - \alpha}\left(e^{-\alpha t} - e^{-\beta t}\right) u(t)$	$\dfrac{1}{(s + \alpha)(s + \beta)}$
$\sin \omega t\, u(t)$	$\dfrac{\omega}{s^2 + \omega^2}$
$\cos \omega t\, u(t)$	$\dfrac{s}{s^2 + \omega^2}$
$\sin (\omega t + \theta) u(t)$	$\dfrac{s \sin \theta + \omega \cos \theta}{s^2 + \omega^2}$
$\cos (\omega t + \theta) u(t)$	$\dfrac{s \cos \theta - \omega \sin \theta}{s^2 + \omega^2}$
$e^{-\alpha t} \sin \omega t\, u(t)$	$\dfrac{\omega}{(s + \alpha)^2 + \omega^2}$
$e^{-\alpha t} \cos \omega t\, u(t)$	$\dfrac{s + \alpha}{(s + \alpha)^2 + \omega^2}$

APPENDIX
B

PHASORS AND PHASOR DIAGRAMS

The concept of the phasor is quite convenient in the analysis of balanced steady-state operation of ac electromechanical devices. Therefore, it is important to be familiar with phasor theory. For this purpose let a steady-state sinusoidal variable be expressed as

$$F_a = \sqrt{2}F \cos \theta_{ef} \qquad (B\text{-}1)$$

where capital letters are used to denote steady-state quantities and F is the rms value of the sinusoidal variation. In the text the subscript s or r is added to denote variables associated with the stator or rotor, respectively. In (B-1),

$$\theta_{ef} = \int_0^t \omega_e(\xi)\, d\xi + \theta_{ef}(0) \qquad (B\text{-}2)$$

where ω_e is the electrical angular velocity and ξ is a dummy variable of integration. For steady-state conditions, (B-2) may be written as

$$\theta_{ef} = \omega_e t + \theta_{ef}(0) \qquad (B\text{-}3)$$

Substituting (B-3) into (B-1) yields

$$F_a = \sqrt{2}F \cos \left[\omega_e t + \theta_{ef}(0) \right] \qquad (B\text{-}4)$$

We know that

$$e^{j\alpha} = \cos \alpha + j \sin \alpha \qquad (B\text{-}5)$$

Thus, (B-4) may also be written as

$$F_a = \text{Re}\,[\sqrt{2}Fe^{j[\omega_e t + \theta_{ef}(0)]}] \tag{B-6}$$

where Re is shorthand for the "real part of." Equations (B-4) and (B-6) are identical. We can rewrite (B-6) as

$$F_a = \text{Re}\,[\sqrt{2}Fe^{j\theta_{ef}(0)}e^{j\omega_e t}] \tag{B-7}$$

By definition, the phasor representing F_a is

$$\tilde{F}_a = Fe^{j\theta_{ef}(0)} \tag{B-8}$$

which is a complex number. Equation (B-7) may now be written as

$$F_a = \text{Re}\,[\sqrt{2}\tilde{F}_a e^{j\omega_e t}] \tag{B-9}$$

A shorthand notation for (B-8) is

$$\tilde{F}_a = F\underline{/\theta_{ef}(0)} \tag{B-10}$$

Equation (B-10) is commonly referred to as the *polar form* of the phasor. The *cartesian form* is

$$\tilde{F}_a = F\cos\theta_{ef}(0) + jF\sin\theta_{ef}(0) \tag{B-11}$$

When using phasors to calculate steady-state voltages and currents we think of the phasors as being stationary at $t = 0$. On the other hand, a phasor is related to the instantaneous value of the sinusoidal quantity it represents. Let us take a moment to consider this aspect of the phasor and, thereby, give some physical meaning to it. We know from analytic geometric that

$$e^{j\omega_e t} = \cos\omega_e t + j\sin\omega_e t \tag{B-12}$$

is a constant-amplitude line of unity length rotating counterclockwise at an angular velocity of ω_e. Now,

$$\sqrt{2}\tilde{F}_{as}e^{j\omega_e t} = \sqrt{2}Fe^{j\theta_{ef}(0)}e^{j\omega_e t}$$
$$= \sqrt{2}Fe^{j[\omega_e t + \theta_{ef}(0)]}$$
$$= \sqrt{2}F\{\cos[\omega_e t + \theta_{ef}(0)] + j\sin[\omega_e t + \theta_{ef}(0)]\} \tag{B-13}$$

is a constant-amplitude line $\sqrt{2}F$ in length rotating counterclockwise at an angular velocity of ω_e with a time zero displacement from the positive real axis of $\theta_{ef}(0)$. The instantaneous value of F_a is the real part of (B-13). In other words, the real projection of the phasor $\tilde{F}_a$ is the instantaneous value of $(1/\sqrt{2})F_a$ at time zero. As time progresses, $\tilde{F}_a$ rotates at ω_e in the counterclockwise direction, and its real projection, in accordance with (B-9), is the instantaneous value of $(1/\sqrt{2})F_a$. Thus, for

$$F_a = \sqrt{2}F\cos\omega_e t \tag{B-14}$$

the phasor representing F_a is

$$\tilde{F}_a = Fe^{j0} = F\underline{/0°} = F + j0 \qquad \text{(B-15)}$$

For
$$F_a = \sqrt{2}F \cos\left(\omega_e t + \tfrac{1}{6}\pi\right) \qquad \text{(B-16)}$$

the phasor is

$$\tilde{F}_a = Fe^{j\pi/6} = F\underline{/30°} = F(0.866 + j0.5) \qquad \text{(B-17)}$$

Finally, for

$$F_a = \sqrt{2}F \sin \omega_e t \qquad \text{(B-18)}$$

the phasor is

$$\tilde{F}_a = Fe^{-j\pi/2} = F\underline{/-90°} = 0 - jF \qquad \text{(B-19)}$$

Although there are several ways to arrive at (B-19) from (B-18), Is it helpful to ask yourself where must the rotating phasor be positioned at time zero so that, when it rotates counterclockwise at ω_e, its real projection is $(1/\sqrt{2})F \sin \omega_e t$? Is it clear that a phasor of amplitude F positioned at $\tfrac{1}{2}\pi$ would represent $-\sqrt{2}F \sin \omega_e t$?

It is often instructive to be able to construct a phasor diagram. For example, let us consider a voltage equation of the form

$$\tilde{V} = (r + jX)\tilde{I} + \tilde{E} \qquad \text{(B-20)}$$

where r is the resistance and X is the reactance. In most cases we will deal with an inductive reactance; however, in a series LC circuit,

$$X = X_L + X_C = \omega_e L + \frac{-1}{\omega_e C} \qquad \text{(B-21)}$$

where L is the inductance and C is the capacitance. The inductive reactance is X_L and X_C is the capacitive reactance. Let us assume that $\tilde{V}$ and $\tilde{I}$ are known and that we are to calculate $\tilde{E}$. The phasor diagram may be used as a rough check on these calculations. Let us construct this phasor diagram by assuming that X is equal to X_L (or $|X_L| > |X_C|$) and $\tilde{V}$ and $\tilde{I}$ are known as shown in Fig. B-1. Solving (B-20) for $\tilde{E}$ yields

$$\tilde{E} = \tilde{V} - (r + jX)\tilde{I} \qquad \text{(B-22)}$$

To perform this graphically, start at the origin in Fig. B-1 and walk to the terminus of $\tilde{V}$. Now, we want to subtract $r\tilde{I}$. To achieve the proper orientation to do this, stand at the terminus of $\tilde{V}$, turn, and look in the $\tilde{I}$ direction which is at the angle ϕ. But we must subtract $r\tilde{I}$; hence, $-\tilde{I}$ is 180° from $\tilde{I}$, so do an about face and now we are headed in the $-\tilde{I}$ direction which is $\phi - 180°$. Start walking in the direction of $-\tilde{I}$ for the distance $r|\tilde{I}|$ and then stop. While still facing in the $-\tilde{I}$ direction, let us consider the next term. We must subtract $jX\tilde{I}$, so let us face in the direction of $-jX\tilde{I}$. We are still looking in the $-\tilde{I}$ direction, so we need only to j ourselves. Thus, we must rotate 90° in the counter-clockwise direction, whereupon we are standing at the end of $\tilde{V} - r\tilde{I}$ looking in

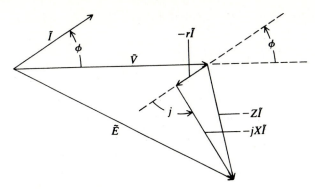

FIGURE B-1
Phasor diagram for (B-22).

the direction of $\phi - 180° + 90°$. Start walking in this direction for the distance of $X|\tilde{I}|$, whereupon we are at the terminus of $\tilde{V} - r\tilde{I} - jX\tilde{I}$. According to (B-22), $\tilde{E}$ is the phasor drawn from the origin of the phasor diagram to where we are. Note that in Fig. B-1 $-Z\tilde{I}$ is $-(r + jX)\tilde{I}$, where Z is the impedance.

The average steady-state power may be calculated by using phasors,

$$P = |\tilde{V}||\tilde{I}| \cos \phi_{pf} \qquad \text{(B-23)}$$

where the power is in watts and the so-called power factor angle is defined as

$$\phi_{pf} = \theta_{ev}(0) - \theta_{ei}(0) \qquad \text{(B-24)}$$

Here, $\tilde{V}$ and $\tilde{I}$ are phasors with the positive direction of $\tilde{I}$ taken in the direction of the voltage drop and $\theta_{ev}(0)$ and $\theta_{ei}(0)$ are the phase angles of $\tilde{V}$ and $\tilde{I}$, respectively.

The reactive power is defined as

$$Q = |\tilde{V}||\tilde{I}| \sin \phi_{pf} \qquad \text{(B-25)}$$

The units of Q are in var (voltampere reactive). An inductance is said to absorb reactive power and thus, by definition, Q is positive for an inductor and negative for a capacitor. Actually, Q is a measure of the exchange of energy stored in the electric (capacitor) and magnetic (inductance) fields; however, there is no average power interchanged between these energy storage devices.

APPENDIX
C

DEFINITION OF BALANCED SETS

Many electromechanical devices are supplied from a balanced two- or three-phase sinusoidal voltage source. Therefore, it is important to be familiar with the properties of balanced sets. By definition, a two-phase set of variables is balanced if the variables are equal-amplitude sinusoidal quantities in time quadrature (90° out of phase). A three-phase set of variables is balanced if the variables are equal-amplitude quantities which are 120° out of time phase with each other.

In the broadest sense of the above definition, two-phase balanced sets may be expressed as

$$f_a = \pm f(t) \cos \theta_{ef} \tag{C-1}$$

$$f_b = \pm f(t) \sin \theta_{ef} \tag{C-2}$$

where
$$\theta_{ef} = \int_0^t \omega_e(\xi) \, d\xi + \theta_{ef}(0) \tag{C-3}$$

In (C-1) and (C-2), f can represent voltage, current, flux linkage, or charge. The amplitude $f(t)$ may be any function of time. However, for steady-state balanced conditions $f(t)$ is a constant. In (C-3), ω_e is the electrical angular velocity and ξ is a dummy variable of integration. Equations (C-1) and (C-2) express four balanced two-phase sets. Like signs of (C-1) and (C-2) define balanced sets where f_a leads f_b by 90°; for unlike signs f_a lags f_b by 90°.

A three-phase balanced set may be expressed as

$$f_a = f(t) \cos \theta_{ef} \tag{C-4}$$

$$f_b = f(t) \cos (\theta_{ef} - \tfrac{2}{3}\pi) \tag{C-5}$$

$$f_c = f(t) \cos (\theta_{ef} + \tfrac{2}{3}\pi) \tag{C-6}$$

This set is referred to as an *abc* sequence, since f_a leads f_b by 120° and f_b leads f_c by 120°. An *acb* sequence is

$$f_a = f(t) \cos \theta_{ef} \tag{C-7}$$

$$f_b = f(t) \cos (\theta_{ef} + \tfrac{2}{3}\pi) \tag{C-8}$$

$$f_d = f(t) \cos (\theta_{ef} - \tfrac{2}{3}\pi) \tag{C-9}$$

For steady-state conditions (C-3) becomes

$$\theta_{ef} = \omega_e t + \theta_{ef}(0) \tag{C-10}$$

Whereupon (C-1) and (C-2) are written as

$$F_a = \pm\sqrt{2}F \cos [\omega_e t + \theta_{ef}(0)] \tag{C-11}$$

$$F_b = \pm\sqrt{2}F \sin [\omega_e t + \theta_{ef}(0)] \tag{C-12}$$

Likewise, (C-4) through (C-6) are written as

$$F_a = \sqrt{2}F \cos [\omega_e t + \theta_{ef}(0)] \tag{C-13}$$

$$F_b = \sqrt{2}F \cos [\omega_e t - \tfrac{2}{3}\pi + \theta_{ef}(0)] \tag{C-14}$$

$$F_c = \sqrt{2}F \cos [\omega_e t + \tfrac{2}{3}\pi + \theta_{ef}(0)] \tag{C-15}$$

where the capital letters are used to denote steady-state quantities and F is the rms value of the sinusoidal variation. When calculating steady-state voltages and currents by using phasors, we need to consider only one phase, since what happens in one phase of a three-phase balanced system happens in the other two phases 120° or 240° later.

APPENDIX
D

MATRIX ALGEBRA

Basic Definitions [1]

A rectangular array of numbers or functions

$$\mathbf{A} = \begin{bmatrix} a_{11} & a_{12} & \cdots & a_{1n} \\ a_{21} & a_{22} & \cdots & a_{2n} \\ \cdots\cdots\cdots\cdots\cdots\cdots\cdots \\ a_{m1} & a_{m2} & \cdots & a_{mn} \end{bmatrix} \tag{D-1}$$

is known as a *matrix* and is denoted in the text by capital boldface letters. The numbers or functions a_{ij} are the elements of the matrix and the subscript i denotes the row and j the column. A matrix with m rows and n columns is of *order* (m, n) or an $m \times n$ (m by n) matrix. If $m = n$, the matrix is a *square matrix*.

If a matrix is an $m \times 1$ matrix, it is a *column vector*. If it is a $1 \times n$ matrix, it is a *row vector*. Generally, lower-case boldface letters are used to denote column or row vectors.

A square matrix in which all elements are zero except those on the main diagonal, $a_{11}, a_{22}, \ldots, a_{nn}$, is a *diagonal matrix*. If all elements of a diagonal matrix are unity, then the matrix is the *identity matrix* and is denoted as $\mathbf{I}$.

When $a_{ij} = a_{ji}$, the matrix is called a *symmetrical matrix*. A *null* or *zero matrix* is one in which all elements are zero.

414

Addition and Subtraction

Two matrices can be added or subtracted only if they are of the same order. Thus, if $\mathbf{A}$ has elements a_{ij} and $\mathbf{B}$ has elements b_{ij}, then, if

$$\mathbf{C} = \mathbf{A} + \mathbf{B} \tag{D-2}$$

$\mathbf{C}$ has elements c_{ij}, where

$$c_{ij} = a_{ij} + b_{ij} \tag{D-3}$$

or if

$$\mathbf{C} = \mathbf{A} - \mathbf{B} \tag{D-4}$$

then

$$c_{ij} = a_{ij} - b_{ij} \tag{D-5}$$

Also, addition is commutative,

$$\mathbf{A} + \mathbf{B} = \mathbf{B} + \mathbf{A} \tag{D-6}$$

and associative,

$$(\mathbf{A} + \mathbf{B}) + \mathbf{C} = \mathbf{A} + (\mathbf{B} + \mathbf{C}) \tag{D-7}$$

Obviously,

$$\mathbf{A} + \mathbf{O} = \mathbf{A} \tag{D-8}$$

where $\mathbf{O}$ is the zero matrix.

Multiplication

If the matrix $\mathbf{A}$ is multiplied by a scalar, every element of the matrix is multiplied by the scalar. For example, $k\mathbf{A}$ means all elements of $\mathbf{A}$ are multiplied by the constant k; $t\mathbf{A}$ means that all elements of $\mathbf{A}$ are multiplied by time t.

To multiply two matrices, say $\mathbf{AB}$, it is necessary that the number of columns of $\mathbf{A}$ equal the numbers of rows of $\mathbf{B}$. If $\mathbf{A}$ is of order $m \times n$ and $\mathbf{B}$ of order $n \times p$, then the order of $\mathbf{AB}$ is $m \times p$. The elements of

$$\mathbf{C} = \mathbf{AB} \tag{D-9}$$

are obtained by multiplying the elements by the ith row of $\mathbf{A}$, by the corresponding elements of the jth column of $\mathbf{B}$, and adding these products. In particular,

$$c_{ij} = a_{i1}b_{1j} + a_{i2}b_{2j} + \cdots + a_{in}b_{nj} \tag{D-10}$$

In (D-9), $\mathbf{A}$ is said to *premultiply* $\mathbf{B}$ whereas $\mathbf{B}$ is said to *postmultiply* $\mathbf{A}$. Let

$$\mathbf{A} = \begin{bmatrix} 1 & 2 & 3 \\ 4 & 5 & 6 \end{bmatrix} \tag{D-11}$$

and

$$\mathbf{B} = \begin{bmatrix} -7 & -8 \\ 9 & 10 \\ 0 & -11 \end{bmatrix} \tag{D-12}$$

Then
$$AB = \begin{bmatrix} 11 & -21 \\ 17 & -48 \end{bmatrix} \tag{D-13}$$

However,
$$BA = \begin{bmatrix} -39 & -54 & -69 \\ 49 & 68 & 87 \\ -44 & -55 & -66 \end{bmatrix} \tag{D-14}$$

We see that, in general, matrix multiplication is not commutative; thus,

$$AB \neq BA \tag{D-15}$$

Multiplying a matrix of $m \times n$ by a column vector of $n \times 1$ yields a column vector of $m \times 1$. Multiplication of a row vector of $1 \times n$ and a column vector of $n \times 1$ yields a function (scalar) which is the sum of the product of specific elements of each vector.

Multiplying the identity matrix by A yields A, that is,

$$AI = IA = A \tag{D-16}$$

We can show that

$$A(BC) = (AB)C \tag{D-17}$$

$$A(B + C) = AB + AC \tag{D-18}$$

$$(B + C)A = BA + CA \tag{D-19}$$

Finally, consider the simultaneous linear equations

$$5x + 3y - 2z = 14 \tag{D-20}$$

$$x + y - 4z = -7 \tag{D-21}$$

$$6x + 3z = 1 \tag{D-22}$$

Let us write the above equations in the form

$$Ax = b \tag{D-23}$$

Here
$$A = \begin{bmatrix} 5 & 3 & -2 \\ 1 & 1 & -4 \\ 6 & 0 & 3 \end{bmatrix} \tag{D-24}$$

$$x = \begin{bmatrix} x \\ y \\ z \end{bmatrix} \tag{D-25}$$

$$b = \begin{bmatrix} 14 \\ -7 \\ 1 \end{bmatrix} \tag{D-26}$$

In this case, A is called the *coefficient matrix*. Care must be taken not to confuse the column vector x and the variable x.

Transpose

The *transpose* of a matrix $\mathbf{A}$ is denoted as $\mathbf{A}^T$. The transpose of $\mathbf{A}$ is obtained by interchanging the rows and columns of $\mathbf{A}$. Thus, if

$$\mathbf{A} = \begin{bmatrix} 1 & 2 & 3 & 4 \\ 5 & 6 & 7 & 8 \end{bmatrix} \tag{D-27}$$

$$\mathbf{A}^T = \begin{bmatrix} 1 & 5 \\ 2 & 6 \\ 3 & 7 \\ 4 & 8 \end{bmatrix} \tag{D-28}$$

The transpose possesses the following properties:

$$(\mathbf{A}^T)^T = \mathbf{A} \tag{D-29}$$

$$(\mathbf{A} + \mathbf{B} + \mathbf{C})^T = \mathbf{A}^T + \mathbf{B}^T + \mathbf{C}^T \tag{D-30}$$

$$(\mathbf{AB})^T = \mathbf{B}^T \mathbf{A}^T \tag{D-31}$$

$$(\mathbf{ABC})^T = \mathbf{C}^T \mathbf{B}^T \mathbf{A}^T \tag{D-32}$$

Partitioning

Partitioning of matrices is used throughout the text. It is helpful in matrix multiplication. For example, let $\mathbf{A}$ and $\mathbf{B}$ be partitioned as

$$\mathbf{A} = \begin{bmatrix} \mathbf{C} & \mathbf{D} \\ \mathbf{E} & \mathbf{F} \end{bmatrix} \tag{D-33}$$

$$\mathbf{B} = \begin{bmatrix} \mathbf{G} & \mathbf{H} \\ \mathbf{J} & \mathbf{K} \end{bmatrix} \tag{D-34}$$

where $\mathbf{C}$ through $\mathbf{K}$ are submatrices. The product of $\mathbf{AB}$ is

$$\mathbf{AB} = \begin{bmatrix} \mathbf{CG} + \mathbf{DJ} & \mathbf{CH} + \mathbf{DK} \\ \mathbf{EG} + \mathbf{FJ} & \mathbf{EH} + \mathbf{FK} \end{bmatrix} \tag{D-35}$$

Determinants

Every square matrix has a scalar associated with it called its *determinant*. In particular, if $\mathbf{A}$ is a square matrix, say,

$$\mathbf{A} = \begin{bmatrix} a_{11} & a_{12} & a_{13} \\ a_{21} & a_{22} & a_{23} \\ a_{31} & a_{32} & a_{33} \end{bmatrix} \tag{D-36}$$

then the determinant of $\mathbf{A}$ is denoted det $\mathbf{A}$ or $|\mathbf{A}|$

$$\det \mathbf{A} = \begin{vmatrix} a_{11} & a_{12} & a_{13} \\ a_{21} & a_{22} & a_{23} \\ a_{31} & a_{32} & a_{33} \end{vmatrix} \tag{D-37}$$

It is important to note the difference between (D-36) and (D-37); (D-36) represents a matrix which is a square array of elements whereas (D-37) represents a scalar associated with the matrix $\mathbf{A}$.

The det $\mathbf{A}$ is determined by obtaining the *minors* and *cofactors*. Given a matrix $\mathbf{A}$, a minor is the determinant of any square submatrix of $\mathbf{A}$. The cofactor of the element a_{ij} is a scalar obtained by multiplying $(-1)^{i+j}$ times the minor obtained from $\mathbf{A}$ by removing the ith row and jth column. To find the determinant of the square matrix $\mathbf{A}$:

1. Pick any one row or any one column of the matrix.
2. For each element in the row or column chosen, find its cofactor.
3. Multiply each element in the row or column chosen by its cofactor and sum the results.

The sum is the determinant of the matrix. For example, find det $\mathbf{A}$, where

$$\mathbf{A} = \begin{bmatrix} 3 & 5 & 0 \\ -1 & 1 & 1 \\ 3 & -6 & 4 \end{bmatrix} \tag{D-38}$$

Expanding in the second column,

$$\det \mathbf{A} = (5)(-1)^{1+2}\begin{vmatrix} -1 & 1 \\ 3 & 4 \end{vmatrix} + (1)(-1)^{2+2}\begin{vmatrix} 3 & 0 \\ 3 & 4 \end{vmatrix} + (-6)(-1)^{3+2}\begin{vmatrix} 3 & 0 \\ -1 & 1 \end{vmatrix}$$

$$= (-5)(-7) + (1)(12) + (6)(3) = 65 \tag{D-39}$$

Adjoint

The *adjoint matrix* of a square matrix $\mathbf{A}$, denoted adjoint $\mathbf{A}$ or $\mathbf{A}^a$, is formed by replacing each element a_{ij} by the cofactor α_{ij} and transposing. Thus, the adjoint of (D-36) is

$$\text{adjoint } \mathbf{A} = \begin{bmatrix} \alpha_{11} & \alpha_{12} & \alpha_{13} \\ \alpha_{21} & \alpha_{22} & \alpha_{23} \\ \alpha_{31} & \alpha_{32} & \alpha_{33} \end{bmatrix}^T$$

$$= \begin{bmatrix} \alpha_{11} & \alpha_{21} & \alpha_{31} \\ \alpha_{12} & \alpha_{22} & \alpha_{32} \\ \alpha_{13} & \alpha_{23} & \alpha_{33} \end{bmatrix} \tag{D-40}$$

Inverse

The *inverse* of a square matrix $\mathbf{A}$ is written as $\mathbf{A}^{-1}$ and is defined as

$$\mathbf{A}^{-1}\mathbf{A} = \mathbf{A}\mathbf{A}^{-1} = \mathbf{I} \tag{D-41}$$

In the text, parentheses are used to avoid confusion with superscripts, that is, the inverse is denoted $(\mathbf{A})^{-1}$. The inverse is defined only for square matrices. In particular,

$$\mathbf{A}^{-1} = \frac{\text{adjoint } \mathbf{A}}{\det \mathbf{A}} \qquad (D\text{-}42)$$

If $\det \mathbf{A}$ is zero, $\mathbf{A}$ does not possess an inverse and is said to be *singular*. Consider a 2×2 matrix.

$$\mathbf{A} = \begin{bmatrix} a_{11} & a_{12} \\ a_{21} & a_{22} \end{bmatrix} \qquad (D\text{-}43)$$

$$\text{adjoint } \mathbf{A} = \begin{bmatrix} a_{22} & -a_{12} \\ -a_{21} & a_{11} \end{bmatrix} \qquad (D\text{-}44)$$

$$\det \mathbf{A} = a_{11}a_{22} - a_{12}a_{21} \qquad (D\text{-}45)$$

Find the inverse of (D-38). The cofactor of a_{11} is

$$\alpha_{11} = (3)(-1)^{1+1} \begin{vmatrix} 1 & 1 \\ -6 & 4 \end{vmatrix} = (3)(10) = 30 \qquad (D\text{-}46)$$

Finding the cofactor of each element and transposing yields the adjoint of $\mathbf{A}$. Thus

$$\text{adjoint } \mathbf{A} = \begin{bmatrix} 30 & 20 & 15 \\ 35 & 12 & 18 \\ 0 & 33 & 32 \end{bmatrix} \qquad (D\text{-}47)$$

The $\det \mathbf{A}$ is given by (D-39), hence,

$$\begin{aligned} \mathbf{A}^{-1} &= \frac{\text{adjoint } \mathbf{A}}{\det \mathbf{A}} \\ &= \frac{1}{65} \begin{bmatrix} 30 & 20 & 15 \\ 35 & 12 & 18 \\ 0 & 33 & 32 \end{bmatrix} \end{aligned} \qquad (D\text{-}48)$$

Derivatives

The derivative of the matrix $\mathbf{A}$, denoted $(d/dt)\mathbf{A}$ or $p\mathbf{A}$, is the derivative of each element of the matrix. The derivative of $\mathbf{A}$, given by (D-1), is

$$p\mathbf{A} = \begin{bmatrix} pa_{11} & pa_{12} & \cdots & pa_{1n} \\ pa_{21} & pa_{22} & \cdots & pa_{2n} \\ \cdots\cdots\cdots\cdots\cdots\cdots \\ pa_{m1} & pa_{m2} & \cdots & pa_{mn} \end{bmatrix} \qquad (D\text{-}49)$$

where p is the operator d/dt.

Matrix Formulation

In the text we deal with equations of the form

$$\mathbf{A}^{-1}\mathbf{y} = \mathbf{A}^{-1}r\mathbf{I}\mathbf{x} + p[\mathbf{A}^{-1}\mathbf{z}] \qquad (D\text{-}50)$$

where $\mathbf{A}$ is a square, nonsingular matrix, the elements of which may be functions of time, and p is the operator d/dt. Solving (D-50) for $\mathbf{y}$ is achieved by premultiplying by $\mathbf{A}$. Thus,

$$\mathbf{A}\mathbf{A}^{-1}\mathbf{y} = \mathbf{A}\mathbf{A}^{-1}r\mathbf{I}\mathbf{x} + \mathbf{A}p[\mathbf{A}^{-1}\mathbf{z}]$$

$$= r\mathbf{A}\mathbf{A}^{-1}\mathbf{x} + \mathbf{A}[p\mathbf{A}^{-1}]\mathbf{z} + \mathbf{A}\mathbf{A}^{-1}[p\mathbf{z}] \tag{D-51}$$

Since $\mathbf{A}\mathbf{A}^{-1} = \mathbf{I}$, (D-51) may be written as

$$\mathbf{y} = r\mathbf{I}\mathbf{x} + \mathbf{A}[p\mathbf{A}^{-1}]\mathbf{z} + p\mathbf{z} \tag{D-52}$$

APPENDIX
E

SOLUTIONS OF LINEAR, ORDINARY DIFFERENTIAL EQUATIONS WITH CONSTANT COEFFICIENTS

We shall consider ordinary differential equations of the form

$$a_n \frac{d^n y}{dt^n} + a_{n-1} \frac{d^{n-1} y}{dt^{n-1}} + \cdots + a_0 y = x \tag{E-1}$$

which can be written

$$(a_n p^n + a_{n-1} p^{n-1} + \cdots + a_0) y = x \tag{E-2}$$

where $p^n = d^n/dt^n$ denotes the nth derivative with respect to time. It is assumed that the input or forcing function x is known for $t > 0$ as are the initial ($t = 0$) values of $y, dy/dt, \ldots, d^{n-1}y/dt^{n-1}$. Our task is to express $y(t)$ for $t > 0$.

We will do this by first considering the homogeneous case where $x = 0$ for $t > 0$. The characteristic equation corresponding to (E-2) may be written as

$$a_n p^n + a_{n-1} p^{n-1} + \cdots + a_0 = 0 \tag{E-3}$$

If we denote $b_1, \ldots, b_n$ as the negative roots of this characteristic equation, the general solution of (E-2) with $x = 0$ for $t > 0$ may be written as

$$y(t) = C_1 e^{-b_1 t} + C_2 e^{-b_2 t} + \cdots + C_n e^{-b_n t} \tag{E-4}$$

The coefficients $C_1, \ldots, C_n$ may be established from the known initial conditions. In particular, if we differentiate (E-4) $n - 1$ times, we obtain

$$y = C_1 e^{-b_1 t} + C_2 e^{-b_2 t} + \cdots + C_n e^{-b_n t}$$

$$\frac{dy}{dt} = C_1(-b_1) e^{-b_1 t} + C_2(-b_2) e^{-b_2 t} + \cdots + C_n(-b_n) e^{-b_n t} \tag{E-5}$$

$$\cdots\cdots\cdots\cdots\cdots\cdots\cdots\cdots\cdots\cdots\cdots\cdots$$

$$\frac{d^{n-1} y}{dt^{n-1}} = C_1(-b_1)^{n-1} e^{-b_1 t} + C_2(-b_2)^{n-1} e^{-b_2 t} + \cdots + C_n(-b_n)^{n-1} e^{-b_n t}$$

Setting $t = 0$ and equating these to the known initial conditions,

$$y(0) = C_1 + C_2 + \cdots + C_n$$

$$\frac{dy}{dt}(0) = C_1(-b_1) + C_2(-b_2) + \cdots + C_n(-b_n) \tag{E-6}$$

$$\cdots\cdots\cdots\cdots\cdots\cdots\cdots\cdots\cdots\cdots\cdots$$

$$\frac{d^{n-1} y}{dt^{n-1}}(0) = C_1(-b_1)^{n-1} + C_2(-b_2)^{n-1} + \cdots + C_n(-b_n)^{n-1}$$

which can be written in matrix form,

$$
\begin{bmatrix} y(0) \\ \dfrac{dy}{dt}(0) \\ \vdots \\ \dfrac{d^{n-1}y}{dt^{n-1}}(0) \end{bmatrix}
=
\begin{bmatrix} 1 & 1 & \cdots & 1 \\ -b_1 & -b_2 & \cdots & -b_n \\ \vdots & & & \vdots \\ (-b_1)^{n-1} & (-b_2)^{n-1} & \cdots & (-b_n)^{n-1} \end{bmatrix}
\begin{bmatrix} C_1 \\ C_2 \\ \vdots \\ C_n \end{bmatrix}
\tag{E-7}
$$

Equation (E-7) can be solved for the coefficients $C_1, \ldots, C_n$.

This procedure is valid even if the roots of the characteristic equation are complex. However, in the case of complex roots we will find it convenient to modify the procedure somewhat. Also, if two or more roots are equal to one another (repeated roots), this procedure also must be modified. In the case of complex or repeated roots, the following procedure may be used [2].

1. For each distinct negative real root b_i we include the term $e^{-b_i t}$ in the general solution.

2. For each distinct complex pair of roots, $-\gamma_i \pm j\beta_i$, we include the terms $e^{-\gamma_i t} \cos \beta_i t$ and $e^{-\gamma_i t} \sin \beta_i t$ in the general solution.

3. For a negative real root b_i of multiplicity k, we include the terms $e^{-b_i t}, te^{-b_i t}, \ldots, t^{k-1}e^{-b_i t}$ in the general solution.

4. For a complex pair of roots, $-\gamma_i \pm j\beta_i$, of multiplicity k, we include the terms

$$e^{-\gamma_i t} \cos \beta_i t, \, te^{-\gamma_i t} \cos \beta_i t, \ldots, t^{k-1}e^{-\gamma_i t} \cos \beta_i t$$

$$e^{-\gamma_i t} \sin \beta_i t, \, te^{-\gamma_i t} \sin \beta_i t, \ldots, t^{k-1}e^{-\gamma_i t} \sin \beta_i t$$

in the general solution.

Note that for an nth-order differential equation, there will be n terms of the form described above.

Lastly, we collect terms and express the complete, general solution as

$$y(t) = C_1 y_1 + \cdots + C_n y_n \tag{E-8}$$

where $y_1, \ldots, y_n$ represent the n terms established above. The coefficients $C_1, \ldots, C_n$ may be obtained, as before, by differentiating (E-8) $n - 1$ times and equating the results at $t = 0$ with the known initial conditions.

The previous approach can be applied, with modification, to inhomogeneous equations wherein the forcing function x in (E-1) is nonzero. We will restrict our attention to two cases: (1) The forcing function is constant for $t > 0$ or (2) the forcing function is sinusoidal for $t > 0$. In either case, we can express the general solution of the inhomogeneous equation (E-1) as

$$y = y_{ss} + y_h$$
$$= y_{ss} + C_1 y_1 + C_2 y_2 + \cdots + C_n y_n \tag{E-9}$$

where y_{ss} represents the steady-state response and y_h represents the general solution of the homogeneous equation given by (E-8). The general solution of the homogeneous equation is sometimes called the *natural response* or the *transient response* and may be denoted y_{tr} instead of y_h. However, its the same animal. We will assume that the general solution of the homogeneous equation has already been established with the coefficients in (E-9) left unspecified. If the forcing function x is constant for $t > 0$, the steady-state component of (E-9) is easily established. In particular, from (E-1),

$$y_{ss} = \frac{x}{a_0} \qquad x \text{ constant} \tag{E-10}$$

If the forcing function is sinusoidal, phasor techniques may be used to establish y_{ss}. In particular, writing (E-2) in phasor form,

$$[a_n(j\omega)^n + a_{n-1}(j\omega)^{n-1} + \cdots + a_0]\tilde{Y}_{ss} = \tilde{X} \qquad x \text{ sinusoidal} \tag{E-11}$$

where $\tilde{X}$ represents the phasor corresponding to the sinusoidal forcing function (ω represents the frequency) and $\tilde{Y}_{ss}$ is the phasor corresponding to the steady-state solution y_{ss}. From (E-11),

$$\tilde{Y}_{ss} = \frac{\tilde{X}}{a_n(j\omega)^n + a_{n-1}(j\omega)^{n-1} + \cdots + a_0} \tag{E-12}$$

from which the steady-state response y_{ss} is readily established. Once y_{ss} is established by using (E-10) for constant x or (E-12) for sinusoidal x, we can calculate the coefficients of (E-9) by differentiating (E-9) $n-1$ times and equating the results at $t = 0$ with the known initial conditions. For apparent reasons, this method of solution is referred to as the *method of undetermined coefficients*.

REFERENCES

1. R. C. Dorf, *Modern Control Systems*, Addison-Wesley Publishing Company, Reading, Mass., 1967.
2. V. B. Haas, "Class Notes," Purdue University, 1978.

INDEX

INDEX